PHASE TRANSITIONS

This accessible text is an introduction to the theory of phase transitions and its application to real materials. Assuming some familiarity with thermodynamics and statistical mechanics, the book begins with a primer on the thermodynamics of equilibrium phase transitions, including the mean-field and Ginzburg–Landau approaches. The general kinetic features and dynamics of phase transitions are explained, ensuring that readers are familiar with the key physical concepts. With the foundations established, the general theory is applied to the study of phase transitions in a wide range of materials, including ferroic materials, caloric materials, liquid crystals and glasses. Non-equilibrium phase transitions, superconductors and quantum phase transitions are also covered. Including exercises throughout and solutions available online, this text is suitable for graduate courses as well as researchers in physics and materials science seeking a primer on popular and emerging research topics.

ANTONI PLANES is Emeritus Professor of Department of Physics of Condensed Matter at the University of Barcelona. His research interests include the study of phase transitions in materials with multiple coupled degrees of freedom, precursor phenomena, avalanche criticality and caloric effects. He is a fellow of the APS and the Royal Academy of Sciences and Arts of Barcelona.

AVADH SAXENA is Senior Scientist and the former Group Leader of the Physics of Condensed Matter and Complex Systems Group at Los Alamos National Laboratory (LANL). His research interests include phase transitions, non-linear science, topological defects and non-Hermitian quantum mechanics. He is a fellow of the APS, a fellow of LANL and an external member of the Royal Academy of Sciences and Arts of Barcelona.

PHASE TRANSITIONS

A Materials Perspective

ANTONI PLANES
Universitat de Barcelona

AVADH SAXENA
Los Alamos National Laboratory

CAMBRIDGE
UNIVERSITY PRESS

Shaftesbury Road, Cambridge CB2 8EA, United Kingdom

One Liberty Plaza, 20th Floor, New York, NY 10006, USA

477 Williamstown Road, Port Melbourne, VIC 3207, Australia

314–321, 3rd Floor, Plot 3, Splendor Forum, Jasola District Centre,
New Delhi – 110025, India

103 Penang Road, #05–06/07, Visioncrest Commercial, Singapore 238467

Cambridge University Press is part of Cambridge University Press & Assessment,
a department of the University of Cambridge.

We share the University's mission to contribute to society through the pursuit of
education, learning and research at the highest international levels of excellence.

www.cambridge.org
Information on this title: www.cambridge.org/9781009549752

DOI: 10.1017/9781009549776

First published 2025

A catalogue record for this publication is available from the British Library

A Cataloging-in-Publication data record for this book is available from the Library of Congress

ISBN 978-1-009-54975-2 Hardback

Additional resources for this publication at www.cambridge.org/9781009549752.

To Teresa and Mònica and to the memory of my parents
Antoni Planes

To my siblings and to the memory of my parents
Avadh Saxena

Contents

Foreword

Phase transitions (aka phase transformations, phase changes, etc.) are fundamental topics, with a long tradition in physics, materials science, chemistry and engineering. The fact that iron must be heated and hammered to yield sturdy tools was known since the dawn of time. The transformation between liquids and solids by freezing is a common natural phenomenon and melting was long known as a widespread physical phenomenon. Great progress was made in the eighteenth century in understanding some of the more basic principles using examples like 'Newton's metal' and the development of steel (Reaumur, Monge, Berthollet and Vandermonde). Another early motivation to study phase transitions, which still goes strong today, is geology and mineralogy. If our planet earth were in thermal equilibrium, it would be a perfect sphere with no local variations of its chemical and physical structure – which is not suitable for most life forms. Fortunately, this is not the case and long-term geological history is understood by the way minerals transform under changing temperatures, pressures and chemical drivers. Phase transitions of minerals are the key indicator for such events, and much research is done in these fields. A new and exciting application of phase transitions is in biological sciences. While giant steps forward were made in the nineteenth century with fundamental contributions by Boltzmann, Ehrenfest, Einstein, Lenz and Ising, modern theories of phase transitions in solids owe much to the contributions of Landau and Lifshitz and the various extensions of their initial approach by Ginzburg and others. The twentieth century saw a great proliferation of the field including the extension to renormalization theory. There are many textbooks available that cater to the broader subject of phase transitions. There has long been a need for a book on phase transitions that straddles physics, materials science, engineering, etc. with ease, ignoring artificial boundaries between these disciplines. Therefore, I was delighted to see the new book *Phase Transitions: A Materials Perspective* which deftly fills this gap. It is easily accessible to advanced undergraduate and graduate students as well as to new and seasoned researchers. It is written by two prominent scientists who have at least four decades of research experience each on the topic of phase transitions in a variety of functional materials both from the vantage of a physicist and a materials scientist. Between the two of them, they have collaborated extensively for more than two decades on these topics. Thus, they bring a unique perspective to the subject matter. The first two chapters lay out the basic physics concepts needed to understand the ensuing materials-specific chapters. A lot of the contents relate to new research during the past two decades such as ferroic glasses, multicaloric effects, avalanches and other non-equilibrium phenomena not covered in one place in other books. The accompanying Exercise book of solutions is a tremendous resource because it is not merely the solutions but it also provides detailed explanations along with supplementary material aimed at helping facilitate various concepts central to understanding phase transitions in a number of materials. Of course, not everything could be captured in the book. My hope is that with the likely success of this edition of the book, the next edition will be able to capture emerging new results on topological materials, phase transitions in selected biological systems, nanoscale effects, phase transitions in quantum materials, etc. I highly recommend this interdisciplinary book to physics, materials science and relevant engineering departments in addition to researchers engaged in this important field worldwide.

Ekhard K. H. Salje, University of Cambridge, June 2024

Preface

The study of phase transitions has been a revered subject for a very long time. It is still a subject of great importance since phase transitions are not only ubiquitous but also technologically important. The first scientific studies go back to the nineteenth century, and today it is a mature discipline. Nevertheless, some aspects are still not completely understood, although most researchers agree that we understand quite well the phase transitions in equilibrium that are driven by thermal fluctuations when disorder does not play a relevant role. However, problems related to materials with interplay between multiple degrees of freedom, disorder or quantum effects are still under debate. As a result, phase transitions are still an important topic of investigation, one which is essential in order to understand specific materials properties and associated functionalities that often emerge through a phase transition. Thus, phase transitions are of utmost importance in the process of designing new advanced materials with improved properties. A key objective of the book is to address new and emerging topics of research in the field of materials to students both in condensed matter physics and materials science and engineering. To that end, we have aimed to write the book in a way that addresses a number of subjects developed in recent years discussed within the general framework of phase transitions, which is well adapted to teaching purposes for these two groups of students as well as for early career researchers.

This book assumes that the reader has been through an introductory course in thermodynamics and statistical mechanics and has some background in materials physics. The main emphasis is on physics rather than materials science but concepts are introduced in a way that is intended to highlight the relevance of these concepts in the field of materials, and thus it is expected that it will help in enhancing research in materials science. Part I of the book will introduce basic concepts in phase transitions, while

Part II aims at applying these concepts to specific classes of materials. It is written and organized in such a way that it addresses the selected topics in a pedagogical and sufficiently rigorous way, rendering the book both insightful and useful for condensed matter physicists but still accessible to materials scientists. Therefore, the book is intended to serve as a natural bridge between the physics and materials communities. While the book will be especially useful for graduate students, it is expected that some introductory parts and certain chapters and, especially, parts of chapters will be of interest to undergraduate students. In addition, one of our objectives is that it will also be useful for early career and even more senior researchers in the topics covered in the book. In this sense, the book discusses certain advanced subjects, which, in some cases, require complex mathematical developments. We have not avoided these developments but have always tried to stress underlying physical aspects and attempted as much as possible to avoid the use of complicated mathematical formalism. In general, the mathematics required to follow the book is not sophisticated. However, some knowledge of variational calculus and group theory would be helpful.

The book puts much emphasis on highlighting the potential of the mean-field approach and the Landau theory when dealing with phase transitions in real materials exhibiting complex behaviour. In these cases the use of more exact techniques, such as the renormalization group approach, is difficult and, in fact, beyond the scope of the book. Regardless of this, the weaknesses arising from mean-field approximations are discussed. From this perspective, an important goal of the book is to bring a number of topics related to phase transitions that have garnered a lot of interest during recent decades among academics in a way that can be taught to graduate and even undergraduate students. In our opinion, this should be an essential part of the aim of researchers when writing a book.

Part I of the book is divided into two chapters. The Chapter 1 represents a formal introduction to general features of equilibrium phase transitions and contains all the basic physics concepts necessary to follow subsequent chapters, where these concepts are applied to specific classes of materials. Chapter 2 deals with general kinetic features and dynamics of phase transitions. Part II is aimed at applying the general theory to the study of phase transitions in selected classes of materials. It starts with a thorough discussion of ferroic and multiferroic materials, which are introduced with the aim of treating the common aspects of the different ferroic classes and of their interplay from, as far as possible, a unified point of view. In any case, specific differences are also taken into account in detail. Ferroic materials are essential to follow most of the subsequent chapters on caloric materials,

materials with disorder and glassy phenomena, as well as non-equilibrium phase transitions. In all of them the same unified point of view is adopted. In particular, in the chapter on materials with disorder the usefulness of this point of view to understanding common features of glassy phenomena in ferromagnetic, ferroelectric and ferroelastic materials is highlighted. Besides ferroic materials, the book also contains chapters dedicated to the study of phase transitions in liquid crystals and superconductors and ends with an introductory chapter on quantum phase transitions. The last two chapters, on superconductors and quantum phase transitions, require some knowledge of quantum mechanics. We have also included an appendix that sets out in detail how symmetry aspects can be taken into account from group theory and explicitly explains how to proceed in order to obtain an adequate Landau free energy based on the symmetry of the parent and product phases, as well as the nature of the order parameter of the system of interest.

All chapters include many relevant references. In general they are not needed to follow the book. They are intended to be useful for those who are interested in further deepening their knowledge of the subject. A short list of exercises is also provided at the end of each chapter. Most of these are quite straightforward; only a few are more challenging. This list has been included with a twofold aim. First, to complement some of the developments of the corresponding chapter and, second, to encourage potential readers to check whether they have correctly understood the contents of the chapter and to train themselves to perform some simple calculations that can permit a deeper understanding of certain problems.

During the process of writing this book and previous research collaborations, we have benefited from discussions with a number a colleagues who have also provided us with some critical comments that have positively contributed to improving the book. In particular, discussions with Teresa Castán, Eduard Vives, Lluís Mañosa, Jordi Ortín, Marcel Porta, Pol Lloveras, Matteo Palasini, Francisco-José Pérez-Reche, Jordi Marcos, Xavier Moya, Ekhard Salje, Mehmet Acet, Alan Bishop, Robert Albers, Turab Lookman, Subodh Shenoy, Mahdi Sanati, Xiaobing Ren, Amar Bhalla, Greg Olson and Peter Littlewood are particularly appreciated. Also, we would like to thank Lucas Maisel, formerly a student at the University of Barcelona, who has helped us with some of the calculations presented in Chapter 8.

Part I
General Aspects

1

Introduction to phase transitions

At the macroscopic scale, matter can be organized into very diverse structures that are distinguished by their properties; they are the so-called phases. When a phase transition takes place, the system undergoes a reorganization and its properties change. Qualitatively, this reorganization can be understood in terms of a competition between energy, essentially associated with the degree of cohesion, and entropy associated with the degree of disorder, which results from the fact that the free energy of a given system in equilibrium must reach a minimum. Therefore, at high temperature, the system prefers to be in a disordered state of high entropy, while below a certain temperature, which may depend on parameters such as pressure and magnetic field, it becomes more favourable to reorganize itself into an ordered and more *compact* phase in order to decrease energy. This is only possible if the constituents of the system exhibit some kind of attractive interaction.

Consequently, phase transitions take place when a certain physical system is subjected to a change in the external conditions that can be achieved by changing the temperature or a field (mechanical, electrical, magnetic, etc.) that is coupled to system properties. The existence of a phase transition is revealed through a singular behaviour of some thermodynamic quantities. In this context, the word singular must be understood not only in the most common sense of unique and somewhat anomalous behaviour, but also, and this is a crucial aspect, in its mathematical sense indicating that these variables have a non-analytical behaviour when the transition takes place. From a formal viewpoint, a singular behaviour is only expected to occur in the thermodynamic limit, which means that phase transitions are intimately related to the thermodynamic behaviour of materials.

These transitions are thermal transitions in the sense that thermal fluctuations are the main microscopic mechanism that induces the change of properties in the system. Thermodynamics is thus essential in order to understand this class of transitions. Nevertheless, it is important to note that phase transitions can also occur in *non-thermal* systems where thermal fluctuations do not play a primary role (or even do not play any role) in driving the transition. In this class of systems, phase transitions are controlled by the variation of a *non-thermal* parameter, which is not directly related to temperature. In analogy with thermal systems, this parameter induces a change in the properties of the system and the transition shows up through a singularity in the variables conjugated to the control parameter or in their derivatives. Examples include dynamic phase transitions, disorder-induced phase transitions and topological (structural) phase transitions. Quantum phase transitions [1] can also be included within this category. This class of transitions occurs at zero temperature induced by a certain non-thermal parameter and is a consequence of competing ground-state phases. The microscopic mechanism at the origin of these transitions is associated with quantum fluctuations, which are a direct consequence of the Heisenberg uncertainty principle. In spite of the fact that these transitions occur in macroscopic systems, they cannot be described by the usual thermodynamic formalism. Nevertheless, the analogy with thermal transitions is an essential point in order to understand this class of systems.

Phase transitions, both thermal and non-thermal, can be classified according to the nature of the singularity that occurs at the transition. Paul Ehrenfest was the first to propose a classification based on this idea in 1933 [2]. He considered the analytical properties of the free energy at the transition point and proposed that a transition belongs to the class of nth-order transitions if the nth-order derivative of the free energy displays a discontinuity at the transition. Therefore, first-order transitions are those in which quantities obtained as first-order derivatives of the free energy show a discontinuity at the transition. Examples of such quantities are mass (or molar) density, magnetization or polarization, which are densities of global quantities associated with the whole system. The discontinuity of these quantities must be understood to be associated with the coexistence of phases with different values of such quantities. If the first derivatives are continuous but the discontinuity occurs in the second-, the third- or the higher-order derivative of the free energy, then the transition is classified as a second-, a third- or a higher-order transition. This is an elegant classification, which has the inconvenience of recognizing only discontinuities rather than more subtle singular behaviour associated with the possible divergent behaviour of response functions at the transition. Therefore, this classification has been proven inappropriate and replaced by a more general one, which proposes only two large classes of transitions: first-order transitions and continuous transitions [3]. The former are those concerned with a discontinuity of properties associated with first-order derivatives of the free energy and thus are characterized by the coexistence of different phases in equilibrium. In the case of continuous transitions, first derivatives of the free energy are continuous, but higher-order derivatives are discontinuous or diverge. Transitions taking place at a critical point belong to this second class.

The order parameter is the essential entity that permits distinguishing the different phases that can occur in a system and, thus, must carry basic information related to the class of singularities occurring at the transition. Therefore, the order parameter is an important quantity in order to analyze and formulate a suitable model to quantitatively describe a phase transition in a given system. Consequently, when studying phase transitions in a given system, we must begin by identifying an appropriate order parameter. The identification must be done on the basis of the phenomenological behaviour of the system considered. Usually it is a quantity or a combination of quantities related to first-order derivatives of the free energy that is able to reflect the discontinuous character of first-order transitions or rather to demonstrate the continuous variation expected across continuous transitions.

In general, it is expected that the chosen order parameter be a quantity that fluctuates in time and space. Actually, these fluctuations provide essential information about relevant response functions of the system. This fact is particularly relevant in the vicinity of critical points where the divergence of response functions is intimately associated with an *anomalous* behaviour of fluctuations. In any case, a large number of aspects related to phase transitions can be explained simply from the behaviour of the statistical mean value of the order parameter. In general, it is convenient to define the order parameter in such a way that this average value vanishes in the disordered phase (usually the high-temperature phase in thermal transitions) and takes a finite value in the ordered phase (low-temperature phase in thermal transitions). In continuous transitions, this mean value will present a continuous variation across the transition, while it will reveal a discontinuity in a first-order transition.

Close to critical points, various systems display critical behaviour, which is also denoted as criticality. This behaviour is characterized by power-law behaviour of thermodynamic quantities, which reflects the absence of characteristic scales and thus scale invariance associated with the anomalous behaviour of the fluctuations [4]. For a given thermodynamic quantity ϕ, its power-law dependence with temperature is expressed as, $\phi \sim |t|^\mu$, where $t = (T - T_c)/T_c$ is a reduced temperature that measures the *distance* in temperature to the critical point T_c. This notation indicates that close to the critical point ϕ has a dominant power-law behaviour with temperature. The exponent μ is thus defined as

$$\mu = \lim_{t \to 0} \frac{\log \phi}{\log |t|}. \tag{1.1}$$

Therefore, the exponent that quantifies this behaviour determines the nature of the singularity occurring at the critical point.[1] Note, in particular, that when the exponent μ is positive, the function ϕ goes to zero at the transition, while it diverges when it is negative. Note also that $\mu = 0$ may correspond to the following three different situations: discontinuity, logarithmic divergence or cusp behaviour of the function ϕ [4].

1.1 Thermodynamics: general features

Thermodynamics is essential in order to understand phase transitions. This section is a general introduction to basic concepts of thermodynamics and its relevance for a proper understanding of phase transitions. We will start with an introduction to the equilibrium thermodynamic description of a complex material constituted of c components that can exchange matter, heat and work[2] with the surroundings. We will assume that the system can exchange ω multiple kinds of work associated with different forces that couple with the properties of the system. The thermodynamic description of such systems requires $c + \omega + 1$ independent extensive variables[3] or degrees of freedom, which include the mol numbers $\{n_i\}$ (or the masses $\{m_i\}$) of the c components, the w generalized displacements $\{\boldsymbol{X}_i\}$ that couple with the forces acting on the system, and entropy S. The internal energy $U = U(\{n_i\}, \{\boldsymbol{X}_i\}, S)$ is thus a function of these independent variables, and its differential change associated with differential exchanges in equilibrium of matter, work, and heat is given by the fundamental thermodynamic equation (also denoted as thermodynamic identity) that combines the first and second laws of thermodynamics [5],

---

[1] It is usual to denote critical exponents with the following common notations: exponent of the heat capacity, α; order parameter, β; susceptibility, γ; critical isotherm, δ; correlation length, ν; correlation function, η.

[2] The definition of work is the same that in mechanics, and, therefore, it is given by the product of a force and a displacement. Forces of different nature can be considered including contact forces or electromagnetic forces among others. In quasi-static trajectories, work can be expressed as the product of an intensive variable and the change of its conjugate extensive variable that characterizes the state of the system. Usually extensive variables are denoted as generalized displacements and intensive variables as generalized forces.

[3] Extensive variables are defined for the whole system such as volume, total magnetic moment or total polar moment (or polarization).

$$dU = TdS + \sum_{i=1}^{\omega} \boldsymbol{y}_i \cdot d\boldsymbol{X}_i + \sum_{j=1}^{c} \mu_j dn_i, \tag{1.2}$$

where the generalized forces $\{\boldsymbol{y}_i\}$, chemical potentials $\{\mu_i\}$, and temperature T are intensive variables thermodynamically conjugated to the extensive variables, $\{\boldsymbol{X}_i\}, \{n_i\}$, and S, respectively, and are given as,

$$\boldsymbol{y}_j = \left(\frac{\partial U}{\partial \boldsymbol{X}_j}\right)_{[\{\boldsymbol{X}_{i\neq j}\},\{n_i\},S]}, \tag{1.3}$$

$$\mu_j = \left(\frac{\partial U}{\partial n_j}\right)_{[\{\boldsymbol{X}_i\},\{n_{i\neq j}\},S]}, \tag{1.4}$$

$$T = \left(\frac{\partial U}{\partial T}\right)_{[\{\boldsymbol{X}_i\},\{n_i\}]}. \tag{1.5}$$

Each pair of conjugated variables $\boldsymbol{X}_i$ and $\boldsymbol{y}_i$ has the same tensorial order so that the tensorial product $\boldsymbol{y}_i \cdot d\boldsymbol{X}_i$ is a scalar that quantifies the reversible work associated with the infinitesimal change of the displacement variable. The existence of interplay between the different degrees of freedom must be taken into account through the explicit dependence of each coordinate on the remaining independent coordinates. It is worth noting that in closed systems, all the n_i are fixed and, thus, the term $\sum_{i=1}^{c} \mu_i dn_i$ will not appear in the fundamental equation. This is the usual situation we will deal with in this book.

In practical situations it is often easier to modify the state of a given system by externally controlling temperature and generalized forces than their corresponding conjugated extensive variables, entropy and generalized displacements, respectively. It is then convenient to use intensive quantities as independent variables, which can be achieved through suitable Legendre transforms. For instance, any Legendre transform that supposes the interchange of entropy by temperature as independent variable defines a free energy. The Helmholtz free energy F is given as

$$F = U - TS. \tag{1.6}$$

In terms of F the fundamental equation reads,

$$dF = -SdT + \sum_{i=1}^{w} \boldsymbol{y}_i \cdot d\boldsymbol{X}_i + \sum_{i=1}^{c} \mu_i dn_i. \tag{1.7}$$

The free energy that has temperature and all generalized forces as independent variables is usually called the Gibbs free energy G and is defined as

$$G = U - TS - \sum_{i=1}^{w} \boldsymbol{y}_i \cdot \boldsymbol{X}_i, \tag{1.8}$$

and its differential change is given by

$$dG = -SdT - \sum_{i=1}^{w} \boldsymbol{X}_i \cdot d\boldsymbol{y}_i + \sum_{i=1}^{c} \mu_i dn_i. \tag{1.9}$$

All the functions that are obtained by means of Legendre transforms of the internal energy are usually denoted as thermodynamic energy potentials or simply thermodynamic potentials. It is worth noting that a set of entropy potentials can also be introduced as Legendre transforms of the entropy. They are also called Massieu–Planck functions.

The complete interchange of extensive by intensive variables gives rise to the Gibbs–Duhem equation,

$$SdT - \sum_{i=1}^{w} \boldsymbol{X}_i \cdot d\boldsymbol{y}_i + \sum_{i=1}^{c} n_i d\mu_i = 0, \tag{1.10}$$

which expresses that intensive variables cannot be all independent. This result is related to the general condition of extensivity that will be discussed in more detail later.

Response functions are quantities that express how a system responds when subjected to external fields that couple to the properties of the system. These quantities provide a good description of the behaviour of given materials and are often the ones of most direct physical interest. Response functions of particular interest are heat capacities and susceptibilities.

Heat capacities measure the thermal response along a given reversible thermodynamic trajectory. The two basic heat capacities are defined respectively at constant generalized displacements and at constant generalized force variables as

$$C_X = T \left(\frac{dS}{dT} \right)_{\{\boldsymbol{X}_i\}}, \tag{1.11}$$

$$C_y = T \left(\frac{dS}{dT} \right)_{\{\boldsymbol{y}_i\}}. \tag{1.12}$$

The susceptibility tensor is a symmetric tensor that quantifies the response of generalized displacements to the changes of generalized forces. The elements of this tensor are defined as

$$\chi_{ij} = \left(\frac{\partial \boldsymbol{X}_i}{\partial \boldsymbol{y}_j}\right)_{T,\{\boldsymbol{y}_{k\neq j}\}} = -\left(\frac{\partial^2 G}{\partial \boldsymbol{y}_i \partial \boldsymbol{y}_j}\right)_{T,\{\boldsymbol{y}_{k\neq j}\}}. \qquad (1.13)$$

Diagonal terms determine the response of a given extensive property to its conjugated field and are simply denoted as susceptibilities. Non-diagonal terms quantify the cross-response to non-conjugated fields and are called cross-susceptibilities.

Maxwell relations are equalities that result from the fact that the second derivatives of the thermodynamic potentials with respect to pairs of their natural variables are independent of the order in which they are carried out. This, for instance, justifies that the susceptibility tensor is symmetric. The most relevant are those involving the entropy. For instance, from dG, equalities of the form

$$\frac{\partial^2 G}{\partial \boldsymbol{y}_i \partial T} = \frac{\partial^2 G}{\partial T \partial \boldsymbol{y}_i} \Rightarrow \left(\frac{\partial S}{\partial \boldsymbol{y}_i}\right)_{T,\{\boldsymbol{y}_{j\neq i}\}} = \left(\frac{\partial \boldsymbol{X}_i}{\partial T}\right)_{\{\boldsymbol{y}_j\}} \qquad (1.14)$$

are obtained. Similar expressions are also obtained from dF, which involve derivatives with respect to the generalized displacements instead of generalized forces.

1.1.1 Equilibrium and stability

General equilibrium conditions are based on the second law of thermodynamics that states that entropy cannot decrease in a system confined by adiabatic walls, which are those that do not allow heat exchange with the surroundings.[4] Therefore, this can be interpreted in the sense that entropy must decrease or remain constant under any virtual displacement consistent with the adiabatic constraint that moves away the system from the equilibrium state. For small displacements about equilibrium, this condition of maximum of the entropy can be expressed as

$$(\delta S)_{[U,\{\boldsymbol{X}_i\},\{n_i\}]} \leq 0, \qquad (1.15)$$

or, alternatively

$$(\delta S)_{[H,\{\boldsymbol{y}_i\},\{n_i\}]} \leq 0, \qquad (1.16)$$

where $H = U - \sum_{i=1}^{n} \boldsymbol{y}_i \cdot \boldsymbol{X}_i$ is the enthalpy. It is easy to show that these inequalities are equivalent to those expressing the minimum of both U and

---

[4] Adiabatic walls are necessarily non-permeable and do not allow exchange of matter. However, we will assume that they can allow exchange of work.

H at equilibrium. That is, $(\delta U)_{[S,\{X_i\},\{n_i\}]} \geq 0$ and $(\delta H)_{[S,\{y_i\},\{n_i\}]} \geq 0$, respectively. It is in fact straightforward to see that all energy potentials show minima at equilibrium, while all entropy potentials show maxima at equilibrium.[5]

It is worth noting that in the preceding inequalities, the symbol δ was used to represent small but not infinitesimal displacements. Therefore, the change corresponding to any thermodynamic potential can be expanded as a power series of its natural variables about equilibrium. For instance, if, for the sake of simplicity, we consider a closed system that can exchange only one kind of work ($\omega = 1$) associated with a scalar generalized displacement, X, then δU can be written as,

$$
\delta U = \left(\frac{\partial U}{\partial S}\right)_X \delta S + \left(\frac{\partial U}{\partial X}\right)_S \delta X
$$
$$
+ \frac{1}{2}\left\{ \left(\frac{\partial^2 U}{\partial S^2}\right)_X (\delta S)^2 + 2\frac{\partial^2 U}{\partial S \partial X}\delta S \delta X + \left(\frac{\partial^2 U}{\partial X^2}\right)_S (\delta X)^2 \right\}
$$
$$
+ \dots . \tag{1.17}
$$

Therefore, to first order in small displacements, we must have $(\delta^{(1)} U)_{S,X} = 0$, which states that U is an extremum in equilibrium. From this condition, considering an arbitrary division of the systems into two parts 1 and 2 such that $U = U_1 + U_2$, $S = S_1 + S_2$ and $X = X_1 + X_2$, it follows that

$$
\delta^{(1)} U = \frac{\partial U}{\partial S_1}\delta S_1 + \frac{\partial U}{\partial S_2}\delta S_2 + \frac{\partial U}{\partial X_1}\delta X_1 + \frac{\partial U}{\partial X_2}\delta X_2, \tag{1.18}
$$

where derivatives must be computed about equilibrium. Taking into account that the displacement is performed at constant S and X, then $\delta S_1 = -\delta S_2$ and $\delta X_1 = -\delta X_2$ must be satisfied, and it is obtained that

$$
\delta^{(1)} U = (T_1 - T_2)\,\delta S_1 + (y_1 - y_2)\,\delta X_1 = 0. \tag{1.19}
$$

From the preceding expression, it is immediately deduced that T and y must be homogeneous along the system in equilibrium.[6] This result can be immediately generalized to the case of open multicomponent systems with multiple work properties. In equilibrium, temperature and all generalized forces and chemical potentials must be homogeneous along the system. These conditions are local conditions that apply to any extremum of the

---

[5] This condition supposes that entropy potentials are concave functions, while thermodynamic potentials are convex functions about equilibrium.

[6] It has been tacitly assumed that force fields that can vary appreciably from point to point in the system are not considered here. A well-known example of such a situation is fluids embedded in a gravitational field that can induce a pressure gradient (this effect is neglected when the vertical dimension of the container of the fluid is small enough).

thermodynamic potentials. To ensure that the equilibrium is stable, the sign of derivatives of higher order must be considered. If second-order derivatives of thermodynamic potentials are non-zero, which occur far from phase transitions, the condition of stability will be given by $\delta^{(2)}U \geq 0$, which states that U is a convex function that shows a minimum in equilibrium. In the same simple case analysed before, this condition can be expressed as

$$\left(\frac{\partial^2 U}{\partial S^2}\right)_X (\delta S)^2 + 2\frac{\partial^2 U}{\partial S \partial X}\delta S \delta X + \left(\frac{\partial^2 U}{\partial X^2}\right)_S (\delta X)^2 \geq 0. \qquad (1.20)$$

Therefore, the preceding quadratic differential form must be positive definite. Since this must hold for arbitrary variations δS and δX, stability conditions can be expressed requiring that all principal minors of the Hessian determinant must be positive. Then, it is obtained that the following inequalities must be satisfied,

$$\left(\frac{\partial^2 U}{\partial S^2}\right)_X = \frac{T}{C_X} \geq 0, \qquad (1.21)$$

$$\left(\frac{\partial^2 U}{\partial S^2}\right)_X \left(\frac{\partial^2 U}{\partial X^2}\right)_S - \left(\frac{\partial^2 U}{\partial S \partial X}\right)^2 = \frac{T}{\chi C_X} \geq 0. \qquad (1.22)$$

Therefore, stability requires that the heat capacity $C_X \geq 0$ and the susceptibility $\chi \geq 0$. Since it can be shown that the heat capacity $C_y \geq C_X$, the condition $C_y \geq 0$ is satisfied as well.

These results can be generalized to multicomponent systems that can exchange multiple kinds of work with the surroundings. In this case, among other conditions involving cross-susceptibilities, it is obtained that heat capacity and all susceptibilities must be positive (or zero). Also, chemical potentials of all c components must satisfy

$$\left(\frac{\partial \mu_i}{\partial n_i}\right)_{S,\{\mathbf{X}\}_j} \geq 0. \qquad (1.23)$$

1.1.2 Phase coexistence: the Clausius–Clapeyron equation

In first-order phase transitions, phases with different properties can coexist in equilibrium, which reflects the fact that all extensive properties show a discontinuity at the transition. In this case, each phase can be treated as an homogeneous thermodynamic subsystem, and thus the whole system is assumed to be constituted of the sum of these subsystems or phases that can exchange heat, work and matter. Instead, in continuous transitions, extensive properties vary continuously across the critical point. Therefore, the singularity that characterizes the phase transition is in this case more subtle than in first-order transitions and determines the critical behaviour that will be discussed in detail afterward.

To analyse phase coexistence a bit deeper, consider the simpler case of a single component substance characterized by a single scalar extensive property X that undergoes a first-order transition. The phase diagram is a chart that shows the regions where the distinct accessible phases can occur in equilibrium as a function of thermodynamic variables. In the case considered, it is customary to represent the lines of coexistence between pairs of phases in a T-y diagram, where y is the force conjugated to X. Indeed, these lines separate the regions of existence of pure phases. This is a simple situation in which the effect of chemical composition need not be taken into account. The prototypical example is an isotropic fluid where vapour, liquid and solid phases can coexist in equilibrium. In this case, the property X is volume and the conjugated thermodynamic force is pressure. Therefore, assume that $y = y(T)$ is a two-phase coexistence line or phase boundary that ends at a critical point (y_c, T_c).[7] Equilibrium conditions must be satisfied along this line, which impose that temperature, generalized force y, and chemical potential μ must be homogeneous along the two phases, 1 and 2, that coexist in equilibrium. Then, the Gibbs–Duhem equation (Eq. 1.10) allows to write

$$s_1 dT + x_1 dy = s_2 dT + x_2 dy, \tag{1.24}$$

where s_i and x_i $(i = 1, 2)$ are entropy and property X per mole unit in each phase. Therefore, the slope of the coexistence line is given by

$$\frac{dy}{dT} = -\frac{\Delta s}{\Delta x}. \tag{1.25}$$

This is the Clausius–Clapeyron equation. In this equation, Δs and Δx quantify, respectively, the discontinuities at the transition. The change in both entropy and volume decreases as the system approaches the critical point and vanishes at this point, where there is no distinction between the two phases. At this point, second derivatives of thermodynamic potentials vanish and the stability conditions must be determined with higher-order derivatives. Since third-order derivatives must also vanish to ensure that the critical point is not an inflection point, stability must be studied with the sign of fourth-order derivatives. In particular, the free energy F must satisfy,

---

[7] The fact that a coexistence line ends at a critical point is not imposed by thermodynamics. As will be discussed later, this is determined by symmetry conditions satisfied by the system.

$$\left(\frac{\partial^2 F}{\partial X^2}\right)_T = 0 \Rightarrow \left(\frac{\partial y}{\partial x}\right)_T = 0, \tag{1.26}$$

$$\left(\frac{\partial^3 F}{\partial X^3}\right)_T = 0 \Rightarrow \left(\frac{\partial^2 y}{\partial X^2}\right)_T = 0. \tag{1.27}$$

This means that the critical point is an inflection point of the equation of state, $y = f(T, x)$, of the system.

It is interesting to discuss the coexistence of phases in multicomponent systems constituted of c components. To this purpose, assume that the system can only exchange one kind of work associated with the scalar property X. Suppose also that at a given temperature T and generalized field y, p phases can coexist in equilibrium. In this case, equilibrium requires that for all components, the chemical potential must be the same for all phases, that is,

$$\mu_i^{(1)} = \mu_i^{(2)} = \dots = \mu_i^{(p)}, \ \forall \ i = 1, \dots, c. \tag{1.28}$$

Chemical potentials can be expressed as functions of T, y and the mole fractions of each component in each phase, $x_i^{(r)} = n_i^r / \sum_i n_i^r$, where r is the index that indicates the phase. The solution of the problem will be given in terms of the composition of each phase at given values of T and y. This requires solving $c(p-1)$ equations with $2 + p(c-1)$ independent variables, where it has been taken into account that for each phase, $\sum_{i=1}^{c} x_i^{(r)} = 1$. When the number of equations exceeds the number of independent variables, no solution can exist, which means that no equilibrium with so many phases is possible. If equations and independent variables are exactly equal, a unique solution exists. Finally, when the number of phases is small and there are more independent variables than phases, equilibrium is possible for a manifold of states. The number of degrees of freedom, f, defined as the difference between the number of independent variables and the number of equations, is given by $f = 2 + c - p$. Then, it follows that the number of phases plus the number of degrees of freedom must exceed the number of components by two for the coexistence of phases in equilibrium to be possible. This result is usually known as the Gibbs phase rule, which provides the limits on the complexity of phase diagrams. Note that for a pure substance $(c = 1)$, the maximum number of phases that can coexist in equilibrium is $p = 3$ with $f = 0$. This in turn corresponds to triple points where T and y are fixed.

Gibbs phase rule can be generalized to systems with multiple properties that can exchange ω kinds of work with the surroundings [6]. In this case, denoting $W = \omega + c$ as the number of independent conjugated thermodynamic pairs in the fundamental equation, it is obtained that $f = W - p + 1$, which leads to the general version of the Gibbs phase rule that states the number of phases plus the number of degrees of freedom must exceed the number of conjugated thermodynamic pairs by one for the coexistence of phases in equilibrium to be possible.

1.2 Extensivity and thermodynamics of small systems

In the preceding introduction to thermodynamics, it has been implicitly assumed that systems are macroscopic and properties such as entropy and generalized displacement satisfy the properties of extensivity, which, strictly speaking, can only be satisfied by infinitely large systems. Actually, this property applies to systems that can be considered as composed of many subsystems all of which can be considered macroscopic as well. Therefore, suppose a system $\mathcal{S}$ which comprises the union of $\mathcal{N}$ subsystems, that is,

$$\mathcal{S} = \mathcal{S}^{(1)} \cup \mathcal{S}^{(2)} \cup ... \cup \mathcal{S}^{(\mathcal{N})}. \tag{1.29}$$

Subsystems should be, in general, separated by walls that impose restrictions on the exchange of heat, work or matter between them. Then, a global property $\boldsymbol{X}$ defined for the whole system $\mathcal{S}$ is said to be additive if it satisfies that

$$\boldsymbol{X} = \boldsymbol{X}^{(1)} + \boldsymbol{X}^{(2)} + ... + \boldsymbol{X}^{(\mathcal{N})}, \tag{1.30}$$

where $\boldsymbol{X}^{(i)}$ is the property $\boldsymbol{X}$ corresponding to the subsystem i. In general, this property applies if subsystems are large enough and surface effects associated with the walls may be safely neglected. In an homogeneous system in equilibrium, a property that satisfies additivity is said to be extensive. If temperature T, generalized forces $\{y_i\}$ and the number of moles of the components $\{n_i\}$ are chosen as independent variables, formally extensivity supposes that the property X scale with system size as

$$\boldsymbol{X}(T, \{\boldsymbol{y}_i\}, \{\lambda n_i\}) = \lambda \boldsymbol{X}(T, \{\boldsymbol{y}_i\}, \{n_i\}), \tag{1.31}$$

which must hold for all λ. The preceding equation states that X is an homogeneous function of degree one. Thus, the Euler theorem enables us to write

$$\boldsymbol{X}(T, \{\mathbf{y}_i\}, \{n_i\}) = \sum_{i=1}^{c} n_i \left(\frac{\partial \boldsymbol{X}}{\partial n_i}\right)_{[T, \{\boldsymbol{y}_i\}, n_{j \neq i}]}. \tag{1.32}$$

Defining partial molar properties x_i as

$$x_i \equiv \left(\frac{\partial \boldsymbol{X}}{\partial n_i} \right)_{[T, \{\boldsymbol{y}_i\}, n_{j \neq i}]}, \tag{1.33}$$

$\boldsymbol{X}$ can be expressed as, $\boldsymbol{X} = \sum_{i=1}^{c} n_i \boldsymbol{x}_i$. This is particularly interesting in the case of the Gibbs potential that has $T, \{\boldsymbol{y}_i\}, \{n_i\}$ as natural variables. In this case, the Gibbs molar properties are precisely the chemical potentials $\{\mu_i\}$. Thus, in this case,

$$G = U - TS - \sum_{i=1}^{\omega} \boldsymbol{y}_i \cdot \boldsymbol{X}_i = \sum_{j=1}^{c} n_i \mu_i. \tag{1.34}$$

The Gibbs–Duhem equation is then obtained by differentiating the preceding expression. This result confirms that Gibbs–Duhem equation is a consequence of extensivity, which means that when extensivity applies, temperature, generalized fields and chemical potentials are intensive local variables independent of the system size.

When dealing with small thermodynamic systems comprising a small number of atoms as, for instance, nanoscale systems, the validity of the standard formulation of thermodynamics must be questioned. First, because surface effects may play a relevant role and thus the property of extensivity might not apply. Second, because the fluctuations of thermodynamic variables about their average value may be large. Both effects have a crucial influence on phase transitions, which can only occur as such in the thermodynamic limit.

A number of approaches have been proposed aimed at generalizing thermodynamics to such situations. Gibbs [7] already discussed the effect of surfaces and surface curvature in thermodynamics and the theory was later extended by Tolman [8]. An interesting approach that takes into account surface and large fluctuation effects has been proposed by Hill [9]. Let us briefly introduce Hill's ideas. Consider a small system with volume V that, for the sake of simplicity, will be considered as the only scalar property relevant to characterize the state of the whole system. Hill considered an ensemble of $\mathcal{N}$ independent replicas of a fixed volume of such a system that can exchange heat and matter with a (macroscopic) reservoir characterized by a temperature T and chemical potential μ. Therefore, internal energy and number of particles of the members of the ensemble must be considered as fluctuating quantities. Hill's basic hypothesis is that for large enough

$\mathcal{N}$, this ensemble follows the laws of macroscopic thermodynamics. Then a *thermodynamic identity* for the ensemble can be written as

$$dU_e = TdS_e - p\mathcal{N}dV + \mu dn_e + \Pi d\mathcal{N}, \tag{1.35}$$

where U_e, S_e and n_e are the internal energy, entropy and number of particles of the whole ensemble of replicas, which are functions of the independent variables, T, V and μ, respectively.[8] The so-called replica energy Π is given by

$$\Pi = \left(\frac{\partial U_e}{\partial \mathcal{N}}\right)_{[T,V,\mu]}. \tag{1.36}$$

The term $\Pi d\mathcal{N}$ can thus be interpreted as the work required to increase the volume of the ensemble by adding a new member. This is in contrast with the term $-p\mathcal{N}dV$ that is related to the work associated with changes in the volume of the ensemble by changing the volume of each replica. Integration of Eq. 1.35 at constant T, V, μ and Π leads to

$$U_e = TS_e + \mu n_e + \Pi\mathcal{N}. \tag{1.37}$$

Average values of the internal energy, entropy and number of particles of a representative member of the ensemble can be defined as

$$U(T,V,\mu) \equiv \frac{U_e(T,V,\mu,\mathcal{N})}{\mathcal{N}}, \tag{1.38}$$

$$S(T,V,\mu) \equiv \frac{S_e(T,V,\mu,\mathcal{N})}{\mathcal{N}}, \tag{1.39}$$

$$n(T,V,\mu) \equiv \frac{n_e(T,V,\mu,\mathcal{N})}{\mathcal{N}}. \tag{1.40}$$

Replacing these expressions into Eq. 1.37, it is obtained that

$$U(T,V,\mu) = TS(T,V,\mu) + \mu n(T,V,\mu) - \pi(T,V,\mu)V, \tag{1.41}$$

which is an equation formally equivalent to Eq. 1.34 that expresses extensivity. Note, however, that in the present equation the generalized field is not the pressure but instead the so-called *integral pressure*, π. Differentiating the preceding equation and combining with Eq. 1.35 after expressing U_e, S_e, and n_e as a function U, S and n for a given value of $\mathcal{N}$, the corresponding Gibbs–Duhem equation for the small system is obtained as

$$d[\pi V] = SdT + pdV + nd\mu, \tag{1.42}$$

[8] It is important to note that this ensemble thermodynamic identity is not equivalent to the thermodynamic identity for macroscopic extensive systems, since the thermodynamics of small systems depends on the choice of environmental control variables. Therefore, changing the set of independent variables cannot be done via Legendre transforms [10].

where πV is often denoted as the subdivision potential. Note that the relationship between p and π is given by

$$p = \left(\frac{\partial[\pi V]}{\partial V}\right)_{T,\mu} = \pi + V\left(\frac{\partial \pi}{\partial V}\right)_{T,\mu}. \qquad (1.43)$$

It is then convenient to define a function $\phi \equiv (\pi - p)V$ that satisfies

$$d\phi = SdT - Vdp + nd\mu, \qquad (1.44)$$

which indicates that in the thermodynamic limit, when π coincides with pressure p the Gibbs–Duhem equation for an extensive system is recovered. An important consequence of this result is the fact that there is no exact analogue of the Clausius–Clapeyron equation (Eq. 1.25 with $x = V/n$ and $y = -p$) for small systems. As a matter of fact, this can be understood by taking into account that isotherms in a small system are smooth analytical curves with no sharply defined end point of the coexistence region.

1.3 Simple microscopic lattice models. The Ising model

Lattice models are simple microscopic models, which are especially adapted to study phase transitions in solid materials.[9] In general, a lattice model is any model defined in a d-dimensional lattice (or network), not necessarily regular, that contains N nodes or lattice sites, so that each site is occupied by a microscopic variable that can be in a number of discrete or continuous microscopic states. These variables are supposed to interact according to given rules that define the hamiltonian of the model. This class of models represents a simplified description of certain physical systems that incorporate only those ingredients that are essential to describe phase transitions that take place associated with given degrees of freedom.

It is common and convenient to describe this class of models using magnetic language. Then, the variables defined on each lattice sites are called spin variables. Among the wide variety of lattice models that have been proposed, the best known and the most paradigmatic is the so-called Ising model. In this model, a classical two-state spin variable S_i is defined at each node of a lattice, which is usually regular, that can take values, $S_i = \pm 1$ corresponding to the spin-up and spin-down states, respectively. For this

---

[9] In spite of being specially adequate for solids, similar models can also be formulated to study fluids. In that case, there is no underlying lattice but it is possible to divide the fluid into cells of a given microscopic volume. It is convenient to choose this volume in such a way that only one particle can occupy one cell. Therefore, cells may be either in an occupied or an empty state.

reason, often the Ising model is classified as a spin-1/2 lattice model. The hamiltonian of the Ising model is

$$\mathcal{H}(\{S_i\}) = -J \sum_{\langle ij \rangle_{nn}} S_i S_j - h \sum_{i=1}^{N} S_i, \qquad (1.45)$$

where the first sum extends over all nearest neighbour pairs of spins given by $\frac{1}{2}zN$. Here z is the coordination number or number of nearest neighbours of a given lattice site, which is an intrinsic property of the lattice. This term determines the interaction between the spin variables and therefore J represents a measure of the interaction energy. $J > 0$ favours neighbouring spins to align parallel to each other and therefore allows a ferromagnetic order to be established. Instead, $J < 0$ favours that the neighbouring spins align antiparallel and, thus, the possibility that an antiferromagnetic order be established. The second sum of the hamiltonian takes into account the possible interaction of the spins with an external field, h. It is the so-called Zeeman term. It is interesting to note that the model does not include a term associated with the kinetic energy of the constituents (spins). Consequently, collective excitations of the lattice are not considered and, thus, the model is only adequate to study static properties. This represents a quite drastic simplification that anyhow captures the essential physics of the problem and, as will be shown, is sufficient to deal with the study of phase transitions, which are a consequence of the interaction between spins. Note that it is possible to extend this class of models to include the possibility of dealing with spin dynamics effects by, for instance, simulating collective excitations by means of a heat bath [11]. Moreover, these models can also be extended with the aim of taking into account more complex effects such as long-range dipolar-like interactions [12].

Within the framework of magnetism, the Ising model can be understood as the limiting case of the Heisenberg model of magnetism. This can be seen taking into account that Dirac showed that in materials with electrons localized in orthogonal orbitals, the effect of the Pauli principle can be taken into account by adding to the hamiltonian a term of the type $-\sum_{i<j} J_{ij}(\frac{1}{2} + 2\mathbf{S}_i \cdot \mathbf{S}_j)$, where the sum extends over all pairs of spins in the system. This term suggests that the spin-dependent contribution to the energy can be written as the following pair interaction quantum operator

$$\mathcal{H}_H = -2 \sum_{i<j} J_{ij} \mathbf{S}_i \cdot \mathbf{S}_j, \qquad (1.46)$$

which is usually known as the Heisenberg hamiltonian. Here, J_{ij} is an exchange energy integral associated with the exchange of electrons between states i and j, and $\mathbf{S}$ is a spin operator related to the magnetic moment $\boldsymbol{\mu} = g\mu_B\mathbf{S}$, where g is the Landé factor[10] and μ_B the Bohr magneton [104]. This model is in fact very accurate in the case of atoms with orbital angular momentum $L = 0$ and is a reasonable approximation in the case of many transition metals. It is, however, a poor approximation in the case of rare earths (except for those for which $L = 0$). In this approach, demagnetization and anisotropy effects are not taken into account.

Often the above model is treated in the classical approximation assuming that magnetic moments are vector quantities that can be continuously oriented in space. In this framework, the magnetic anisotropy can be taken into account assuming that the term $J_{ij}\mathbf{S}_i \cdot \mathbf{S}_j$ can be written in the form $J_{ij}^x S_{i_x} S_{j_x} + J_{ij}^y S_{i_y} S_{j_y} + J_{ij}^z S_{i_z} S_{j_z}$, where S_{i_α} ($\alpha = x, y, z$) are the Cartesian coordinates of the classical spin vector $\boldsymbol{S}_i$. If it is assumed, for example, that the exchange interaction in the z-direction is different from the interaction in the x-y plane, the hamiltonian of the system can be written in the form

$$\mathcal{H} = -J \sum_{\langle ij \rangle_{nn}} aS_{i_z}S_{j_z} + b(S_{i_x}S_{j_x} + S_{i_y}S_{j_y}). \tag{1.47}$$

Indeed, the Ising limit corresponds to the strong uniaxial limit for which $a \simeq 1$ and $b \simeq 0$. In this case, the spin variable can be treated as a scalar that can take only two values, ± 1. Another interesting model corresponds to the limit of strong planar anisotropy. In this case, it is assumed that $a \simeq 0$, and $b \simeq 1$. This model is called the XY-model or planar Heisenberg model.

1.3.1 The Ising model for non-magnetic systems

A particularly interesting aspect of the Ising model is the fact that, through a reinterpretation of spin variables, it can be easily adapted to describe systems characterized by a two-state microscopic property that can be of a very different nature than that of the spin giving rise to the magnetic moment. A well-known example is that of binary alloys. In that case, lattice sites are occupied by atoms of species A or B so that a spin variable can be defined in each lattice site that takes values ± 1 when the site is occupied by an atom A or B, respectively.

Suppose, in general, a system is such that each lattice site is characterized by a constituent that can be in two states denoted as 1 and 2. Of course, this can correspond to a spin that can be in a state up or in a state down, or to an atom that can be of the species A or B, among other examples. In all

---

[10] The Landé factor is given as, $g = 1 + \frac{J(J+1)+S(S+1)-L(L+1)}{2J(J+1)}$, where S is the spin, L the orbital angular moment and J the total angular moment. Note that when $L = 0$, $g = 2$.

these systems, if pair interactions between nearest neighbour constituents are assumed, the configuration energy of the system can be written in the form

$$E_c = N_{11}\varepsilon_{11} + N_{22}\varepsilon_{22} + N_{12}\varepsilon_{12}, \tag{1.48}$$

where $\varepsilon_{\alpha\beta}$ $(\alpha, \beta = 1, 2)$ are interaction energies of pairs α-β. Introducing occupation variables $P_i^{(\alpha)}$ defined as $P_i^{(\alpha)} = 1$ if the constituent in the lattice site i is in the state α and 0 otherwise, the configurational energy can be expressed as

$$E_c = \sum_{\langle ij \rangle_{nn}} \sum_{\alpha, \beta} \varepsilon_{\alpha\beta} P_i^{(\alpha)} P_j^{(\beta)}. \tag{1.49}$$

Now a spin variable S_i can be defined in each lattice site in such a way that $S_i = +1$ if the constituent in the lattice site i is in the state 1 and $S_i = -1$ if it is in the state 2. Then, the occupation variables can be given as, $P_i^{(1)} = \frac{1}{2}(S_i + 1)$ and $P_i^{(2)} = \frac{1}{2}(S_i - 1)$. Replacing these expressions in Eq. 1.49, the following Ising-like hamiltonian is obtained

$$\mathcal{H}(\{S_i\}) = E_0 - \epsilon \sum_{\langle ij \rangle_{nn}} S_i S_j - V \sum_i S_i. \tag{1.50}$$

In the preceding equation $E_0 = \frac{1}{4}Nz(\varepsilon_{11} + \varepsilon_{22} + 2\varepsilon_{12})$, $\epsilon = -\frac{1}{4}(\varepsilon_{11} + \varepsilon_{22} - 2\varepsilon_{12})$ and $V = \frac{1}{2}z(\varepsilon_{22} - \varepsilon_{11})$, where z is the lattice coordination number, N is the number of lattice sites and E_0 is a constant. The second term of the hamiltonian is the relevant term that describes the interaction between the constituents. Here, ϵ is the parameter that quantifies this interaction and must be identified with the magnetic exchange energy J in the case of magnetic systems. It is interesting to note that, as in the magnetic case, this parameter represents, in general, an exchange energy since it can be interpreted as the difference between having nearest neighbour constituents in the same state (1-1 or 2-2), or in different states (1-2 or 2-1). Therefore, if $\epsilon > 0$, each constituent will prefer to be surrounded by nearest neighbour constituents in the same state and, thus, the model will be suitable to describe ferromagnetism in the case of magnetism or phase separation for atom-like constituents. On the other hand, in the case $\epsilon < 0$, each constituent will prefer to be surrounded by constituents in a different state and the model will be suitable to describe antiferromagnetism in the case of magnetic systems and order–disorder transitions in other cases. The third term, which must not be identified with the Zeeman term in the magnetic Ising model, is also a constant. It must be zero in the case of magnetic systems since, for symmetry reasons, it is expected that a pair of spin-up and a pair of spin-down constituents have the same interaction, and can be non-zero in other cases. Besides this difference, it is important to remark that magnetic systems comprising spins and atom-like systems are not strictly equivalent since in the former case the number of constituents in one state is not conserved, while it is conserved in the second case. For instance, the number of A and B atoms are conserved quantities in a binary alloy, while the number of spin-up and spin-down constituents are not.[11] We will see that, strictly speaking, from a Statistical Mechanics point of view, the two models are only equivalent when the magnetic model is treated in the canonical ensemble and the alloy (and, in general, atom-like mixtures) in the grand-canonical ensemble. In this case, it is necessary to add to the hamiltonian of the alloy a term of the type $-\mu \sum_i S_i$, where μ is the difference of chemical potentials of the two species A and B. This is the term that plays the same role as the Zeeman term in the magnetic Ising model.

The procedure discussed to obtain an Ising-like hamiltonian in systems with two-state constituents can be easily generalized to systems comprising n-state constituents. For instance, in the case of a three-state system a spin variable that takes values $S_i = +1, 0, -1$ corresponding to states 1, 2 and 3 respectively, can be defined such that, $P_i^{(1)} = \frac{1}{2}(S_i^2 + S_i)$, $P_i^{(2)} = \frac{1}{2}(1 - S_i^2)$ and $P_i^{(3)} = \frac{1}{2}(S_i^2 - S_i)$. Then, the corresponding spin-1 hamiltonian reads

$$\mathcal{H}(\{S_i\}) = -\epsilon \sum_{\langle ij \rangle_{nn}} S_i S_j + K \sum_{\langle ij \rangle_{nn}} S_i^2 S_j^2 + L \sum_{\langle ij \rangle_{nn}} S_i^2 S_j$$

$$+ M \sum_{i=1}^{N} S_i^2 + V \sum_{i=1}^{N} S_i + V_0. \qquad (1.51)$$

This model is often called the Blume–Emery–Griffiths model since these authors proposed for the first time a reduced version of the spin-1 model to study He^3-He^4 mixtures [14] (see Exercise 1.10). In the preceding hamiltonian, $\epsilon = -\frac{1}{4}(\varepsilon_{11} + \varepsilon_{33} - 2\varepsilon_{13})$, $K = \frac{1}{4}(\varepsilon_{11} + \varepsilon_{33} + 2\varepsilon_{23}) + \varepsilon_{22} - \varepsilon_{12} - \varepsilon_{32}$, $L = \frac{1}{2}(\varepsilon_{11} - \varepsilon_{33}) - (\varepsilon_{12} - \varepsilon_{32})$, $M = \frac{1}{3}(\varepsilon_{12} + \varepsilon_{32} - 2\varepsilon_{22})$, $V = \frac{1}{2}z(\varepsilon_{12} - \varepsilon_{32})$ and $V_0 = \frac{1}{2}Nz\varepsilon_{22}$, where z and N are the lattice coordination number and number of constituents, respectively.

---

[11] A spin-up can flip to spin-down while an atom of a given species cannot turn into an atom of another species.

1.4 Statistical mechanics of lattice models and the mean-field approximation

Consider an Ising model for a system with N constituents in contact with a thermal bath at temperature T. In the canonical ensemble, the partition function is given by

$$Q(\beta) = \sum_{\{S_i\}} e^{-\beta \mathcal{H}(\{S_i\})}, \qquad (1.52)$$

where $\beta = 1/k_B T$ (k_B is the Boltzmann constant) and $\mathcal{H}(\{S_i\})$ is the Ising hamiltonian (with or without an external applied field). The sum must be performed over all the spin configurations. This number of configurations depends obviously on whether there is a conservation of the number of constituents in a given state or not. The relevance of this conservation is perhaps better reflected if occupation numbers are used to express the partition function. To be specific, let us first consider the case of a magnetic system without conservation of the numbers N_+ and N_- of spins in the up and down states, respectively, which correspond to states 1 and 2. The Ising hamiltonian (Eq. 1.45) can then be expressed as

$$\mathcal{H} = -J(N_{++} + N_{--} - N_{+-}) - h(N_+ - N_-), \qquad (1.53)$$

where N_{++}, N_{--} and N_{+-} are the number of nearest neighbour pairs of up-up, down-down and up-down spins. Taking into account that the total number of pairs is $\frac{1}{2} z N$, and that $N_+ + N_- = N$, it is straightforward to see that $\mathcal{H}$ can be expressed in terms of two independent occupation numbers, N_+ and N_{++} as

$$\mathcal{H} = -J \left(4N_{++} - 2z N_+ + \frac{1}{2} z N \right) - h(2N_+ - N). \qquad (1.54)$$

Note that N_+ and N_{++} can be given in terms of the probabilities of finding a spin-up and a pair of nearest neighbours of spin-up, given, respectively, as, $p_+ = N_+/N$ and $p_{++} = 2N_{++}/z N$.

Then, the canonical partition function can be given by

$$Q(\beta, h) = e^{\beta N(\frac{1}{2} z - h)} \sum_{N_+=0}^{N} e^{-\beta(zJ - h)N_+} \sum_{\{N_{++}\}} g(N_+, N_{++}) e^{4\beta J N_{++}}, \quad (1.55)$$

where the second sum must be performed over all pairs N_{++} consistent with the number N_+ of spins in the state up. Here, $g(N_+, N_{++})$ is the number of different configurations that can be established with the same numbers N_+ and N_{++}.

Consider now the case of a binary alloy as an example of a system with conservation of the number of constituents in a given state. The states 1 and 2 are now denoted as A and B corresponding to the two species of atoms. In this case, the canonical partition function can be written as

$$Q(\beta) = e^{(\frac{1}{2}\beta zN - 2z\epsilon N_A)} \sum_{\{N_{AA}\}} g(N_A, N_{AA})e^{4\epsilon JN_{AA}}, \qquad (1.56)$$

where, for the sake of simplicity, a symmetric alloy has been considered for which, $\varepsilon_{AA} = \varepsilon_{BB}$. Note that in this case there is no sum over N_A, which is a fixed quantity. Indeed, the sum over pairs N_{AA} must be performed consistently with the number N_A of A-atoms. We can overcome the conservation restriction by computing the partition function in the grand canonical ensemble, which is given as

$$\mathcal{Q}(\beta, \mu) = e^{\beta N(\frac{1}{2}z - \mu)} \sum_{N_A=0}^{N} e^{-2\beta(z\epsilon - \mu)N_A} \sum_{\{N_{AA}\}} g(N_A, N_{AA})e^{4\epsilon JN_{AA}}, \qquad (1.57)$$

where μ is the difference between chemical potentials of A and B atoms. Comparing expressions given by Eqs. 1.55 and 1.57, it is clear that the grand canonical partition function of the binary alloy is equivalent to the canonical partition function of the magnetic system. In addition, this corroborates that μ plays the role of the magnetic field in the case of the binary alloy.

1.4.1 Mean-field approximation

As we have already discussed, lattice models represent very simplified representations of physical systems that undergo phase transitions. Despite the crude approximation to reality, the calculation of the corresponding partition function is very difficult and has only been obtained in very few cases. In the case of the Ising model, exact solutions are only known for $1d$ and $2d$ models [15]. In general, it will be necessary to use some kind of approximation to find a solution that at least provides us with a qualitatively reasonable description of the thermodynamic behaviour of the system under consideration. The most common approximation is the mean-field approximation, which, in general, can be understood as a self-consistent, variational approach in which correlations are treated in an approximate manner. As pointed out by Kadanoff [16], mean-field theory provides a partial, and partially imprecise, answer to the problem of phase transitions, but it is important to remark that *partially imprecise* means that it is partially right. There are several ways to perform this class of approximations, which are not all equivalent.[12]

---

[12] As we will discuss later, all of them give the same description of critical behaviour.

One of the most convenient methods is the Bragg–Williams approach, introduced by Bragg and Williams to study the order–disorder transition in the β-(Cu-Zn) brass [17]. In its original version, they introduced the method using occupation variables. We will begin by discussing it starting from the Ising hamiltonian expressed in spin variables [18]. The partition function is then given by Eq. 1.52. We will begin by assuming that $J > 0$ and define the magnetization m per spin of a given configuration as

$$m \equiv \frac{1}{N} \sum_{i=1}^{N} S_i = \frac{N_+}{N} - \frac{N_-}{N} = 2p_+ - 1, \tag{1.58}$$

which is a convenient order parameter for a ferromagnetic system. We can now rewrite the partition function in the form

$$Q(\beta, h) = \sum_{\{m'\}} \sum_{\{r\}} e^{-\beta E_r}, \tag{1.59}$$

where the second sum over $\{r\}$ is performed over all spin configurations with the same value of m'. Therefore, E_r are the energies of these configurations. If $g(m')$ is the number of configurations corresponding to a given value of m', the average energy of these configurations can be obtained as

$$\bar{E}(m') = \frac{1}{g(m')} \sum_{\{r\}} E_r. \tag{1.60}$$

Then, the exponential term of the partition function can be written as $e^{-\beta E_r} = e^{-\beta \bar{E}} e^{-\beta(E_r - \bar{E})}$, and expanding the second exponential in power series it is obtained that

$$\begin{aligned} e^{-\beta E_r} &= e^{-\beta \bar{E}} \left\{ 1 - \frac{E_r - \bar{E}}{k_B T} + \frac{1}{2!} \left(\frac{E_r - \bar{E}}{k_B T} \right)^2 - \frac{1}{3!} \left(\frac{E_r - \bar{E}}{k_B T} \right)^3 + \dots \right\} \\ &= e^{-\beta \bar{E}} \sum_{j=0}^{\infty} \left(\frac{-1}{k_B T} \right)^j \frac{(E_r - \bar{E})^j}{j!}. \end{aligned} \tag{1.61}$$

Defining the moment of j-order of the energy distribution as

$$M_j \equiv \frac{1}{g(m')} \sum_{\{r\}} (E_r - \bar{E})^j, \tag{1.62}$$

the term $e^{-\beta E_r}$ can be expressed as

$$e^{-\beta E_r} = g(m') e^{-\beta \bar{E}} \sum_{j=0}^{\infty} \left(\frac{-1}{k_B T} \right)^j \frac{M_j}{j!}. \tag{1.63}$$

The logarithm of the sum in the preceding expression can be given as $\ln \sum_{j=0}^{\infty}(M_j x^j)/j! = \sum_{n=1}(B_n x^n)/n!$, where $x \equiv (-1/kT)$. Differentiating both sides of this equation with respect to x yields

$$\sum_{n=1}^{\infty}\sum_{j=0}^{\infty}\frac{n}{n!j!}B_n M_j x^{n+j-1} = \sum_{j=0}^{\infty}\frac{jM_j}{j!}x^{j-1}. \tag{1.64}$$

Then, equating equal powers in x on both sides of the obtained equation leads to

$$B_1 M_0 = M_1,$$
$$B_1 M_1 + B_2 M_0 = M_0,$$
$$B_3 M_0 + 2B_2 M_1 + B_1 M_2 = M_3,$$
$$\dots, \tag{1.65}$$

and taking into account that $M_0 = 1$ and $M_1 = 0$, it is obtained that $B_1 = 0$, $B_2 = M_2$, $B_3 = M_3$, $B_4 = M_4 - 3M_2^2$, ... Therefore, the partition function can be written as

$$Q_N = \sum_{\{m'\}} g(m') \exp\left[-\frac{\bar{E}}{k_B T} + \frac{M_2}{2(k_B T)^2} + \frac{M_3}{3!(k_B T)^3} + \dots\right]. \tag{1.66}$$

Now, applying the saddle point method, the preceding sum can be approximated by its maximum term, which corresponds to $m' = m$. This value represents the magnetization per spin in the equilibrium state of the system at T and h. It is obtained from the condition of minimum of the following free energy function

$$F(m') = \bar{E} - \frac{M_2}{2k_B T} - \frac{M_3}{3!(k_B T)^2} - \dots - Nk_B T \ln g(m'). \tag{1.67}$$

Therefore, the approach must be considered as a variational method and thus, m is a solution of the equation

$$\left(\frac{\partial F}{\partial m'}\right)_{m'=m} = 0. \tag{1.68}$$

In the Bragg–Williams approximation, all moments of order higher than the first are omitted, which supposes that correlations between neighbouring spins are neglected. Therefore, taking into account this fact, the mean value $\bar{E}$ is given by

$$\bar{E}(m') = -\frac{1}{2}zJNm'^2 - Nhm', \tag{1.69}$$

which is simply obtained by substituting the spin variables S_i for their mean values in the Ising hamiltonian. This equation assumes the self-consistent condition, which imposes that the mean spin value is the same for all lattice sites.

On the other hand, $g(m)$ is the number of spin configurations that have the same order parameter m and, thus, the same average energy. Since the Bragg–Williams approximation neglects correlations between neighbouring spins, it is simply given by, $g(m') = N!/(N_+!N_-!)$. Taking into account that $N_+ = N(1+m')/2$ and $N_- = N(1-m')/2$, the following free energy function per spin is obtained:

$$f(m', T, h) = -\frac{1}{2}zJm'^2 - hm' + k_B T \left(\frac{1+m'}{2} \ln \frac{1+m'}{2} + \frac{1-m'}{2} \ln \frac{1-m'}{2} \right),$$
(1.70)

where the Stirling approximation has been used to compute $\ln g(m')$. After minimization, it is obtained that the equilibrium magnetization per spin must satisfy the following equation:

$$m = \tanh \left(\frac{zJm + h}{k_B T} \right). \tag{1.71}$$

The solution $m = m(T, h)$ of the preceding equation can be obtained numerically. For m close to zero, the hyperbolic tangent can be expanded in a power series of m. It is then easy to see that for $h = 0$, the solution is $m = 0$ above a temperature $T_c = zJ/z$. Below this temperature, this solution corresponds to a maximum of the free energy function and is thus an unstable solution. Two stable symmetric solutions occur at $\pm m_0 \neq 0$. For $h \neq 0$, a non-zero solution is obtained at all finite temperatures, which has the same sign as that of h. Thus, a phase transition cannot occur in the presence of an external field (that breaks the symmetry of the up and down spins). Therefore, the present model displays a continuous phase transition at the critical point, $h_c = 0$ and $T_c = zJ/k$ from a high-temperature paramagnetic phase ($m = 0$) to a low-temperature ferromagnetic phase ($m \neq 0$). It is easy to show that close to the critical point thermodynamic quantities show a power-law behaviour, characterized by the following exponents: $\alpha = 0$ (corresponding to a discontinuity of the heat capacity), $\beta = 1/2$, $\gamma = 1$, $\delta = 3$. These are the mean-field critical exponents that are obtained in any mean-field theory.

Equation 1.71 plays the role of the equation of state of the system. It is worth remarking that this equation of state is formally equivalent to the paramagnetic equation of state of a spin-1/2 system subjected to an effective or mean-field $h_{eff} = zJm + h$, which is a function of the mean magnetization, m. Therefore, it can be assumed to result from an effective hamiltonian of the type, $\mathcal{H}_{eff} = -\sum_i h_{eff}(m)S_i$. This point of view precisely represents the approach proposed within the molecular field theory. This approach was first introduced by Weiss [19] as an extension of the Langevin theory of paramagnetism aimed at taking into account the interaction between magnetic moments. Indeed, the obtained results are equivalent to those obtained within the Bragg–Williams approximation. It is interesting to note that the effective field can simply be obtained as the mean value of the local field h_i acting on a spin S_i, which is given as

$$h_i = -\frac{\partial \mathcal{H}(\{S_i\})}{\partial S_i} = J \sum_{j\ nn\ i} S_j + h, \tag{1.72}$$

where the sum must be performed over the z nearest neighbours of S_i. As first proposed by Bethe [20], this procedure can be generalized by treating exactly the interactions of a given spin with its nearest neighbour spins exactly and the interactions of these spins with the rest through a mean-field. From this point of view, the Bragg–Williams theory corresponds to a first-order mean-field theory, while the Bethe approximation is of second order.

The mean-field treatment discussed so far is self-consistent in the sense that it is imposed that all spins see the same effective field, which reduces the N-body original problem to a one-body or to a few-body problem, as happens in the Bethe approximation. Therefore, these approximations suppose neglecting the fluctuations of h_i that extend beyond the length scale associated with a single atom in the case of the Bragg–Williams or Weiss approximation, and to the length scale of two atoms in the case of the Bethe approximation. This is an important fact given that the fluctuations of h_i is the mechanism that couples the state of a given spin with the state of its neighbouring spins. Therefore, neglecting fluctuations results in the fact that correlations between spins are not treated correctly. Therefore, although the solution obtained is qualitatively correct, it will be inadequate when the correlations extend over great distances, as happens at critical points. This is essentially evidenced by the fact that the critical exponents of the theory are independent of space and order parameter dimensions and, in general, differ from those measured experimentally. Actually, in mean-field theories, the lattice only plays a role through the number of neighbours. This means, for instance, that assuming interactions between nearest neighbour constituents, in this approximation, the same result is obtained if the system is defined on a simple cubic $3d$ lattice ($z = 6$) or in a triangular $2d$ lattice ($z = 6$). In the particular case of the Bragg–Williams approximation, it is found that a $1d$

chain exhibits a phase transition, although this is known to be an incorrect result.[13] Notwithstanding, the mean-field exponents are correct in systems with long-range interactions between the constituents and, as we will see later when the space dimension is large enough (often this dimension is greater than 3).

In principle, one might think that it is possible to find a better approximation by considering more terms in the series of moments given by Eq. 1.62 that determines the partition function. Actually, high-order moments have been estimated by Kirkwood [21, 22], which show that taking them into account only allows to improve the prediction of the critical temperature, but the obtained critical behaviour is exactly the same than that predicted by the Bragg–Williams approximation. The problem comes from the fact that the series of moments converges very weakly and, considering only a finite number of terms in the expansion, supposes again to treat the fluctuations (and consequently the correlations) in an approximate form, which always brings to the same mean-field critical behaviour. In fact, the Kirkwood approach with a finite number of moments represents a kind of high-order mean-field approximation, which, apart from criticality, gives results which are not strictly the same as those obtained in other mean-field approximations such as the Bethe approximation.

Generalization to q-state spin lattice models

The Bragg–Williams method introduced above can be easily generalized to systems with a spin variable that can take more than two states. Consider that the spin variables S_i defined at each lattice point can take values $\{S^\alpha\}$ where $\alpha = 1, 2, ..., q$. In the Bragg–Williams approximation, spin variables in the hamiltonian must be replaced by their mean value, which, in this general case, will be given by

$$m = \langle S_i \rangle = \sum_{\alpha=1}^{q} p(S^\alpha) S^\alpha, \tag{1.73}$$

---

[13] This can be seen using the Peierls argument. This supposes a ferromagnetic Ising chain of N sites with free boundary conditions for which the ground-state energy is $E_0 = -(N-1)J$. Consider now configurations corresponding to lower-energy excitations with the first l spins of the chain up and the rest $N - l$ down. The excess of energy of such states is $E - E_0 = 2J$. Therefore, at a temperature T, the free energy change due to these excitations is $\Delta F = 2J - kT \ln(N - 1)$. In the limit $N \to \infty$, $\Delta F < 0$ for all T. This means that at any finite temperature, the disordered $m = 0$ phase is stable. Therefore, there cannot be a phase transition to a ferromagnetic state in $1d$ with finite-range interactions. A generalization of this argument to higher dimensions shows that a $2d$ Ising model should show a phase transition at a finite temperature.

where $p(S^\alpha)$ is the probability that the spin takes the value S^α, which is the same for all lattice sites, *i.*[14] The corresponding number of spin configurations for a given m is now $g(m) = N! / \prod_{\alpha=1}^{q} [Np(S^\alpha)]!$ Therefore, for a hamiltonian which has the Ising form with nearest neighbour interactions between q-state spins, the corresponding free energy function per particle will be of the form:

$$f(\{p(S^\alpha)\}) = -\frac{1}{2}zJ[\sum_{\alpha=1}^{q} p(S^\alpha)S^\alpha]^2 + k_B T \sum_{\alpha=1}^{q} p(S^\alpha) \ln p(S^\alpha), \quad (1.74)$$

which is equivalent to that given by Eq. 1.70. In this case, equilibrium probabilities should be found by minimizing the free energy function with respect to the probabilities $\{p(S^\alpha)\}$ under the normalization constraint

$$\sum_{\alpha=1}^{q} p(S^\alpha) = 1, \quad (1.75)$$

which must be done using the method of Lagrange undetermined multipliers. In some cases, due to symmetry reasons, some probabilities must be equal, which may simplify the problem. Finally, note that this method can easily be extended to the case of systems with continuous spin variables (see Exercise 1.3 as an example).

The infinite range Ising model: gaussian integral method

The infinite-range Ising model is interesting since it can be exactly solved and, in the thermodynamic limit, the solution coincides with the corresponding mean-field solution. To address this problem, consider a ferromagnetic Ising model defined on a lattice with N sites where the exchange pair interaction is J/N for all pairs, which can be interpreted as if the range of the interactions would extend to infinity. Indeed, the model may look quite artificial, but it is interesting in the following two aspects. First, because it highlights that the mean-field approach is essentially correct in real systems where long-range effects are relevant,[15] and second because the formal treatment used to obtain the solution is important in the study of spin glasses, as will be seen in Chapter 7.

In the absence of an applied field, the hamiltonian of the system considered is thus the following:

---

[14] Note that in the case of the two-sates Ising model, we have defined $\langle S_i \rangle = (\sum_i S_i)/N = (N_+/N)(+1) + (N_-/N)(-1) = p(S = +1) - p(S = -1) = p_+ - p_- = 2p_+ - 1$, where $p_+ = N_+/N$ and $p_- = N_-/N$.
[15] This is important, for instance, to justify that in systems where elasticity plays an important role or in magnetic systems with dipolar interactions mean-field provides an excellent solution.

$$\mathcal{H}(\{S_i\}) = -\frac{J}{2N} \sum_{i \neq j} S_i S_j, \tag{1.76}$$

that can be written in the form

$$\mathcal{H}(\{S_i\}) = -\frac{J}{2N} \left[\left(\sum_{i=1}^{N} S_i \right)^2 - 2 \sum_{i=1}^{N} S_i^2 \right], \tag{1.77}$$

where the term, $\sum_{i=1}^{N} S_i^2 = N$. Therefore, the partition function is given by

$$Q(\beta) = e^{-\beta J} \sum_{S_i = \pm 1} \exp \left\{ \frac{\beta J}{2N} \left(\sum_i S_i \right)^2 \right\}. \tag{1.78}$$

Now we take into account the gaussian integral identity

$$e^{ax^2/2} \equiv \sqrt{\frac{aN}{2\pi}} \int_{-\infty}^{\infty} dm'\, e^{-Nam'^2/2 + \sqrt{N}am'x}. \tag{1.79}$$

Then, the partition function can be rewritten in the form

$$Q(\beta) = e^{-\beta J} \sum_{S_i = \pm 1} \sqrt{\frac{\beta J N}{2\pi}} \int_{-\infty}^{\infty} dm'\, e^{-N\beta J m'^2/2 + \beta J \sum_i S_i}. \tag{1.80}$$

Therefore, the summation can be performed independently from the integral and it gives

$$Q(\beta) = e^{-\beta J} \sqrt{\frac{\beta J N}{2\pi}} \int_{-\infty}^{\infty} dm'\, e^{-N\beta J m'^2/2 + N \ln[2\cosh(\beta J m')]}. \tag{1.81}$$

It is clear that the problem has been reduced to a single integral problem. While the integral cannot be evaluated, taking into account that the argument of the exponential in the integrand is proportional to N, we can apply the saddle point method that provides an excellent approximation in the thermodynamic limit. Therefore, in this limit

$$Q(\beta) \simeq e^{-\beta J} \sqrt{\frac{\beta J N}{2\pi}} e^{-\beta N f(m)}, \tag{1.82}$$

where $f(m) \equiv J\beta m^2/2 - \ln[2\cosh(\beta J m)]$ and m is the value of the variable m' that maximizes $f(m')$. Thus, it is a solution of $(\partial f/\partial m')_{m'=m} = 0$, which leads to

$$m = \tanh\left(\frac{Jm}{k_B T}\right). \tag{1.83}$$

This is the same equation as Eq. 1.71 of the mean-field approximation in the case $h = 0$, with zJ replaced by $N(J/N)$. Therefore, the coordination number z in the infinite-range model is the number of spins $N - 1$ that in the thermodynamic limit is given by N. Note as well that the variable m, which has been introduced as an *artificial* variable to use the gaussian integral method, can now be identified with magnetization.

The preceding result shows that the mean-field solution is the exact solution in the case of the infinite-range Ising model. As a matter of fact, this is not a surprising result. In fact, if the term $\left(\sum_{i=1}^{N} S_i\right)^2$ in the hamiltonian Eq. 1.77 is rewritten in the form $-(J/2)\sum_{i=1}^{N} S_i(N^{-1}\sum_{j=1}^{N} S_j)$, where, in the thermodynamic limit, the sum in the parenthesis is the magnetization m. Therefore, if this term is replaced by its average value m, the problem reduces to a single-body problem, which is precisely the solution given by Eq. 1.83.

1.4.2 Antiferromagnetic case: sublattices

Let us consider now the same Ising hamiltonian as given in Eq. 1.45, but now with the exchange parameter $J < 0$. In this case, the exchange will favour an antiparallel alignment of the spins. We will assume that the lattice has a symmetry that allows a subdivision into two sublattices so that all the z nearest neighbours of a spin that belong to a sublattice are located on the other sublattice. This condition ensures that a perfect antiferromagnetic order can be established. This subdivision is possible in simple cubic (*sc*) and in body-centred cubic (*bcc*) lattices, but not in face-centred cubic (*fcc*) or triangular lattices. In these cases, geometrical frustration effects may arise that prevent that the perfect long-range antiferromagnetic order can be established [23].[16]

In the case of a *bcc* lattice, the two sublattices are interpenetrating *sc* lattices constituted of the vertex and central nodes of the lattice, respectively. This division is shown in Figure 1.1. Then the problem can be solved using the Bragg–Williams solution that we have obtained in the ferromagnetic case. If we denote the two sublattices as α and β sublattices, respectively, we can define the spin magnetization of a sublattice or staggered magnetization as,

---

[16] In a triangular lattice with nearest neighbour spins antiferromagnetically coupled, once two spins of the equilateral triangle are aligned antiparallel, the third one is frustrated because its two possible orientations, up and down, give the same energy. Thus, the third spin cannot mutually minimize its interaction energy with the other two spins. This is the geometrical frustration effect that leads to multiple ground states with the same energy, and long-range order is suppressed.

$$m_s = \frac{2}{N} \sum_{i \in s} S_i, \tag{1.84}$$

where the subindex $s = \alpha, \beta$ indicates the sublattice. Note that the two sublattices have the same number of lattice sites, $N_\alpha = N_\beta = N/2$. Here m_s plays the role of the order parameter in this case since, in the absence of an applied field, its mean value is expected to vanish at high temperature and to be non-zero at low temperature. Therefore, in the absence of an external

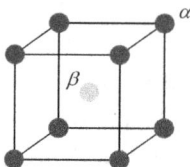

Figure 1.1 Unit cell of the *bcc* lattice with the division in α and β *sc* sublattices.

field h, the effective fields acting on the spins of the two sublattices are

$$h^\alpha_{eff} = zJm_\beta, \tag{1.85}$$

$$h^\beta_{eff} = zJm_\alpha. \tag{1.86}$$

Therefore, the equations of state for the two sublattices are

$$m_\alpha = \tanh\left(\frac{zJ}{k_B T}m_\beta\right), \tag{1.87}$$

$$m_\beta = \tanh\left(\frac{zJ}{k_B T}m_\alpha\right). \tag{1.88}$$

Expanding these equations in a power series of m_β and m_α, respectively, it is obtained that, at high temperature, $m_\alpha = m_\beta = 0$ is the solution of these equations. Below a temperature T_c, the following stable solutions are possible, $m_\alpha = m_\beta \neq 0$ and $m_\alpha = -m_\beta \neq 0$. In the first case $T_c = zJ/k$, while $T_c = -zJ/k$ in the second case. The former solution is only physically meaningful for $J > 0$, and corresponds to the ferromagnetic case. The second solution is the antiferromagnetic solution that requires that $J < 0$. It is worth noting that the critical temperature of the model, $T_c = z|J|/k$, is the same in ferromagnetic and antiferromagnetic cases. The order parameters of the ferromagnetic and antiferromagnetic transitions can be, respectively, defined as $m = (m_\alpha + m_\beta)$ and $\eta = (m_\alpha - m_\beta)$, which show exactly the same temperature dependence in the absence of an applied field. Thus, the mean-field critical exponents of antiferromagnetic and ferromagnetic transitions are exactly the same.

An interesting issue to be considered is the effect of a magnetic field on the transition. When a field is applied, the condition $m_\alpha = -m_\beta$ is no longer satisfied. However, the presence of the field does not prevent, as in the ferromagnetic case, that the system may experience symmetry breaking at a finite temperature. Therefore, contrary to the case of ferromagnetism, the transition from the paramagnetic phase to the antiferromagnetic phase may continue to exist in the presence of an applied magnetic field. Actually, this is easily understood from a thermodynamic viewpoint taking into account that the magnetic field is not thermodynamically conjugated to the antiferromagnetic order parameter. As a matter of fact, the field conjugated to the staggered magnetization would be a field that points along opposite directions in the α and β sublattices. Indeed, this field cannot be realized in practice.

1.4.3 Multiple variables: the Deformable Ising model

The goal of this section is to discuss how Ising models, which are adequate to deal with systems that undergo a phase transition between a disordered and an ordered configurational or magnetic phases can be extended to address secondary degrees of freedom that may affect the properties of the phase transition. Thus, we will consider that these degrees of freedom are not responsible for the phase transition but may have an influence on the phase stability. As an interesting example of this situation. We will discuss how to take into account lattice vibrations in the Ising model.

Both, in alloys and magnets (with localized magnetic moments), atoms are not fixed at lattice sites as considered so far but rather oscillate around these positions and, in spite of the fact that lattice vibrations are usually of small amplitude, these oscillations can affect phase stability [24]. To deeply understand this effect, it is instructive to consider a phase transition between ordered (o) and disordered (d) configurational or magnetic phases and, for each phase, to decompose the free energy into configurational (c) and vibrational (v) parts, that is,

$$F^i = F_m^i + F_v^i = (E_m^i + E_v^i) - T(S_m^i + S_v^i), \tag{1.89}$$

where the superscript $i = o, d$ stands for the phase considered. At high enough temperature E_v^i is simply determined by the equipartition theorem and is independent of the phase considered. Thus, in these conditions, the influence of vibrations on phase stability is determined by the change of vibrational entropy between the two phases. This is a reasonable result since the vibrational entropy can be understood as a measure of the average stiffness of the system. The softer is the system, the larger should be the oscillations and thus the entropy. Consequently, a phase with a large vibrational entropy should be stabilized with respect to harder phases. From this point of view, entropy plays a very relevant role in soft matter [25].

The effect of vibrations on the spin variables can be introduced through the bond proportion model [24], which assumes that the force constant between pairs of nearest neighbour atoms i and j in microscopic states S_i and S_j, is of the form

$$\Phi_{ij}^{cc'} = (\phi_0)_{ij}^{c_i c_j}(1 + \lambda S_i S_j), \tag{1.90}$$

where $(\phi_0)_{ij}^{cc'}$ are constants, and the indices c, c' denote cartesian coordinates (x, y, z). Here, λ is the parameter that couples configurational and vibrational degrees of freedom. Note that $\lambda = 0$ corresponds to the case in which vibrational and configurational or magnetic degrees of freedom are not coupled. Force constants are obtained as second derivatives of the potential energy with respect to lattice displacements about equilibrium lattice positions as done usually in the harmonic treatment of lattice vibrations (see, for instance, Ref.[26]). Then the hamiltonian of the model includes an Ising configurational or magnetic term and a term accounting for the kinetic and harmonic potential energies of lattice vibrations.

For the sake of generality, it is convenient to assume that the lattice can be subdivided into two equivalent sublattices α and β as done in the preceding section. This method is useful since it permits to consider systems with ferro- and antiferromagnetic interactions simultaneously [27]. In the mean-field approximation, the force constant parameters should be expressed as $\Phi_{ij}^{cc'} = (\phi_0)_{ij}^{cc'}(1 + \lambda m_\alpha m_\beta)$. Therefore, in this approximation, the canonical partition function will be

$$Q(\beta) = Q_m Q_v = \sum_{\{m'_\alpha, m'_\beta\}} g(m'_\alpha, m'_\beta) e^{\frac{1}{2}\beta z N m'_\alpha m'_\beta} \prod_{k=1}^{3N} \frac{1}{\beta \hbar \omega_k(m'_\alpha, m'_\beta)}, \tag{1.91}$$

where we have not included an external field coupled to spin variables. In the preceding expression

$$g(m'_\alpha, m'_\beta) = \frac{N_s!}{\left(\frac{1+m'_\alpha}{2}N_s\right)!\left(\frac{1-m'_\alpha}{2}N_s\right)!} \times \frac{N_s!}{\left(\frac{1+m'_\beta}{2}N_s\right)!\left(\frac{1-m'_\beta}{2}N_s\right)!}, \tag{1.92}$$

with $N_s = N/2$ being the number of lattice sites in each sublattice. Here, $\omega_k(m_\alpha, m_\beta)$ denotes the normal vibrational frequency modes. Using the saddle point method, the variational free energy function per particle is given by

$$f = f_m + f_v, \tag{1.93}$$

where

$$
\begin{aligned}
f_m &= -\frac{1}{2} z J m_\alpha m_\beta + \frac{k_B T}{2} \left\{ \frac{1+m_\alpha}{2} \ln \frac{1+m_\alpha}{2} + \frac{1-m_\alpha}{2} \ln \frac{1-m_\alpha}{2} \right\} \\
&+ \frac{k T_B}{2} \left\{ \frac{1+m_\beta}{2} \ln \frac{1+m_\beta}{2} + \frac{1-m_\beta}{2} \ln \frac{1-m_\beta}{2} \right\},
\end{aligned} \tag{1.94}
$$

and

$$f_v = k_B T \sum_{k=1}^{3N} \ln[\beta \hbar \omega_k(m_\alpha, m_\beta)]. \tag{1.95}$$

The square of the frequencies of the normal modes are the eigenvalues of the dynamical matrix, with elements $D_{cc'}(\boldsymbol{q})$ proportional to $\sum_{\{i,j\}} \Phi_{ij}^{cc'} \exp[i\boldsymbol{q} \cdot (\boldsymbol{R}_j - \boldsymbol{R}_i)]$, where $\boldsymbol{R}_i$ and $\boldsymbol{R}_j$ are vector positions in the lattice of atoms i and j, respectively. In the present mean-field approximation,

$$D_{cc'}(\boldsymbol{q}, m_\alpha, m_\beta) = (1 + \lambda m_\alpha m_\beta) D_{cc'}(\boldsymbol{q}, m_\alpha = 0, m_\beta = 0). \tag{1.96}$$

Therefore, the normal frequencies depend on the staggered magnetizations as

$$\omega^2(\boldsymbol{q}, m_\alpha, m_\beta) = (1 + \lambda m_\alpha m_\beta) \, \omega^2(\boldsymbol{q}, m_\alpha = 0, m_\beta = 0), \tag{1.97}$$

and thus the vibrational contribution to the free energy is

$$f_v = \frac{3}{2} k_B T \ln(1 + \lambda m_\alpha m_\beta) + \frac{k_B T}{N} \sum_{k=1}^{3N} \ln[\beta \hbar \omega_k(m_\alpha = 0, m_\beta = 0)], \tag{1.98}$$

where the second term on the right-hand side member of the equation is independent of the order parameter. Minimization of the complete free energy function given by the sum $f_m + f_v$ with respect to m_α and m_β leads to the equations of state

$$
\begin{aligned}
m_\alpha &= \tanh\left(\frac{zJ}{k_B T} m_\beta + \frac{3\lambda m_\alpha}{1 + \lambda m_\alpha m_\beta} \right), \\
m_\beta &= \tanh\left(\frac{zJ}{k_B T} m_\alpha + \frac{3\lambda m_\beta}{1 + \lambda m_\alpha m_\beta} \right).
\end{aligned} \tag{1.99}
$$

Indeed, these equations reduce to Eqs. 1.87 and 1.88 when $\lambda = 0$, as expected. Numerical analysis of these equations in the case of a magnetic system [29] leads to the phase diagram of the system for $J > 0$ and $J < 0$. For low enough coupling parameter λ, both para-ferromagnetic (P-F) and para-antiferromagnetic (P-A) transitions are found to be second order. Transition temperatures can be found analytically and are given by, $T_c^{P-F}(\lambda) = zJ/k(1 + 3\lambda)$, for $J > 0$, and $T_c^{P-A}(\lambda) = -zJ/k(1 - 3\lambda)$, for $J < 0$. The line of second-order transitions ends at $\lambda \, \text{sgn}(J) = 1/3$, which is a tricritical point.[17] For higher values of coupling, the transition is first order.

In general, for negative values of $\lambda \, \text{sgn}(J)$, the transition temperature is higher than in the absence of coupling since the low-temperature ferromagnetic phase is elastically softer than the paramagnetic phase and thus vibrational entropy increases its range of stability. For $\lambda \, \text{sgn}(J) < -1/3$ the vibrational entropy difference between ordered and disordered magnetic phases cannot be balanced by the corresponding magnetic entropy difference and the ordered phase is stable at all temperatures. For positive $\lambda \, \text{sgn}(J)$, the paramagnetic phase is softer than the ordered low-temperature phase and, consequently, the transition temperature is lower than in the absence of coupling. For $\lambda \, \text{sgn}(J) > 1/3$, due to the large entropy difference between both phases, the phase transition becomes first order. It is interesting to remark that for $J > 0$ (<0), an antiferromagnetic (ferromagnetic) phase can be stabilized at high enough temperature by vibrational entropy. This allows that a triple point where ferro-, antiferro- and paramagnetic phases coexist occurs at $\lambda \, \text{sgn}(J) = 0.75$. The phase diagram is shown in Figure 1.2.

It is worth noting that, beyond the magnetic/configurational properties, the model also enables us to study the elastic properties of the lattice in the different magnetic phases. In particular, the elastic constants C_{ijkl} of the system can be obtained from the slopes at the origin of the dispersion curves $\omega(\mathbf{q}, m_\alpha, m_\beta)$. Therefore, they are found to show the following dependence on the staggered magnetizations

$$C_{ijkl} = \frac{1 + \lambda m_\alpha m_\beta}{1 + \lambda \text{sgn}(J)} C_{ijkl}(0), \qquad (1.100)$$

where $C_{ijkl}(0)$ are the elastic constants at $T = 0$. This expression corroborates that lattice hardening or softening occurs in the cases previously discussed.

---

[17] Tricritical points will be discussed later on in Section 1.5.2.

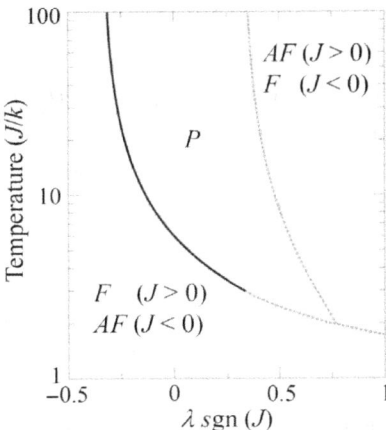

Figure 1.2 Phase diagram of the deformable Ising model. The regions where ferromagnetic (F), antiferromagnetic (AF) and paramagnetic (P) phases occur as a function of the coupling constant for positive and negative values of the exchange parameter are indicated. Continuous and dashed lines stand for continuous and first-order transitions, respectively.

1.5 The Landau approach

In this section, we will introduce basic aspects of the Landau theory.[18] We will start by putting the theory into a proper perspective. The Landau theory is a general phenomenological theory suitable to address and understand phase transitions. It is based on the idea that a phase transition is associated with a symmetry breaking that entails a change in system properties. The symmetry breaking is reflected by the behaviour of the order parameter, which is the basic quantity of the theory. As usual, the (average) order parameter represents a property of the system that vanishes in the high-temperature phase and is different from zero below the transition. The crucial hypothesis of the theory assumes that near the transition, the free energy function can be expanded in a power series of the order parameter and its spatial derivatives. Actually, this means admitting that the free energy function is an analytical function of the order parameter across the transition. Therefore, the theory is, in principle, only appropriate to analyse the behaviour of the studied system in the neighbourhood of a phase transition, where the order parameter is small. It is somewhat surprising that a polynomial free energy describes the singular behaviour expected in the vicinity of any phase transition. In fact, in the theory, the singularity is taken into account by the value of the order parameter resulting from

---

[18] For a more comprehensive introduction to the Landau theory, Ref. [28] is recommended.

the minimization of the free energy function, which is the value expected in the thermodynamic equilibrium state. Therefore, the free energy function corresponding to this value of the order parameter determines the thermodynamic free energy. The symmetry properties of the considered system are taken into account in the series expansion of the free energy function which may contain only the terms allowed by the symmetry operations of the high-temperature phase. This ensures that the Landau free energy is invariant under the symmetry operations of this phase. As a phenomenological approach, the Landau theory deals only with macroscopic quantities and avoids requiring the microscopic hamiltonian to treat the system of interest. Then, when the theory is combined with the thermodynamic formalism, it gives rise to a very general and powerful method for the study at the thermodynamic scale of systems undergoing phase transitions. It allows to relate measurable quantities with the parameters of the theory that appear in the series expansion of the free energy. These parameters are phenomenological but it is interesting to note that they can also be obtained from first principles calculations based, for instance, on the density functional theory. From this point of view, the Landau theory can be understood as a bridge that allows connecting microscopic models with the thermodynamic description.

In his first version, Landau proposed a theory for homogeneous systems undergoing a continuous phase transition [62]. Some years later, Devonshire realised that the theory may also be used to deal with first-order transitions [31]. The generalization to inhomogeneous systems was proposed by Ginzburg and Landau to study the phase transitions from a normal conductor to a superconductor. The inhomogeneity requires including gradient terms of the order parameter in the free energy expansion. This generalized theory is often called the Ginzburg–Landau theory [32].

To formally introduce the Landau theory, let us consider a macroscopic system in a d-dimensional space. Instead of looking at the microscopic lattice scale, as done in the Ising model, let us consider a coarse-graining of the system that consists of dividing the lattice into cells of linear size ℓ, as illustrated in Figure 1.3. An order parameter density, $m(\boldsymbol{r})$, is then defined in each grain located at position $\boldsymbol{r}$. For the sake of simplicity, we will suppose that the order parameter is a scalar quantity. This approach assumes that there are no fluctuations of $m(\boldsymbol{r})$ with wavelength smaller than ℓ.[19] Therefore, the procedure defines a field theory model and the corresponding partition function can be expressed in the form

---

[19] This means that $m(\boldsymbol{r})$ has no Fourier components with a wave number greater than a cutoff $\sim 1/\ell$.

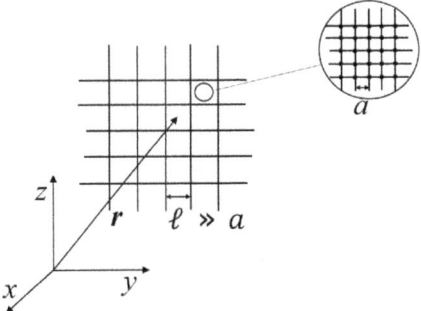

Figure 1.3 Schematic illustration of the coarse-graining of the lattice.

$$Q_{CG}(\beta, h) = e^{-\beta F(\beta, h)} = \int [Dm'] e^{-\beta \mathcal{H}_{CG}(\{m', h\})}, \qquad (1.101)$$

where h is an inhomogeneous external field conjugate to the order parameter, and $\mathcal{H}_{CG}(\{m', h\})$ is the coarse-graining hamiltonian given by the functional integral

$$\mathcal{H}_{CG}(\{m', h\}) = \int d^d r \; f[m'(\boldsymbol{r}), h(\boldsymbol{r})]. \qquad (1.102)$$

The integration measure $[Dm']$ in Eq. 1.101 includes a summation over all functions $\{m'(\boldsymbol{r})\}$, which reflects all possible configurations of the order parameter field. Note that f in Eq. 1.102 is usually referred to as the Landau free energy density. For small $m'(\boldsymbol{r})$, close to a phase transition, it can be expanded in powers of $m'(\boldsymbol{r})$ and its derivatives, that is,

$$f[m'(\boldsymbol{r}), h(\boldsymbol{r})] = f_0 - h(\boldsymbol{r})m'(\boldsymbol{r}) + \frac{1}{2}b|\nabla m'(\boldsymbol{r})|^2 + \dots$$
$$+ \frac{1}{2}a_2 m'^2(\boldsymbol{r}) + \frac{1}{3}a_3 m'^3(\boldsymbol{r}) + \frac{1}{4}a_4 m'^4(\boldsymbol{r}) + \dots \quad (1.103)$$

In the preceding expression, all coefficients are phenomenological and may depend on temperature and other parameters such as pressure or magnetic field. In general, they are simply chosen based on the purposes of the model. Here, f_0 is the non-singular part of the free energy, which is often omitted. Usually, a_2 is assumed to be proportional to $t = (T - T_c)/T_c$, where T_c is a critical temperature, and the remaining parameters are supposed to be temperature independent. The coefficient of the linear term in $m'(\boldsymbol{r})$ must obviously be $h(\boldsymbol{r})$, to ensure that the thermodynamic expression $h = -\partial F/\partial m$

is satisfied. Some terms of the expansion may vanish due to symmetry conditions. In fact, the expansion must be invariant under all symmetry operations of the high-temperature phase. The general procedure to determine the expansion based on the symmetries of the system is explained in detail in the Appendix. As a simple example, we can consider systems that are invariant under the reversal (or sign change) of the order parameter. In this case the cubic term, a_3 must vanish. This may correspond, for instance, to the case of a magnetic system in the absence of an applied external field. The number of terms taken in the expansion is also determined by purely phenomenological criteria. Actually, the model must be as complex as necessary and as simple as possible.[20] If, for instance, the system undergoes a continuous phase transition, the simplest model is the one in which the expansion is limited to fourth order with $a_4 > 0$, and only a_2 is assumed to be a function of temperature such that it changes its sign from positive at high temperature to negative at low temperature at the critical temperature T_c. Note that, in this case, the coefficient of higher-order term must be positive to ensure that the function shows minima associated with stable phases for both $a_2 > 0$ and $a_2 < 0$. Concerning the gradient terms, in general, terms of order higher than the second are not considered. The second-order gradient term introduces a characteristic length scale in the system such as domain wall width.

The Landau theory for homogeneous systems results from a saddle point approximation consisting of replacing in the partition function given by Eq. 1.101 the integrand by its maximum value. If we suppose that h is a uniform field, the functional form of $m'(r)$ that maximizes the integrand of the partition function is indeed a function $m(\mathrm{t}, h)$, which must be a solution of

$$\left[\frac{\partial f}{\partial m'}\right]_{m'=m} = a_2 m + a_3 m^2 + a_4 m^3 + \ldots - h = 0. \qquad (1.104)$$

The solution corresponding to thermodynamic equilibrium must satisfy the stability condition $[\partial^2 f/\partial m'^2]_{m'=m} \geq 0$. In the case of a system that is invariant under reversal of the order parameter, $a_3 = 0$, and supposing that $a_2 = a_0 \mathrm{t}$, the free energy function F is given as

$$\frac{F(m', T, h)}{V} = f_0 + \frac{1}{2}a_0 \mathrm{t} m'^2 + \frac{1}{4}a_4 m'^4 - hm'. \qquad (1.105)$$

The function $f(m)$ is depicted in Figure 1.4 for $h = 0$ and temperatures $T > T_c$, $T = T_c$ and $T < T_c$.

---

[20] This is the so-called *Ockham's razor* criterion that is usually stated in the form, *Entia non sunt multiplicanda praeter necessitatem* (entities should not be multiplied beyond necessity).

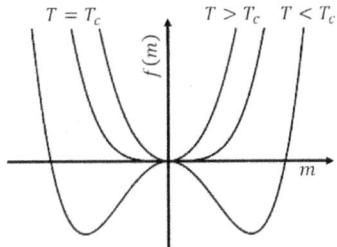

Figure 1.4 Free energy function $f(m)$ for temperatures $T > T_c$, $T = T_c$ and $T < T_c$, and $h = 0$.

It is easy to show that the same free energy function can be obtained from the Bragg–Williams mean-field free energy function corresponding to a ferromagnetic Ising model (that will be denoted here as F_{BW}) given in Eq. 1.70, by expanding the entropy term in a power series about $m' = 0$. That is, close to T_c, the Bragg–Williams free energy function can be expressed as

$$\frac{F_{BW}}{N} = kT_c \ln 2 + \frac{1}{2} z J t m'^2 + \frac{1}{4} \frac{4kT_c}{3} m'^4 + \dots . \qquad (1.106)$$

Therefore, identification of coefficients yields, $a_0 \to zJ$, $T_c \to zJ/k$ (which is, indeed, the critical temperature in the Bragg–Williams approximation), $a_4 \to 4kT_c/3$ and $f_0 \to kT_c \ln 2$. The fact that the same free energy function is obtained in both the Bragg–Williams and Landau formalism is not surprising since the ferromagnetic Ising model is also invariant under the change $S_i \to -S_i$, which implies that the free energy must be invariant under reversal of the order parameter sign. This result corroborates that the Landau approach represents a mean-field treatment. In any case, it is important to remark that although the Landau and mean-field approaches lead to the same free energy close to the critical point, the Landau approach must be considered as a more general theory based solely on the symmetry properties of the considered system in which the coefficients of free energy function are purely phenomenological and not predetermined by the parameters of a given microscopic hamiltonian.

Equilibrium solutions of the order parameter are solutions of

$$a_0 t m + a_4 m^3 = h, \qquad (1.107)$$

which represents the equation of state of the system. Therefore, along the critical isotherm (t = 0), $m \propto h^{1/3}$.

For $h = 0$, the order parameter has the following behaviour:

$$m = \begin{cases} 0 & \text{for } t > 0, \\ \pm(a_0/a_4)|t|^{1/2} & \text{for } t < 0. \end{cases} \tag{1.108}$$

Therefore, in the absence of an applied field, the model displays a continuous transition at $t = 0$ ($T = T_c$). At this point, the coefficient of the harmonic term of the free energy function changes from positive at $t > 0$ to negative at $t < 0$. It is interesting to note that the equilibrium order parameter does not reach a saturation value in the limit $T \to 0$, which is in contradiction with the third law of thermodynamics since the latter requires that the order parameter become independent of temperature near absolute zero. This behaviour reflects the fact that the Landau theory is only adequate to describe the behaviour of the studied system close to the transition temperature. In any case, it has been shown [33] that the saturation effect of the order parameter can be incorporated in the theory assuming a temperature dependence of the term a_2 of the form

$$a_2(T) \propto \left(\coth \frac{\theta_s}{T} - \coth \frac{\theta_s}{T_c} \right), \tag{1.109}$$

where θ_s is a characteristic temperature. With this choice the order parameter shows the same square root dependence on temperature close to T_c and reaches saturation at low temperature.

The susceptibility χ can be obtained by differentiating the equation of state with respect to h. It is given by

$$\chi = \frac{1}{a_0 t + 3a_4 m^2} = \begin{cases} \frac{1}{a_0} t^{-1} & \text{for } t > 0, \\ \frac{1}{2a_0}|t|^{-1} & \text{for } t < 0. \end{cases} \tag{1.110}$$

The heat capacity can be computed from the entropy, $s = -\frac{\partial f}{\partial T} = -\frac{1}{2}\frac{a_0}{T_c}m^2$, as $\frac{C}{T} = \frac{\partial s}{\partial T}$. The obtained result is

$$C = \begin{cases} 0 & \text{for } t > 0, \\ \frac{a_0^2}{2T_c^2 a_4} T & \text{for } t < 0. \end{cases} \tag{1.111}$$

From this expression, it is deduced that the heat capacity shows a discontinuity, $\Delta C = \frac{a_0^2}{2T_c a_4}$, at T_c.

Therefore, as expected, the model reproduces a second-order transition at $T = T_c$ ($t = 0$), with mean-field critical exponents, $\alpha = 0$, $\beta = 1/2$, $\gamma = 1$ and $\delta = 3$.

1.5.1 Ginzburg–Landau approach: spatial fluctuations

In some cases, it is important to take into account the existence of spatial fluctuations that give rise to inhomogeneities in the system. In this case, gradient terms must remain in the free energy density expansion given by Eq. 1.103 and, using the saddle-point approximation, it results that the function $m'(\mathbf{r})$ in Eq. 1.101 should be the one that minimizes the following free energy functional

$$F[\{m'(\mathbf{r})\}, \{h(\mathbf{r})\}, T]$$
$$= \int d^d r \left[\frac{1}{2} a_0 t m'^2(\mathbf{r}) + \frac{1}{4} a_4 m'^4(\mathbf{r}) - h(\mathbf{r}) m'(\mathbf{r}) - \frac{1}{2} b |\nabla m'(\mathbf{r})|^2 \right], \quad (1.112)$$

where the case of a uniaxial ferromagnet that is invariant under reversal of the order parameter has been considered and only the lowest order gradient term has been kept. Indeed, the integral is over the whole hypervolume V. The function $m(\mathbf{r})$ that minimizes this functional is the solution of the associated Euler–Lagrange equation[21] given by

$$\left[\nabla^2 - \frac{a_0 t}{b} \right] m(\mathbf{r}) - \frac{a_4}{b} m^3(\mathbf{r}) = -\frac{1}{b} h(\mathbf{r}). \quad (1.113)$$

Indeed, this equation reduces to the usual equation of state of the Landau theory in the homogeneous case when $h(\mathbf{r}) = h$ and $\nabla m(\mathbf{r}) = 0$.

The preceding Eq. 1.113 can be used to compute the correlation function $G(\mathbf{r})$ defined as

$$G(\mathbf{r}) = \langle m(\mathbf{r}) m(0) \rangle - \langle m(\mathbf{r}) \rangle \langle m(0) \rangle, \quad (1.114)$$

where $\langle \cdot \rangle$ indicates averages over the distribution of functions $\{m'(\mathbf{r})\}$ of the coarse-grained system. Therefore,

$$\langle m(\mathbf{r}) \rangle = \frac{1}{Q_{CG}(\beta, h)} \int [Dm'] m'(\mathbf{r}) e^{-\beta \mathcal{H}_{CG}(\{m', h\})}, \quad (1.115)$$

and, thus, the correlation function can be obtained as

$$\beta G(\mathbf{r}) = \frac{\delta \langle m(\mathbf{r}) \rangle}{\delta h(0)}. \quad (1.116)$$

The derivative on the right-hand side member of the preceding equation can be obtained as follows. Consider an homogeneous system in equilibrium at temperature T in the absence of an applied field and suppose that it is perturbed by a local field, $h(\mathbf{r}) = h_0 \delta(\mathbf{r})$, where $\delta(\mathbf{r})$ is the Dirac-delta

---

[21] It must be taken into account that m vanishes over the boundary of the volume V.

distribution. As a consequence of the local perturbation, the magnetization will change as

$$m(\boldsymbol{r}) = m(T) + \varphi(\boldsymbol{r}), \tag{1.117}$$

where $m(T)$ is given by Eq. 1.108, which depends on whether $T > T_c$ or $T < T_c$. Obviously,

$$\frac{\delta\langle m(\boldsymbol{r})\rangle}{\delta h(0)} = \frac{\varphi(\boldsymbol{r})}{h_0}. \tag{1.118}$$

The response $\varphi(\boldsymbol{r})$ can be obtained by replacing $m(\boldsymbol{r})$ into the Euler–Lagrange Eq. 1.113. Neglecting non-linear terms in φ the following differential equation for φ is then obtained

$$\left[\nabla^2 - \frac{a_0 t}{b} - \frac{3a_4}{b}m^3\right]\varphi(\boldsymbol{r}) - \frac{a_0 t}{b}m - \frac{a_4}{b}m^3 = -\frac{1}{b}h_0\delta(\boldsymbol{r}). \tag{1.119}$$

Substituting m for the equilibrium values corresponding to $T > T_c$ and $T < T_c$, respectively, it turns out that $\varphi(\boldsymbol{r})$ must be the solution of

$$[\nabla^2 - \xi^{-2}]\varphi(\boldsymbol{r}) = -4\pi a\delta(\boldsymbol{r}), \tag{1.120}$$

where $4\pi a = h_0/b$ and the correlation length, ξ, is given by

$$\xi(t) = \begin{cases} \left[\dfrac{b}{a_0 t}\right]^{1/2} & \text{for } t > 0, \\[2ex] \left[\dfrac{b}{2a_0|t|}\right]^{1/2} & \text{for } t < 0. \end{cases} \tag{1.121}$$

Assuming spherical symmetry, in $3d$ the solution is,[22]

$$\varphi(r) = \frac{h_0}{4\pi b}\frac{1}{r}e^{-r/\xi(t)}. \tag{1.122}$$

In arbitrary dimension d, it can be shown that for $r \ll \xi$, $\varphi(r) \propto r^{2-d}$, while for $r \gg \xi$, $\varphi(r) \propto e^{-r/\xi}$. Therefore, the order of magnitude of the correlation function is correctly given by

$$G(r) \approx \frac{e^{-r/\xi}}{r^{d-2}}. \tag{1.123}$$

---

[22] It is convenient to find the solution of the differential equation in Fourier k-space. In arbitrary dimension d, writing, $\varphi(\boldsymbol{r}) = \frac{1}{(2\pi)^d}\int e^{-i\boldsymbol{k}\cdot\boldsymbol{r}}\varphi(\boldsymbol{k})d^dk$ and $\nabla^2\varphi(\boldsymbol{r}) = -\frac{1}{(2\pi)^d}\int e^{-i\boldsymbol{k}\cdot\boldsymbol{r}}k^2\varphi(\boldsymbol{k})d^dk$, and remembering that $\delta(\boldsymbol{r}) = \frac{1}{(2\pi)^d}\int e^{-i\boldsymbol{k}\cdot\boldsymbol{r}}d^dk$, the equation can be written in Fourier space as, $[k^2 + \xi^{-2}]\varphi(\boldsymbol{k}) = 4\pi a$. Therefore, $\varphi(\boldsymbol{k}) = \frac{4\pi a}{[k^2+\xi^{-2}]}$ and, $\varphi(\boldsymbol{r})$ will be obtained as the inverse Fourier transform of $\varphi(\boldsymbol{k})$. It can be shown that $\varphi(r) \propto (\xi r)^{(2-d)/2}K_{(d-2)/2}(r/\xi)$, where $K_{(d-2)/2}$ is a modified Bessel function of the second kind.

Taking into account that the correlation length $\xi(t)$ diverges at the critical point ($t = 0$) as $\xi \sim |t|^{-\nu}$, with $\nu = 1/2$, the correlation function must be of the form[23]

$$G(r) \sim r^{-(2-d+\eta)}, \tag{1.124}$$

with $\eta = 0$.

Although Ginzburg–Landau theory takes into account the spatial fluctuations of the order parameter, within the approximations considered in this section it leads to the mean-field critical behaviour. This fact is a consequence of the approximate treatment of the correlation function that assumes a linear behaviour of the response to a local perturbation field and, thus is not adequate to describe the behaviour of long-range (distance) correlations. Therefore, the exponents $\alpha = 0$, $\beta = 1/2$, $\gamma = 1$, $\delta = 3$, $\nu = 1/2$, and $\eta = 0$, represent the set of mean-field critical exponents, which are independent of space dimension and tensor-rank nature of the order parameter. From this point of view, it is expected that a mean-field approximation provides a good thermodynamic description close to critical points when the following inequality is satisfied

$$\frac{\int_{\Omega(\xi)} d^d r \, G(\boldsymbol{r})}{\int_{\Omega(\xi)} d^d r \, m^2} \ll 1, \tag{1.125}$$

which is the so-called Ginzburg criterion. Integrals are performed over a d-dimensional hypersphere of radius equal to the correlation length, ξ. Given that $m^2 \sim |t|^{2\beta}$, and $G(r) \sim \exp(-r/\xi)/r^{(d-2)}$, carrying out the integrals in spherical coordinates, the Ginzburg criterion can be expressed as

$$|t|^{(2-d)\nu-2\beta} \ll 1, \tag{1.126}$$

which requires that

$$d > 2 + \frac{2\beta}{\nu}. \tag{1.127}$$

This inequality enables us to define an upper critical dimension $d_c \equiv 2 + 2\beta/\nu$, which is the space dimension above which mean-field behaviour provides a correct description of critical behaviour. Replacing mean-field values, $\beta = 1/2$ and $\nu = 1/2$, we obtain that $d_c = 4$, which is the space dimension above which the mean-field description is correct. It is worth noting that these results are valid in systems with a short-range interaction. In systems with long-range interactions, as shown above, mean-field provides the correct critical behaviour [4].

---

[23] This approximation is equivalent to the Ornstein–Zernike approximation, which is often used to study density correlations in fluids [4].

The existence of an upper critical dimension can be understood in the sense that for a given symmetry, the number of interacting entities of a given spin is larger in higher dimension, which would yield a larger field acting on each entity. This effect is expected to reduce fluctuations, which should result in a better reliability of the mean-field treatment in which fluctuations are ignored or treated in an approximate manner.

Note that a lower critical dimension is in general also defined as the space dimension above which symmetry breaking can occur at a finite temperature. This lower critical dimension depends on space and order parameter dimensions. For the Ising model, the lower critical dimension is $d = 1$. When the order parameter has a continuous symmetry, the Mermin-Wagner theorem [34] has to be taken into account. This theorem states that continuous symmetries cannot be spontaneously broken at finite temperature in systems with sufficiently short-range interactions in dimensions $d \leq 2$. Actually, this result supposes that in this class of systems, long-range fluctuations can be created with little energy cost and since they increase the entropy they are favoured and, consequently, avoid that the long-range order can be established in $2d$. Thus, in models such as the XY or the continuous Heisenberg model, the lower critical dimension is $d = 2$.

1.5.2 First-order transitions and tricritical points

The Landau theory can also be used to account for first-order phase transitions. Two general cases must be considered. First, in systems that are invariant under reversal of the order parameter, a first-order phase transition can occur if the fourth-order term of the free energy expansion is negative. In this case, the expansion must be, at least, extended up to the sixth-order if $a_6 > 0$. In the homogeneous case, the free energy density function in the absence of an applied field is thus of the form

$$f(m', T) = f_0 + \frac{1}{2}a_0 t m'^2 - \frac{1}{4}|a_4|m'^4 + \frac{1}{6}a_6 m'^6. \qquad (1.128)$$

The second case corresponds to systems that are not invariant under reversal of the order parameter. In the simplest situations, the free energy function is then of the form

$$f(m', T) = f_0 + \frac{1}{2}a_0 t m'^2 + \frac{1}{3}a_3 m'^3 + \frac{1}{4}a_4 m'^4, \qquad (1.129)$$

where, again, no external field has been considered. In this case, $a_4 > 0$ and a_3 must be either positive or negative. Free energy functions given by Eqs. 1.128 and 1.129 are shown in Figures 1.5 and 1.6, respectively, for selected values of the temperature.

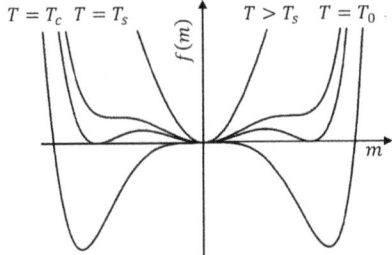

Figure 1.5 Free energy function given by Eq. 1.128 for temperature $T > T_s$, $T_s > T > T_0$, $T = T_0$, and $T = T_c$. T_0 is the temperature at which a first-order transition between $m = 0$ and $m \neq 0$ can occur in equilibrium (the case of three degenerate minima). T_s is the metastability limit of the low-temperature phase.

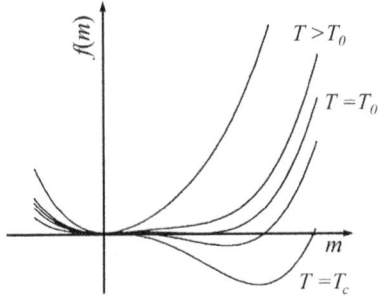

Figure 1.6 Free energy function given by Eq. 1.129 for temperature $T > T_0$, $T = T_0$ (two degenerate minima), $T < T_0$ and $T = T_c$.

In both cases, at a temperature $T_0 > T_c$ a first-order transition can occur in equilibrium. Therefore, at this temperature,

$$f(m \neq 0, T_0) = f(m = 0, T_0), \qquad (1.130)$$

which is the condition that determines the equilibrium coexistence of the low-temperature ordered phase with $m \neq 0$ and high-temperature disordered phase with $m = 0$. The equilibrium values of the order parameter must be solutions of the equation

$$\left[\frac{\partial f(m', T)}{\partial m'} \right]_{m'=m} = 0. \qquad (1.131)$$

The equilibrium temperatures T_0 of the first-order transition can be obtained by solving the two preceding equations. For both models, they are given as,

$T_0 - T_c$

$$= \begin{cases} \frac{3}{16} \frac{a_4^2}{a_0 a_6} & \text{systems invariant under reversal of the order parameter,} \\ \frac{2}{9} \frac{a_3^2}{a_0 a_4} & \text{systems non-invariant under reversal of the order parameter.} \end{cases}$$

$$(1.132)$$

It is important to note that in both cases, the first-order transition may still occur in the presence of an external field h thermodynamically conjugate to the order parameter.

Tricritical points

Usually, the negative character of the fourth order term in Eq. 1.128 is a consequence of the coupling of the primary order parameter, m, with a secondary parameter. This can, for instance, be the case of some antiferromagnetic material subjected to an applied magnetic field, which plays the role of a secondary field. For low values of the magnetic field, these systems undergo a continuous transition from a high-temperature paramagnetic phase to a low-temperature antiferromagnetic phase. However, when the magnetic field is high enough, the transition is first order. The magnetic field is not thermodynamically conjugate to the staggered magnetization, which is the order parameter but, since it induces a non-zero magnetization that couples to the order parameter, it can affect transition features. In the limit of high magnetic anisotropy, it can be shown that this effect can be effectively taken into account considering that the expansion coefficients of the free energy given by Eq. 1.128 are functions of the magnetization and, thus, of the applied magnetic field [35]. Due to this dependence, for low fields a_4 is positive, but changes to negative for high enough fields and the transition changes from continuous to first order. Then, an interesting situation occurs when both, $a_2 = a_0 t$, and a_4 go to zero simultaneously. The points in the phase diagram where this happens are denoted as tricritical points. They are those points that separate a line of first order from a line of continuous transitions and are controlled by the secondary field and temperature [18]. Actually, if Δ is this secondary field, while $a_4(\Delta) > 0$, a continuous transition occurs at $a_2(T_c, \Delta < \Delta_t) = 0$, while for $a_4(\Delta) < 0$, a first-order transition takes place at $a_2(T_0, \Delta) = 3a_4(\Delta > \Delta_t)/16a_6$. Therefore, assuming that a_4 is not temperature dependent, the tricritical point, (T_t, Δ_t) is given by the following conditions:

$$a_2(T_t, \Delta_t) = 0,$$
$$a_4(\Delta_t) = 0.$$

The generic phase diagram of this class of systems is shown in Figure 1.7.

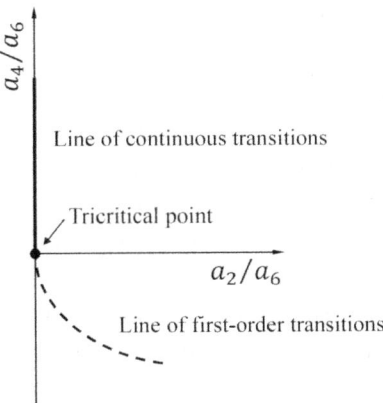

Figure 1.7 Generic phase diagram of a system described by the free energy 1.128 that shows a tricritical point. The coefficient a_4 is supposed to be a function of a secondary field.

It is interesting to note that when the tricritical point is approached in such a way that a_4 goes to zero faster than a_2, the system shows a tricritical behaviour. It is easy to show that the tricritical behaviour is characterized by mean-field exponents, $\alpha_t = -1$, $\beta_t = 1/4$, $\gamma_t = 1$ and $\nu_t = 1/2$, which are different from mean-field critical exponents. Note that in this case, Ginzburg criterion given by Eq. 1.127 leads to an upper critical dimension $d_t = 3$. This means that, except for logarithmic corrections that are practically unmeasurable, mean-field description of tricritical behaviour is essentially correct.

1.6 Scaling and renormalization

All mean-field theories give the same set of independent mean-field critical exponents, regardless of space and order parameter dimension, the latter determined by the ground-state degeneracy. Nevertheless, these results are in contradiction with many experiments that suggest that both space and order parameter dimension have a strong influence on criticality and, thus, on critical exponents. Furthermore, it is also known that critical exponents satisfy a number of equalities that suggest that they are not independent. In other words, critical behaviour seems to be determined by a small number of exponents. We are interested in justifying these results on the basis of plausible physical arguments. We will first discuss the hypothesis of homogeneity of the singular part of free energy that will enable us to obtain the so-called scaling relations, which are the relations that must be satisfied by the set of critical exponents. Next we will justify the hypothesis of homogeneity taking into account that criticality is characterized by the divergence of the correlation length. This point of view can in fact be justified more rigorously by means of the formalism of the renormalization group theory. However, in the present book, we will not discuss this formalism that has been developed in great detail in several recent books [36, 37].

As in the previous section, we will continue considering a system described by a scalar order parameter m with states of equilibrium controlled by two parameters, t and h that measure the proximity to the critical point. In the previous section, we have seen that within the Landau approach, the equation of state of such a system near the critical point is of the form

$$h \simeq a_0 m(t + \frac{a_4}{a_0} m^2 + ...), \tag{1.133}$$

from which the set of mean-field critical exponents can be obtained as shown in the preceding section. Arbitrary values of the exponents might be obtained assuming that the equation of state in the neighbourhood of the critical point is of the form

$$h \simeq m(t + cm^{1/\beta} + ...)^{\gamma}, \tag{1.134}$$

which means that the equation of state should be of the more general form, $h = m\Psi(t, m^{1/\beta})$, where Ψ should be an homogeneous function of degree γ of t and $m^{1/\beta}$. This suggests that the singular part of the free energy should also be an homogeneous function that under a change of scale λ that modifies the proximity to the critical point, behaves, for all λ, as

$$f(t, h) = \lambda f(\lambda^s t, \lambda^r h). \tag{1.135}$$

This is the hypothesis of homogeneity of the free energy that was first introduced by Widom [38]. Thermodynamic relations permit to find the connection between the exponents r and s and the critical exponents. This connection can be determined as follows. From the free energy, close to the critical point, the order parameter m is expected to scale as

$$m(t, h) = -\frac{\partial f(t, h)}{\partial h} = -\lambda \frac{\partial f(\lambda^s t, \lambda^r h)}{\partial(\lambda^r h)} \frac{\partial(\lambda^r h)}{\partial h} = \lambda^{r+1} m(\lambda^s t, \lambda^r h).$$
$$\tag{1.136}$$

Proceeding in a similar way, it can be obtained that the susceptibility scales as

$$\chi(t, h) = \frac{\partial m(t, h)}{\partial h} = \lambda^{2r+1} \chi(\lambda^s t, \lambda^r h), \tag{1.137}$$

and the heat capacity as

$$C(t, h) = -T_c \frac{\partial^2 f(t, h)}{\partial t^2} = \lambda^{2s+1} C(\lambda^s t, \lambda^r h). \tag{1.138}$$

Choosing $h = 0$, $\lambda = |t|^{-1/s}$ and $t = 0$, $\lambda = |h|^{-1/r}$, it is obtained that

$$\begin{cases} m(t,0) = m(-1,0)(-t)^{(r+1)/s}, \\ m(0,h) = m(0,\pm1)|h|^{-(r+1)/r}, \\ \chi(t,0) = \chi((\pm1,0)|t|^{(2r+)/s}, \\ C(t,0) = C(\pm1,0)|t|^{(2s+1)/s}. \end{cases} \qquad (1.139)$$

Note that the coefficients are all estimated far from the critical point. Therefore, critical exponents should be related to r and s as

$$\begin{cases} \alpha = \frac{2s+1}{s}, \\ \beta = -\frac{r+1}{s}, \\ \gamma = \frac{2r+1}{s}, \\ \delta = -\frac{r}{s+1}. \end{cases} \qquad (1.140)$$

The following scaling relations are immediately obtained

$$\alpha + 2\beta + \gamma = 2, \qquad (1.141)$$

$$\beta(\delta - 1) = \gamma. \qquad (1.142)$$

Kadanoff [39] showed that the hypothesis of homogeneity of the singular part of the free energy can be justified by taking into account that close to the critical point the properties of the system are determined by the fact that the correlation length ξ of the order parameter is very large and diverges at T_c. To be specific, let us consider an Ising model defined on a d-dimensional hypercubic lattice with lattice parameter a_0. The states of the system are controlled by the two variables t and h that measure the separation from the critical point. Therefore, for $|t| \ll 1$ and $|h| \ll 1$, we expect that ξ is much larger than the characteristic microscopic length, a_0. Then, as illustrated in Figure 1.8, the idea is to define a lattice of blocks of size ba_0 and replace the spins within the original lattice with a new spin variable that can take the same values as the original spins, S_i. This is justified by taking into account that, in general, each block may be in 2^n states with $n = b^d$ but, close to the critical point, since $\xi/a_0 \gg b$ short-range correlations will be very intense and many of the 2^n states will have to be suppressed. In other words, near the critical point, it is expected that homogeneous or quasi-homogeneous states will be dominant, which justify that the state of the block can be determined by a spin variable S_I, where I is an index that locates the blocks. The state of the system constituted of spin blocks will be controlled by variables t' and h' that measure the distance of the block system from the critical point. We expect that these parameters depend on

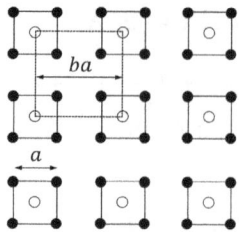

Figure 1.8 Illustration of the Kadanoff construction. The original spins (solid symbols) sit on the lattice of lattice parameter a. They are grouped into blocks constituted of four spins that define the new lattice of parameter ba where b is the scaling parameter. In this figure $b = 2$.

t, h and on the linear dimension b of the blocks. The simplest relationship between the original variables and block variables that are consistent with the symmetry conditions, $h' \to -h'$ when $h \to -h$, $t' \to t'$ when $h \to -h$ and $t' = h' = 0$ when $t = h = 0$, is

$$h' = b^x h, \tag{1.143}$$

$$t' = b^y t, \tag{1.144}$$

where x and y are positive exponents to ensure that the block system is further away from the critical point than the original system, as expected. Indeed, the functional dependence of the singular part of the block-free energy on t' and h' must be the same as that of the singular part of the original free energy on t and h. Given that there are b^d spins per block, it will turn out that

$$f(t, h) = b^{-d} f(t^y, h^x). \tag{1.145}$$

This equation determines how f changes under scale changes and, thus, justifies Widom's homogeneity hypothesis. Now, λ must be identified with b^{-d}, which leads to, $s = -y/d$ and $r = -x/d$. Note that the present scaling expression is interesting as it explicitly includes the space dimension d.

Given that the characteristic length of the block system is b times greater than the characteristic length of the original system, we expect that the correlation length decreases by the same factor, that is,

$$\xi(t, h) = b \, \xi(t^y, h^x). \tag{1.146}$$

Then choosing $b = |t|^{-1/y}$ and $h = 0$, from Eqs. 1.145 and 1.146, we obtain,

$$f(t, 0) = |t|^{d/y} f(\pm 1, 0), \tag{1.147}$$

$$\xi(t, 0) = |t|^{-1/y} \xi(\pm 1, 0). \tag{1.148}$$

From the second of the preceding equations, it results that the critical exponent $\nu = 1/y$ and taking into account that $C \sim \partial^2 f / \partial t^2$, the following scaling relation is obtained:

$$d\,\nu = 2 - \alpha, \tag{1.149}$$

which is denoted as the hyperscaling relation since it includes explicitly the space dimension d.

Scaling laws are interesting since they can be experimentally verified, which provides a useful method to determine critical exponents. The most common way to analyse experiments is as follows. Consider, for instance, the case of the magnetization, $m = -\partial f / \partial h$, which scales as, $m(t, h) = |t|^{-(x-d)/y} m(\pm 1, h/|t|^{x/y})$, by choosing $b = |t|^{-1/y}$. Taking into account that $d\,x/y = \beta$ and $x/y = \beta + \gamma$, the following scaling equation results,

$$m(t, h) = |t|^{\beta} \mathcal{M}_{\pm} \left[\frac{h}{|t|^{\beta+\gamma}} \right], \tag{1.150}$$

where $\mathcal{M}_+$ is the scaling function for $t > 0$ and $\mathcal{M}_-$ the corresponding function for $t < 0$. The preceding equation suggests that if we represent measurements performed close to the critical point T_c as $h/|t|^{(\beta+\gamma)}$ versus $m/|t|^{\beta}$, results should collapse in the two universal curves, Φ_+ and Φ_+ corresponding to the data obtained for $t > 0$ and $t < 0$, respectively. An example of such a representation has been reported for the ferromagnetic compound CrB_3 in Ref. [40]

Let us conclude this section with a brief glance at the renormalization group. This approach is based on and provides rigour to the ideas of Kadanoff's block transformation. It consists of a reduction of the degrees of freedom followed by a proper rescaling of characteristic lengths. This rescaling is performed by introducing an effective hamiltonian for the block of spins system, which leads to the same behaviour as the original hamiltonian for magnetization and spin correlations on distances large in comparison to the block size. To be specific, suppose that the hamiltonian contains two coupling parameters, K_1 and K_2, that, in the case of an Ising model are given by $K_1 = -J/kT$ and $K_2 = -h/kT$. After a block transformation of scale b, the new parameters will be K_{1_b} and K_{2_b}. If $\mathbf{K}$ and $\mathbf{K}_b$ are vectors of components (K_1, K_2) and (K_{1_b}, K_{2_b}), respectively, the block transformation equation is of the form, $\mathbf{K}_b = \mathcal{T}\,\mathbf{K}$, where $\mathcal{T}$ is a transformation operator that incorporates both the reduction of degrees of freedom and rescaling. Since the transformed hamiltonian is formally equivalent to the original, the block transformation process may be applied as many times as required. After applying the transformation s times, we will have:

$$\mathbf{K}_{sb} = \mathcal{T}\,\mathbf{K}_{s-1}. \qquad (1.151)$$

If the system is not critical, the correlation length will be finite and, through successive block transformations, the effective correlation length will decrease, reflecting a departure from the critical point. If, on the other hand, the system is at a critical point, the correlation length will diverge and through the transformation process a fixed point $\mathbf{K}_c$ will be reached at which the operator $\mathcal{T}$ may satisfy,

$$\mathbf{K}_c = \mathcal{T}\,\mathbf{K}_c. \qquad (1.152)$$

The transformation $\mathcal{T}$ is what is in fact called the renormalization group even though it has semigroup properties from a mathematical point of view. It is essential that the transformation operator $\mathcal{T}$ is not singular, at least not close to the critical coupling $\mathbf{K}_c$ [41]. Critical behaviour can be determined from flow properties of the effective hamiltonian coupling parameters about the non-trivial fixed points of the transformation (see, for instance, a simple example in Ref. [42]). It is worth remarking that application of this real space renormalization group approach requires performing drastic approximations and the obtained results are often quite poor. It must be noted that, in contrast, the momentum space approach is much more powerful, especially when applied within the framework of continuum field theory models, and provides very accurate results [43].

Finally, it should be noted that a large number of experimental results also point out that critical behaviour is independent of a large number of specific microscopic features such as lattice symmetry or the nature of the interaction (while remaining short range). Renormalization group shows the two parameters that determine asymptotic phenomena such as critical behaviour are space and microscopic spin variable dimension. These two properties allow to classify physical systems in universality classes. $3d$ systems with a scalar order parameter, such as fluids, uniaxial or Ising magnets, and binary mixtures among others belong to the same Ising universality class.

Exercises

1.1 Consider a fluid consisting of N spherical molecules that are located in a container of volume V. The interaction among the molecules is approximated by an effective potential energy of the form:

$$u(r) = \begin{cases} \infty & \text{for } r < r_0, \\ -\bar{u} & \text{for } r \geq r_0. \end{cases}$$

This represents a mean-field approximation that supposes that molecules can be treated as independent molecules, since each molecule sees the same effective potential created by the rest of the molecules. Show that in this approximation, the gas can be described by a van der Waals equation of state, $p = \frac{RT}{v-b} - \frac{a}{v^2}$, where $v = V/n$ is the molar volume and $n = N/N_A$ the mole number (and N_A is the Avogadro's number). Here a and b are related to the potential energy parameters as, $a = -\bar{u}NV/n^2$ and $b = V_{exc}/n$, where V_{exc} is an excluded volume associated with the space occupied by the molecules.

1.2 The axial (or anisotropic) next-nearest neighbour Ising model, which is usually known as the ANNNI model, is a variant of the Ising model in which competing ferromagnetic and antiferromagnetic exchange interactions couple spins at nearest and next-nearest neighbour sites along one of the crystallographic axes of the lattice. The model is considered a suitable prototype to deal with complicated spatially modulated magnetic superstructures that occur in certain alloys. The hamiltonian of the model is

$$\mathcal{H}_A = -J_0 \sum_{\langle nn \rangle_{xy}} S_i S_j - J_1 \sum_{\langle nn \rangle_z} S_i S_j - J_2 \sum_{\langle nnn \rangle_z} S_i S_j,$$

where the spin variables S_i can take values ± 1. The first sum in the right-hand side extends over all nearest neighbour pairs in the x-y plane, the second sum is over nearest neighbour pairs along the axial direction z and the third sum is over next-nearest neighbour pairs also along the z-direction. Find the ground state of this model for positive and negative values of the exchange parameters, J_0, J_1 and J_2.

1.3 The classical Heisenberg model is an isotropic magnetic spin model defined on a lattice with the hamiltonian,

$$\mathcal{H}_{HC} = -J \sum_{\langle ij \rangle_{nn}} \boldsymbol{S}_i \cdot \boldsymbol{S}_j - \boldsymbol{H} \cdot \sum_i \boldsymbol{S}_i,$$

where S_i is the spin located at lattice site i that has modulus $|S_i| = S$, and can rotate continuously over space. The first sum in the hamiltonian extends over nearest neighbours and H is a magnetic field applied along the z-direction. In the ferromagnetic case, corresponding to a magnetic exchange $J > 0$, show that:

(*a*) In the mean-field approximation, the equation of state of the model is:

$$m = \mathcal{L}\left(\frac{SH_{eff}}{k_BT}\right),$$

where m is the average magnetization per lattice site, $H_{eff} = zJm + H$ with z being the number of nearest neighbours in the lattice and $\mathcal{L}$ the Langevin function defined as $\mathcal{L}(x) \equiv \coth x - \frac{1}{x}$.

(*b*) For $H = 0$, the model has a critical point at $T_c = \frac{zJS}{3k_B}$.

1.4 Consider the free energy function given by Eq. 1.67 and suppose that the moment expansion is truncated at second order instead of first order as done in the Bragg–Williams approximation. Taking the expression of $M_2 = \frac{1}{8}NzJ^2(1 - m^2)^2$ proposed in Ref. [21], and supposing that $g(m)$ is the same as in the Bragg–Williams approximation, show that

(*a*) The critical temperature of the model, T_c, is given by

$$T_c = \frac{zJ}{2k_B}\left(1 + \sqrt{1 - \frac{1}{z}}\right). \tag{1.153}$$

(*b*) The critical exponents of the model are mean-field exponents.

1.5 Consider a ferromagnetic Ising model defined on a lattice with z nearest neighbours per site. Find a solution of this model within the Bethe approximation, which consists of treating the interaction of any given spin, or *central* spin S_0, with its nearest neighbours exactly and the interaction of these nearest neighbours with the remaining spins of the lattice through a mean-field, h_{eff}. In this approximation, the hamiltonian of any group of $z + 1$ spins formed by the central spin and its nearest neighbours is of the form,

$$\mathcal{H}_{z+1} = -hS_0 - (h + h_{eff})\sum_{j=1}^{z} S_j - J\sum_{j=1}^{z} S_0S_j,$$

where J is the exchange energy that will be considered positive and h is an external field. Show that:

(*a*) Defining $\tilde{h} = h/k_BT$, $\tilde{h}_{eff} = h_{eff}/k_BT$ and $\tilde{J} = J/k_BT$, the partition function of the group of $z+1$ spins can be written as, $Q_{z+1} = Q_{z+1}^+ + Q_{z+1}^-$ where

$$Q_{z+1}^{\pm} = e^{\pm\tilde{h}} \left[2\cosh(\tilde{h} + \tilde{h}_{eff} \pm \tilde{J}) \right]^z.$$

(*b*) The self-consistency condition reads,

$$e^{2\tilde{h}_{eff}} = \left[\frac{\cosh(\tilde{h} + \tilde{h}_{eff} + \tilde{J})}{\cosh(\tilde{h} + \tilde{h}_{eff} - \tilde{J})} \right]^{z-1},$$

which is obtained by imposing that $\langle S_0 \rangle = \langle S_j \rangle$, where $\langle S_0 \rangle$ is the mean value of the central spin and $\langle S_j \rangle$ is the mean value of one of its nearest neighbours.

(*c*) From the self-consistency equation, a critical point exists for $h = 0$ at the temperature

$$T_c = \frac{1}{\ln \frac{z}{z-2}} \frac{2J}{k_B}.$$

1.6 Consider an alloy A_xB_{1-x} constituted of N_A A-atoms and N_B B-atoms localized on the sites of a *bcc*-lattice (see Figure 1.1) of N-sites. x is the fraction N_A/N and $1 - x = N_B/N$, and the concentration of the alloy is defined as $c \equiv 2x - 1$. Therefore, for a stoichiometric alloy, $x = 0.5$ and $c = 0$. Suppose that the configurational energy of the alloy is, $E_c = N_{AA}v_{AA} + N_{BB}v_{BB} + N_{AB}v_{AB}$, where $N_{\mu\nu}$, with μ, ν standing for A and B, are the number of nearest neighbour pairs of atoms μ and ν and $v_{\mu\nu}$ is the corresponding interaction energy between these pairs of atoms. Show that:

(*a*) Defining a spin variable S_i that takes the values $+1$ when the site i is occupied by an atom A and -1 when it is occupied by an atom B, the configurational energy of the alloy can be expressed as an Ising hamiltonian of the form,

$$\mathcal{H} = E_0 - J \sum_{\langle ij \rangle_{nn}} S_iS_j - h \sum_i S_i,$$

where, $E_0 = \frac{1}{4}zN(v_{AA} + v_{BB} + 2v_{AB})$, $J = -\frac{1}{4}\epsilon = \frac{1}{4}(v_{AA} + v_{BB} - 2v_{AB})$, $h = \frac{1}{2}z(v_{BB} - v_{AA})$. Note that, as in the Ising model, J represents an exchange energy. Note, however, that compared with the magnetic Ising model, the term proportional to h is now a constant and the first sum over nearest neighbour pairs is constrained from the fact that N_A and N_B are constants.

For $\epsilon > 0$ $(J < 0)$ the model is adequate to describe an alloy that undergoes an order–disorder transition from a high-temperature disordered phase with A and B atoms randomly distributed over the lattice according to their fraction and a low temperature ordered phase where A-atoms have a tendency to occupy one of the two sublattices, α and β, of the bcc-lattice (indicated in Figure 1.1) and the B-atoms occupy the other sublattice. In the case $x \geq 0.5$, defining an order parameter for this transition as, $S = (N_A^\alpha - N_A^\beta)/N_A$, where $N_\mu^{\alpha,\beta}$ is the number of μ-atoms in sublattice α or β and within the mean-field approximation show that:

(b) The order parameter is a solution of the transcendental equation,

$$zx\epsilon S - \frac{1}{2}k_BT \ln \frac{(1+S)[1-x(1-S)]}{(1-S)[1-x(1+S)]} = 0.$$

(c) There is a line of critical points given by:

$$T_c(x) = \frac{z\epsilon x(1-x)}{k_B}.$$

1.7   Suppose the same binary alloy A_xB_{1-x} considered in the preceding exercise. Compute the structure factor,

$$\mathbf{F}(Q) = \sum_r f_r e^{iQ \cdot r},$$

where r denotes position vectors of lattice sites and f_r is the scattering factor of the atom located at position r. Taking into account that the detected intensity, I, in a scattering experiment is proportional to the square of the structure factor, show that it will be given as:

$$I \propto \mathbf{F}^2 = (f_A - f_B)S^2,$$

where f_A and f_B are the scattering factors for atoms A and B, respectively, and S is the order parameter defined in the preceding exercise. This result shows that the intensity vanishes at $T > T_c$ while superstructure peaks will grow at $T \leq T_c$. Show that these peaks occur at reciprocal space positions $\mathbf{Q} = \frac{2\pi}{a}(n_x + n_y + n_z)$ such that the sum of the natural numbers n_x, n_y, n_z is odd.

1.8 Consider the same model for a binary alloy as in Exercise 1.6 but now with $\epsilon < 0$ ($J > 0$). In that case, atoms of a given species prefer to have nearest neighbours of the same species. The model is thus adequate for dealing with phase separation in alloys. Show that in the mean-field approximation:

(a) The free energy function of the alloy can be written as:

$$\mathcal{F}(T, N_A, N_B) = E_0 - \frac{1}{2} z\epsilon \frac{N_A N_B}{N_A + N_B}$$
$$+ k_B T \left(N_A \ln \frac{N_A}{N_A + N_B} + N_B \ln \frac{N_B}{N_A + N_B} \right).$$

(b) Imposing the equilibrium coexistence condition, $\mu_A = \mu_B$, of A-rich and B-rich separated phases, with μ_A and μ_B being the chemical potentials of these phases, leads to the coexistence curve:

$$\frac{1}{2}(v_{AA} - v_{BB}) + \frac{1}{2}z\epsilon c + \ln \frac{1+c}{1-c} = 0,$$

where c is the alloy concentration.

(c) For a symmetric alloy with $v_{AA} = v_{BB}$, a critical point exists at:

$$T_c = \frac{z|\epsilon|}{4k_B}.$$

1.9 Consider a system described by a Landau free energy of the form:

$$\mathcal{F}(\phi, \Delta, T) = a_0(\Delta, T) + \frac{1}{2}a_2(\Delta, T)\phi^2 + \frac{1}{4}a_4(\Delta, T)\phi^4$$
$$+ \frac{1}{6}a_6(\Delta, T)\phi^6,$$

where ϕ is the main order parameter assumed to be scalar and Δ is an external field thermodynamically conjugated to a secondary order parameter, $x = \frac{1}{N}\frac{\partial \mathcal{F}}{\partial \Delta}$, where N is the number of constituents. This model can be used to describe an antiferromagnetic material subjected to an applied magnetic field or to a mixture of ^{3}He-^{4}He. In the first case, ϕ is the antiferromagnetic order parameter and Δ an applied magnetic field that controls the net magnetization, while in the second case, ϕ is the order parameter for the normal to superfluid transition and Δ the difference of chemical potentials of ^{3}He and ^{4}He constituents that controls the molar fraction of ^{3}He in the mixture.[24] Consider that as temperature is lowered, the coefficient $a_4 \geq 0$ while $\Delta \leq \Delta_t$, and changes sign for $\Delta > \Delta_t$. Show that:

---

[24] For more details, see Ref. [18].

(a) The model is adequate to describe a line of critical points, $T_c(\Delta)$ given by the condition $a_2(\Delta, T_c) = 0$ that ends at a tricritical point given by the condition $a_2(\Delta_t, T_t) = a_4(\Delta_t, T_t) = 0$. For $\Delta > \Delta_t$ and $a_4 < 0$, the model predicts a line of first-order transitions, $T_0(\Delta)$, given by the condition,

$$a_2(\Delta, T_0) - \frac{3a_4^2(\Delta, T_0)}{16a_6(\Delta, T_0)} = 0,$$

which requires that $a_2(\Delta, T_0) \neq 0$ and $a_6 > 0$.

(b) Close to the tricritical point, for $\Delta > \Delta_t$, the system undergoes a phase separation at which the variable x, of the coexisting phases, shows a discontinuity given by:

$$\delta x = x_1 - x_2 \simeq \frac{1}{2N} \frac{\partial a_2(T, \Delta)}{\partial \Delta} \phi^2 \simeq \frac{3|a_4|}{8a_6 N} \frac{\partial a_2(T, \Delta)}{\partial \Delta}.$$

(c) When the tricritical point is approached, the order parameter ϕ behaves as

$$\phi(T) \sim (T_t - T)^\beta,$$

where, if a_4 approaches zero faster than a_2, $\beta = 1/4$ (tricritical behaviour) while, if a_2 approaches zero faster than a_4, $\beta = 1/2$ (critical behaviour).

1.10 Consider the following simplified version of the Blume–Emery–Griffiths model defined on a lattice with N sites,

$$\mathcal{H}_{BEG}(\{S_i\}) = -\epsilon \sum_{\langle ij \rangle_{nn}} S_i S_j + M \sum_{i=1}^{N} S_i^2,$$

where the variables S_i can take values, -1, 0, $+1$, and $\epsilon > 0$, which supposes that the exchange is ferromagnetic. In the Bragg–Williams mean-field approximation, show that:

(a) The free energy function can be expressed in the form:

$$\mathcal{F} = -\frac{1}{2} N z \epsilon (p^+ - p^-)^2 + N M (p^+ + p^-)$$
$$+ N k_B T (p^+ \ln p^+ + p^- \ln p^- + p^0 \ln p^0),$$

where p^+, p^-, and p^0 are the probabilities that any i-site is occupied by a $S_i = +1$, $S_i = -1$, or $S_i = 0$ spin, respectively.

(b) Defining, $m \equiv \langle S \rangle = p^+ - p^-$, $x \equiv 1 - \langle S^2 \rangle = p^0$, minimization of $\mathcal{F}$ under the condition, $p^+ + p^- + p^0 = 1$, leads to:

$$p^+ = \frac{1}{Z} \exp\{\beta(z\epsilon m - M)\}, \quad p^- = \frac{1}{Z} \exp\{-\beta(z\epsilon m + M)\}, \quad p^0 = \frac{1}{Z},$$

where $Z = 1 + 2e^{-\beta|M|} \cosh(\beta z\epsilon m)$. Then, the free energy function can be written as

$$\mathcal{F} = \frac{1}{2}z\epsilon m^2 - k_B T \ln\left[1 + 2e^{-\beta|M|} \cosh(\beta z\epsilon m)\right]. \quad (1.154)$$

(c) Expansion of this free energy in a power series of m to the sixth order leads to a free energy of the general type given in Exercise 1.9 with, $a_0(T, M) = -k_B T \ln(1 + 2e^{-\beta M})$, $a_2(T, M) = z\epsilon \left(1 - \frac{z\epsilon}{\eta k_B T}\right)$ and $a_4(T, M) = \frac{z\epsilon}{2\eta^2}(\beta z\epsilon)^3 \left(1 - \frac{\eta}{3}\right)$, where $\eta = 1 + \frac{1}{2}e^{\beta M}$ and M must be identified with the field Δ.

(d) The tricritical point of this model occurs at $x_t = \frac{2}{3}$.

1.11 Consider a $2d$ triangular lattice with an Ising spin $S_i = \pm 1$ defined on each of the N sites. In addition to the two-spin ferromagnetic interaction suppose that spins also interact via a three-spin interaction so that the hamiltonian is of the form:

$$\mathcal{H} - J_1 \sum_{\langle ij \rangle_{nn}} S_i S_j - J_2 \sum_{\langle ijk \rangle_{nn}} S_i S_j S_k,$$

where the first sum extends over all pairs of nearest neighbour spins and the second term extends over all the spins sitting on equilateral triangles of nearest neighbour sites.

(a) Use the Bragg–Williams mean-field approximation to determine the free energy function per spin, $f(T, m)$, where m is the mean value of the spin per site $\langle S_i \rangle$ and show that the equation of state of the system is given by

$$m = \tanh[6\beta(J_1 m + J_2 m^2)].$$

(b) Show that for $J_2 \neq 0$, the model may describe a first-order transition.

(c) Plot the phase diagram of this model for both positive and negative values of J_2 in a diagram giving kT/J_1 vs. J_2/J_1.

2

Dynamics of phase transitions

In the preceding chapters, we have only considered equilibrium aspects of phase transitions. In practice, phase transitions rarely occur in strict equilibrium conditions and, in fact, non-equilibrium features are especially relevant in the case of first-order transitions. Often, due to dynamic constraints, metastable or even unstable states are reached and the decay to equilibrium involves a phase transition. The dynamics of phase transitions is precisely concerned with the transient dynamical evolution of a system that relaxes toward equilibrium from an initial metastable (or unstable) state [44, 45, 46]. This problem involves dynamical processes such as nucleation and subsequent growth, spinodal decomposition, ordering and coarsening processes. Prototypical examples are the order–disorder transitions [47, 48] and phase separation processes [49, 50] in an AB binary alloy quenched from an initial temperature above its critical ordering temperature T_c, to a final temperature below T_c. Anyhow, beyond alloys these concepts are essential to understand the behaviour of a broad class of materials, undergoing first-order phase transitions. In general, in any of these systems, immediately after the quench, the system is in a non-equilibrium state with an ordered state characteristic of the high-temperature phase. A fluctuation into the low-temperature phase is then necessary to begin the formation of ordered regions [51]. For metastable states, the decay to equilibrium is caused by rare localized amplitude fluctuations (nucleation process), while the decay of an unstable state originates in the amplification of initially small long wavelength fluctuations (spinodal decomposition or ordering).

In this chapter, we will first study a simple model that illustrates the importance of dynamical effects on phase transitions and its relation to hysteresis effects. In the next section, we will formally introduce how to establish dynamic equations in systems undergoing phase transitions, and later on we will discuss mechanisms such as nucleation or spinodal decomposition employed to initiate the decay to the equilibrium state. Finally, we will also consider coarsening processes expected to control the dynamics at the late times.

2.1 Metastable and unstable states in the framework of Landau theory

Metastable states are thermodynamic states corresponding to minima of the free energy but not to the lowest minimum, which corresponds to the equilibrium stable state. Therefore, metastability corresponds to a situation of precarious stability that can, in a sense, be *easily* disturbed. Unlike an equilibrium state, a given disturbance can cause the system to decay into a lower free energy state. Within the mean-field or Landau description, metastable states correspond to the continuation of isotherms beyond the situation of two-phase equilibrium coexistence. To illustrate the concept and relevance of metastable states and metastability, let us consider a simple uniaxial magnetic material with a scalar order parameter as described by the free energy given in Eq. 1.105, that is,

$$f(m, \epsilon) = f_0(T) + \frac{1}{2}a_2(T)m^2 + \frac{1}{4}a_4m^4 - hm, \qquad (2.1)$$

where, $a_2(T) = a_0(T - T_c)/T_c$. We may suppose that initially the system is in equilibrium at a temperature $T < T_c$ subjected to a strong applied positive field. When the field is decreased keeping the temperature constant, a transition is expected to occur at $h = 0$ at which the magnetization should switch from positive to negative. Nevertheless, for the transition to occur, large enough thermal fluctuations are needed to provide the dynamical mechanism necessary to overcome the energy barrier that separates the states with positive and negative magnetization. Therefore, it may happen that as the field is further increased, the magnetization does not switch and the system remains in a metastable minimum, as illustrated in Figure 2.1. The decay of the metastable state to an equilibrium state will essentially depend on temperature and on the rate at which the field is swept. In any case, for a certain value of the field the barrier will disappear, and the system will decay into the equilibrium state. This point corresponds to the limit of metastability and defines the so-called spinodal point, which corresponds to an unstable state. A schematic illustration of these regions in the m-T phase diagram is depicted in Figure 2.2. It is important to indicate that, in fact, the distinction between metastable and unstable regions is sharp only in a mean-field picture [52]. In real systems, a gradual crossover between both mechanisms is expected. Nevertheless, the concepts of metastability and spinodal retain some validity as a limiting behaviour.

Once the system reaches the equilibrium state corresponding to a strong applied negative magnetic field, if the field is now increased it is expected that the system displays the same transition behaviour as described above. Therefore, as illustrated in Figure 2.1, decreasing and increasing field paths will be different and, as a consequence of metastability, the transition will occur with hysteresis, which reflect the existence of energy dissipation associated with non-equilibrium effects.

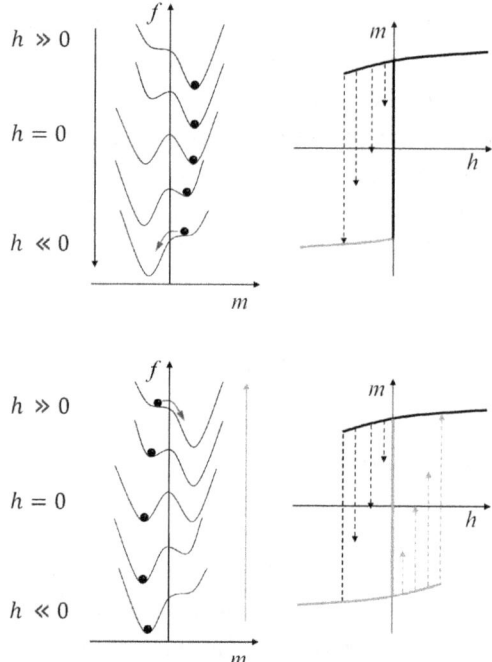

Figure 2.1 Top panels illustrate the evolution of a system from an equilibrium state with $m > 0$ corresponding to a positive and high field as the field decreases. When it becomes negative, the system can be in a metastable state with $m > 0$ until the spinodal limit is reached. In this range of fields, the system can jump to the stable state with $m < 0$ driven by thermal fluctuations. The path followed by the system in the m-h diagram is shown. In this diagram dashed arrows indicate the decay to the equilibrium state. Bottom panels illustrate the case in the same situation but with an increasing field starting from an equilibrium state with negative and high field. The complete loop defines the hysteresis cycle.

It is interesting to analyse in more depth the importance of the field rate on the decay to equilibrium. This can be viewed as a mean first-passage

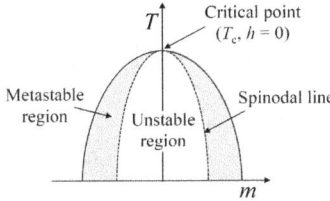

Figure 2.2 Schematic phase diagram in the m-T space showing the metastable and unstable regions.

time problem [53] from an initial metastable state to a more stable state in a system driven at finite rate by an external field. Actually, this can be understood by considering the competition between the sweep rate of the driving parameter and the passage or decay rate from the metastable state. From this point of view, hysteresis is a consequence of the persistence of metastable states due to the fact that sweep rate is larger than the decay rate [54].

Since the energy barrier to be overcome decreases as the field is swept, at some value h_d the metastable state will decay. Owing to the randomness of thermal fluctuations, h_d will be a random variable, which means that for a sequence of repetitive initialization of a given sweep process this field will take values according to a certain distribution $\{h_d\}$. To a first approximation, this distribution may be conveniently characterized by its mean value $\langle h_d \rangle$ and its standard deviation, $\sigma = \sqrt{\langle h_d^2 \rangle - \langle h_d \rangle^2}$. Formally, these two quantities can be related to the metastable state lifetime that can be estimated from a first-passage time formulation [55]. It is reasonable to assume that the probability $P(t)$ for the system to remain in the metastable state after a time t is given by the following generalized Poisson process:

$$\frac{dP(t)}{dt} = -\eta(t)P(t), \qquad (2.2)$$

where $\eta(t)$ is a transition probability, which is assumed to be a function of time. This equation does not satisfy detailed balance (that we will discuss in detail in the next section), but this is the expected microscopic behaviour, at least, close to the metastability limit where reverse jumps must be very unlikely.

Let us consider the process with $v_h \equiv dh/dt \gg 0$ and assume that two limiting fields, $h_i \leq h \leq h_f$, exist such that, $\eta = 0$ for $h \leq h_i$ and $\eta = \infty$ for $h \geq h_f$. These two fields define the range where the system should decay from the metastable to the more stable state. Assuming that the field is linearly swept with time ($v_h = $ constant), then

$$\langle h_d \rangle = h_i + v_h \langle t \rangle, \tag{2.3}$$

$$\sigma^2 = v_h^2 \langle t^2 \rangle + h_i^2 + 2 v_h h_i \langle t \rangle - \langle h_d \rangle^2, \tag{2.4}$$

where $\langle t \rangle$ and $\langle t^2 \rangle$ are time and square time means of the lifetime distribution. They can be expressed as

$$\langle t \rangle = \frac{1}{v_h} \int_{h_i}^{h_f} \exp\left[W(h)\right] dh, \tag{2.5}$$

$$\langle t^2 \rangle = \frac{2}{v_h^2} \left\{ v_h h_i - \int_{h_i}^{h_f} h \exp\left[W(h)\right] dh \right\}, \tag{2.6}$$

where $W(h) = -\frac{1}{v_h} \int_{h_i}^{h_f} \eta(u) du$.

The transition probability $\eta(t)$ is related to the energy barrier that separates the initial metastable state and the final state. This barrier can be obtained from the free energy function 2.1. We consider that initially, $T < T_c$ and $h > 0$. Then, at $h = 0$, the barrier that separates the states with $m_0(T) = \pm\sqrt{|a_2(T)|/a_4}$ is given by $\Delta(0) = f(m_0^2) - f_0(T) = [a_2(T)]^2/4a_4$. At the spinodal point, $h_s = 2b(|a_2(T)|/3a_4)^{3/2}$, the barrier disappears, $\Delta(h_s) = 0$. We then assume that

$$\eta[h(T)] = \begin{cases} 0 & \text{for } h > 0, \\ \frac{\omega}{\Delta(h)} & \text{for } h \leq 0, \end{cases} \tag{2.7}$$

where ω is an energy per unit time related to thermal fluctuations that, indeed, must depend on temperature. In this case, the fields h_i and h_f must be identified with $h = 0$ and h_s, respectively. Introducing the explicit dependence of Δ on h derived from Eq. 2.1, for $h \leq 0$ the following transition probability is obtained:

$$\eta(h) \simeq \frac{\omega}{\Delta(0)} \frac{h_s}{|h_s - h|}, \tag{2.8}$$

where ω has been assumed constant. Note that in this model, η displays a discontinuity at $h = 0$.

Now, using the general expression for $W(h)$, it is obtained that

$$W(h) = \frac{\omega h_s}{v_h \Delta(0)} \ln \frac{|h_s - h|}{h_s}. \tag{2.9}$$

It is convenient to introduce $\tau_h \equiv h_s/v_h$ and $\tau_{tf} \equiv \Delta(0)/\omega$, which are characteristic time scales associated with field-sweep rate (driving rate) and thermal fluctuations, respectively. Then, $\langle t \rangle$ and $\langle t^2 \rangle$ can be computed, and from Eqs 2.3 and 2.4, it follows that

$$\langle h_d \rangle = \frac{h_s}{1 + \tau_h/\tau_{tf}}, \tag{2.10}$$

$$\sigma^2 = \langle h_d \rangle \sqrt{\frac{\tau_h/\tau_{tf}}{2 + \tau_h/\tau_{tf}}}. \tag{2.11}$$

The following two limits are especially interesting,

- for $\tau_h/\tau_{tf} \to 0$, $\langle h_d \rangle \to h_s$ and $\sigma \to 0$,
- for $\tau_h/\tau_{tf} \to \infty$, $\langle h_d \rangle \to 0$ and $\sigma \to 0$.

The first case corresponds to the athermal case, which can be considered as a particular case of a non-thermal transition. In this situation, thermal fluctuations are not operative and the energy barrier can only be overcome when the spinodal field h_s is reached. In this case, $\langle t \rangle = h_s/v_h$ is simply the time needed to sweep the field from 0 to h_s. In contrast, the second case corresponds to the strong thermally activated situation in which, due to the large amplitude of thermal fluctuations, the transition occurs in equilibrium at $h = 0$. Indeed, no hysteresis is expected to occur in such a situation. Usual thermally activated situation occurs in the intermediate regime between these two extreme limits. In this case $\sigma \neq 0$ and the transition occurs with hysteresis.

In the case of transitions that approach the athermal limit to a high degree, it is convenient to introduce a new time scale τ_d. This is a relaxation time that characterizes the decay of the metastable state. Often, this time is assumed to be negligible, but in real materials displaying athermal behavior, it is a finite time that can be important when it becomes comparable with τ_h. This is important when the system is swept at a high rate and the condition of quasistatic driving, which assumes that the system spends an overwhelming majority of time in a static situation, is not satisfied. This enables to distinguish the different situations for the transition,

Equilibrium: $\tau_d \ll \tau_{tf} \ll \tau_h$,

Thermally activated (non-equilibrium): $\tau_d \ll \tau_h \sim \tau_{tf}$,

Athermal (quasistatic): $\tau_d \ll \tau_h \ll \tau_{tf}$,

Athermal (rate dependent): $\tau_d \sim \tau_h \ll \tau_{tf}$.

As we will see in Chapter 8, these situations are commonly observed in transitions taking place in real materials.

2.2 Dynamic equations: formal treatment

The aim of this section is to formally establish general dynamic equations to address kinetic problems in systems undergoing phase transitions. To be specific, let us consider a microscopic lattice system with spin-like variables $\{S_i\}$, described by a hamiltonian $\mathcal{H}(\{S_i\})$. In Chapter 1, we have seen that this class of models is, in principle, only adequate to study static properties. We are, however, interested in introducing equations that allow studying kinetic problems in such a class of systems. The usual way to proceed is to formulate an equation of motion for the probability density, $P(\{S_i\}, t)$, which represents the probability that the lattice is in a given configuration at time t. The simplest alternative is based on the following master equation,

$$\frac{\partial P(\{S_i\}, t)}{\partial t} = \sum_{\{S_i\}} [w(\{S_i\}, \{S_i'\}) P(\{S_i'\}, t) - w(\{S_i'\}, \{S_i\}) P(\{S_i\}, t)],$$

(2.12)

where the sum is over all the accessible configurations $\{S_i\}$, and $w(\{S_i\}, \{S_i'\})$ is the transition rate induced by thermal fluctuations from configuration $\{S_i'\}$ to configuration $\{S_i\}$. Indeed, the probability P must satisfy the normalization condition

$$\sum_{\{S_i\}} P(\{S_i\}) = 1.$$

(2.13)

It is clear that the master equation simply expresses a balance where the first term in the left-hand side represents the increase per unit time of the probability $P(\{S_i\}, t)$ due to transitions from configuration $\{S_i'\}$ to configuration $\{S_i\}$, and the second term the decrease due to transitions from $\{S_i\}$ to $\{S_i'\}$. All the physics is included in the transition probability. The main assumption is that the process is Markovian. Therefore, the following two properties must be satisfied. (*i*) The transition probability $w(\{S_i\}, \{S_i'\})$ does not depend on time and (*ii*) it only depends on the configurations $\{S_i\}$ and $\{S_i'\}$, which supposes that it does not depend on the history preceding the state $\{S_i\}$. In other words, this means that memory effects are not considered within the present assumptions.

The transition probability w can be supposed to originate from the interaction of variables S_i with a heat bath, and thus associated with thermal fluctuations. In solids, for instance, this is provided to a large extent by phonon modes. We will assume that, in general, this class of excitations equilibrates very rapidly in comparison to the change of local spin variables.

For steady states, and in particular for equilibrium, $\partial P(\{S_i\}, t)/\partial t = 0$ which leads to

$$\sum_{\{S_i\}} w(\{S_i\}, \{S_i'\}) P(\{S_i'\}, t) = \sum_{\{S_i\}} w(\{S_i'\}, \{S_i\}) P(\{S_i\}, t). \qquad (2.14)$$

Moreover, in the long time limit, the equilibrium stationary solution $P_e(\{S_i\})$ must be proportional to the Boltzmann factor. This is, for instance satisfied by imposing the sufficient, but not necessary, detailed balance condition

$$\frac{w(\{S_i\}, \{S_i'\})}{w(\{S_i'\}, \{S_i\})} = \frac{P_e(\{S_i'\})}{P_e(\{S_i\})} = e^{-\beta(\mathcal{H}(\{S_i'\}) - \mathcal{H}(\{S_i\}))}. \qquad (2.15)$$

Specific features of the dynamics must be considered in the expression of the transition probability w. In particular, this transition probability should depend on whether variables $\{S_i\}$ satisfy a conservation law or not. In the conserved case, for instance, if the variable S_i, located in a given lattice site, changes by an amount ζ, neighbouring spins must also change so that $\sum_i S_i$ remains constant. In the case of microscopic lattice models, local changes ζ can only be discrete. However, within a coarse grained framework as introduced in Section 2.2, we may assume infinitesimal changes so that the process can be assumed as continuous. In this case, the detailed balance condition should read

$$\frac{w(\{\phi(\boldsymbol{r})\}, \{\phi'(\boldsymbol{r})\})}{w(\{\phi'(\boldsymbol{r})\}, \{\phi(\boldsymbol{r})\})} = e^{-\beta[F(\{\phi(\boldsymbol{r})\}) - F(\{\phi'(\boldsymbol{r})\})]}, \qquad (2.16)$$

where $\phi(\boldsymbol{r})$ is the order parameter density obtained by averaging the microscopic variables over the cells of the coarse grained system, and $F(\{\phi(\boldsymbol{r})\})$ is a Landau free energy functional. Therefore, we can write

$$\begin{aligned} w(\{\phi(\boldsymbol{r})\}, \{\phi'(\boldsymbol{r})\}) \\ = \Omega(\{\phi(\boldsymbol{r})\}, \{\phi'(\boldsymbol{r})\}) \exp\left\{\frac{1}{2}\beta[F(\{\phi'(\boldsymbol{r})\}) - F(\{\phi(\boldsymbol{r})\})]\right\}, \end{aligned} \qquad (2.17)$$

where Ω must be a symmetric function in the initial and final states that carries specific details of the interaction between microscopic variables with the heat bath. We expect that Ω can thus be expressed only as a function of the local change ζ. Since cells in the coarse grained system comprise a large number of lattice sites, ζ must be small compared with the local value of the order parameter and, consequently, Ω must be a sharply peaked function of ζ that should be characterized by its moments.

As an example, we can glance at how to proceed in the case of a system with dynamics constrained by a conservation law. Taking into account

Eq. 2.17 and still assuming that the system is constituted of discrete cells, the master equation can be rewritten in the form [56],

$$\frac{\partial P(\{\phi(r)\}, t)}{\partial t} = \frac{1}{2} \sum_{\{r,r'\}} D_{rr'} \int_{-\infty}^{\infty} d\zeta \, \Omega(\zeta)\Pi(\zeta), \qquad (2.18)$$

where $D_{rr'} = 1$ if cells located at r and r' are nearest neighbours and 0 otherwise, that is,

$$\Pi(\zeta) = P(\cdots, \phi(r) + \zeta, \phi(r') - \zeta, \cdots, t)e^{(\beta \Delta F/2)}$$
$$- P(\cdots, \phi(r), \phi(r'), \cdots, t)e^{(-\beta \Delta F/2)}, \qquad (2.19)$$

and

$$\Delta F = F(\cdots, \phi(r) + \zeta, m(r') - \zeta, \cdots) - F(\cdots, \phi(r), \phi(r'), \cdots). \qquad (2.20)$$

The right-hand side of Eq. 2.18 can now be expanded in a power series of ζ. Then the different moments of $\Omega(\zeta)$ occur in the expansion. Since $\Omega(\zeta)$ is a symmetric function and it is expected that higher order (even) moments are small, the expansion can be truncated at second order, which means that,

$$\frac{\partial P(\{\phi(r)\}, t)}{\partial t} = \frac{\Gamma_m}{4} \sum_{\{r,r'\}} D_{rr'} \left(\frac{\partial}{\partial \phi(r)} - \frac{\partial}{\partial \phi(r')} \right)$$
$$\left[\beta \left(\frac{\partial F}{\partial \phi(r)} - \frac{\partial F}{\partial \phi(r')} \right) P + \left(\frac{\partial P}{\partial \phi(r)} - \frac{\partial P}{\partial \phi(r')} \right) \right], \qquad (2.21)$$

where Γ_m, given as,

$$\Gamma_m = \int_{-\infty}^{\infty} \zeta^2 \Omega(\zeta) d\zeta, \qquad (2.22)$$

is expected to be proportional to a fluctuation frequency, which determines the jump rate per microscopic site. Taking into account the definition of $D_{rr'}$, Eq. 2.21 can be written in the alternative form

$$\frac{\partial P(\{\phi(r)\}, t)}{\partial t} = \frac{\Gamma_m}{2} \sum_{\{r\}} \frac{\partial}{\partial \phi(r)}$$
$$\left[\beta \left(2d \frac{\partial F}{\partial \phi(r)} - \sum_{r'n.n.r} \frac{\partial F}{\phi(r')} \right) P + \left(2d \frac{\partial P}{\partial \phi(r)} - \sum_{r'n.n.r} \frac{\partial P}{\phi(r')} \right) \right], \qquad (2.23)$$

where $\sum_{r'n.n.r}$ denotes a sum over the $2d$ nearest neighbour cells of the cell located at r. It is convenient to express the preceding equation in the form

$$\frac{\partial P(\{\phi(r)\}, t)}{\partial t} = -\sum_{\{r\}} \frac{\partial J(\{\phi(r)\})}{\partial \phi(r)}, \qquad (2.24)$$

where $\boldsymbol{J}$ is a probability-current vector given by,

$$\boldsymbol{J}(\{\phi(\boldsymbol{r})\}) = -\sum_{\{\boldsymbol{r}'\}} \Gamma_{\boldsymbol{r}\boldsymbol{r}'} \left(\beta \frac{\partial F}{\partial \phi(\boldsymbol{r}')} P + \frac{\partial P}{\partial \phi(\boldsymbol{r}')} \right), \qquad (2.25)$$

with $\Gamma_{\boldsymbol{r}\boldsymbol{r}'} = (\Gamma_m/2)[2d\,\delta_{\boldsymbol{r}\boldsymbol{r}'} - \sum_{\boldsymbol{r}'n.n.\boldsymbol{r}''} \delta_{\boldsymbol{r}'\boldsymbol{r}''}]$, where δ is the Kronecker delta.

In the strict continuity limit the sum over cells must be replaced by an integral and normal derivatives by functional derivatives. In this limit, $\Gamma_{\boldsymbol{r}\boldsymbol{r}'} \propto \nabla^2(\boldsymbol{r} - \boldsymbol{r}')$. Then, the preceding dynamic equation can be written in the form of a Fokker–Planck equation, which is a continuity equation in ϕ-space. This equation has the form

$$\frac{\partial P[\phi(\boldsymbol{r}), t]}{\partial t} = -\int d^d r \frac{\delta \boldsymbol{J}[\phi(\boldsymbol{r}, t)]}{\delta \phi(\boldsymbol{r})}, \qquad (2.26)$$

where $\boldsymbol{J}$ is now given as

$$\boldsymbol{J}[\phi(\boldsymbol{r}, t)] = \Gamma \nabla^2 \left[\beta P[\phi(\boldsymbol{r}), t] \frac{\delta F[\phi(\boldsymbol{r})]}{\delta \phi(\boldsymbol{r})} + \frac{\delta P[\phi(\boldsymbol{r}, t)]}{\delta \phi(\boldsymbol{r})} \right]. \qquad (2.27)$$

In the preceding equation, Γ_c is a parameter proportional to the microscopic jump rate Γ_m that determines the time scale of the dynamic process. The Laplacian in the right-hand side term of the preceding equation expresses the local conservation of the order parameter. This term is expected to play a relevant role at the spatial scale of the correlation length.

In systems with no conservation of the order parameter one can proceed in a similar way, which leads to the following Fokker–Planck equation,

$$\frac{\partial P[\phi(\boldsymbol{r}), t]}{\partial t} = \Gamma_c \int d^d r \frac{\delta}{\delta \phi(\boldsymbol{r})} \left[\frac{\delta P}{\delta \phi(\boldsymbol{r})} + P \frac{\delta F}{\delta \phi(\boldsymbol{r})} \right]. \qquad (2.28)$$

2.2.1 Langevin formulation: models A and B

An alternative and, perhaps, more usual way to establish dynamic equations is by writing down a set of phenomenological equations of motion for a small number of macroscopic variables that are chosen in such a way that they describe correctly the macroscopic properties of the system considered. Let us denote these macroscopic variables $\{\eta_i\}$ and suppose that other degrees of freedom have a much faster evolution so that their effect on the macrovariables can be accounted for by a random noise term. Then, equations of motion are assumed to have the form of a Langevin equation as

$$\frac{\partial \eta_i}{\partial t} = \Phi(\{\eta_i\}, t) + \xi_i(t), \qquad (2.29)$$

where, in general, Φ is a non-linear function. The functions $\xi_i(t)$ correspond to random noise terms. It can be demonstrated that the probability distribution $P(\{\eta_i\}, t)$ of variables η_i satisfies a Fokker–Planck equation provided that the noise terms have zero mean and are delta-correlated,

$$\langle \xi_i(t) \rangle = 0, \tag{2.30}$$

$$\langle \xi_i(t)\xi_j(t') \rangle = 2D_i\delta_{ij}\delta(t - t'), \tag{2.31}$$

with D_i independent of variables η_i.

It is easy to see that the first moment of Fokker–Planck Eqs. 2.26 and 2.28 is of the form

$$\frac{\partial \langle \eta_i \rangle}{\partial t} = \langle \Phi(\{\eta_i\}, t) \rangle, \tag{2.32}$$

where, in the case considered above, the order parameter $\phi(\mathbf{r}, t)$ must be identified with the macrovariables η_i and in the non-conserved case the function Φ must be identified with $\delta F/\delta \phi$, and with $\nabla^2(\delta F/\delta \phi)$ in the conserved case. This leads to two interesting dynamical models, usually called model A and model B in the literature [57]. The corresponding Langevin equations are

$$\frac{\partial \phi(\mathbf{r}, t)}{\partial t} = -\Gamma_{nc}\frac{\delta F[\phi(\mathbf{r})]}{\delta \phi(\mathbf{r})} + \xi(\mathbf{r}, t), \tag{2.33}$$

in the non-conserved case (model A), and

$$\frac{\partial \phi(\mathbf{r}, t)}{\partial t} = -\Gamma_c\nabla^2\frac{\delta F[\phi(\mathbf{r})]}{\delta \phi(\mathbf{r})} + \xi(\mathbf{r}, t), \tag{2.34}$$

in the conserved case (model B). In the preceding equations, F is a free energy functional that adequately describes the system considered.

It is important to note that model A is a purely relaxational model such that ϕ evolves toward its equilibrium value corresponding to the minimum of the functional F. Note, for instance, that in this case if the noise term is neglected,

$$\frac{\partial F}{\partial t} = \int d^d r \frac{\delta F}{\delta m}\frac{\partial m}{\partial t} = \int d^d r \frac{\delta F}{\delta m}\left(-\Gamma_{nc}\frac{\delta F}{\delta m}\right)$$

$$= -\Gamma_{nc}\int d^d r \left(\frac{\delta F}{\delta m}\right)^2 \leq 0, \tag{2.35}$$

which means that F is strictly a non-increasing function of time. Therefore, in the absence of fluctuations associated with noise, this model excludes thermally activated processes. Instead, the evolution toward equilibrium described by model B strongly differs from a pure relaxational one due to the local conservation constraint. In any case, an inequality similar to Eq. 2.35 can also be found in the conserved case by replacing Γ_{nc} by $\Gamma_c\nabla^2$.

Example 1: Motion of a planar interface in a magnetic material

As an example of application of Eq. 2.33, we may consider a uniaxial magnetic material described by the free energy functional,

$$F[m(\boldsymbol{r}), T] = \int d^3r \left\{ f[T, m(\boldsymbol{r})] - hm(\boldsymbol{r}) + \frac{1}{2}b[\nabla m(\boldsymbol{r})]^2 \right\}, \qquad (2.36)$$

where the local free energy density is, $f[T, m(\boldsymbol{r})] = \frac{1}{2}a_2(T)m^2(\boldsymbol{r}) + \frac{1}{4}a_4 m^4(\boldsymbol{r})$, which, as required, is invariant under reversal of the magnetization. The magnetization m is the order parameter, which is indeed a non-conserved quantity. Let us consider the problem of the dynamics of a flat, infinite, interface that separates two magnetic domains with opposite magnetization at a temperature $T < T_c$, subjected to a magnetic field, h. For symmetry reasons, in the absence of an applied field, the magnetization only varies along the x-axis so that, for $x \to \pm\infty$, $m(x) \to \pm m_0$, where $m_0 = [|a_2(T)|/a_4]^{1/2}$ is the equilibrium magnetization of the homogeneous system at temperature T. Since the magnetization is a non-conserved order parameter, the interface dynamics induced by the field will be described by model A. Neglecting the noise term, the equation of motion of the magnetization is given as

$$\frac{1}{\Gamma_{nc}}\frac{\partial m}{\partial t} = b\frac{\partial^2 m}{\partial x^2} - \frac{d}{dm}[f(m) - hm]. \qquad (2.37)$$

In the absence of an applied magnetic field f has two symmetric minima at $\pm m_0$ and no dynamics is possible if fluctuations associated with the noise term are not considered. In this case, the stationary profile of the interface is a solution of

$$b\frac{d^2 m(x)}{dx^2} = \frac{df}{dm}. \qquad (2.38)$$

It is easy to check that the solution of this differential equation is

$$m(x) = m_0 \tanh\left[\frac{x}{\sqrt{2}\xi(T)}\right], \qquad (2.39)$$

where $\xi(T) = [b/|a_2(T)|]^{1/2}$. The magnetization profile is depicted in Figure 2.3.

In the presence of an external magnetic field, we are interested in a moving solution of Eq. 2.37 of the form, $m(x, t) = m_d(x - vt)$. Then, taking into

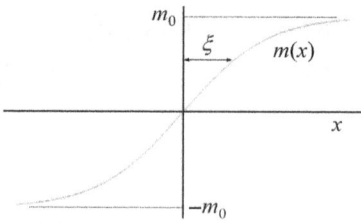

Figure 2.3 Steady state profile of a two-phase interface.

account that, $\partial m_d/\partial x = -v\partial m_d/\partial t$, replacing this solution in Eq. 2.37 the dynamic equation can be expressed as

$$b\frac{d^2 m_d}{dx^2} = \frac{d}{dm_d}[f(m_d) - hm_d] - \frac{v}{\Gamma_{nc}}\frac{dm_d}{dx}. \tag{2.40}$$

This equation has the form of a one-dimensional mechanical equation of motion of a particle of mass b. According to this analogy, x must be identified with time and m_d with the displacement. Then, $-[f(m_d)-hm_d]$ represents a potential energy and the last term of the right-hand side can be interpreted as an effective friction opposing the motion of the interface. In the absence of an applied field, which corresponds to the steady state case, friction vanishes and the corresponding total energy must be conserved, which supposes that, $\frac{1}{2}b(dm/dx)^2 - f(m) = -f(m_0)$. This is an interesting equation that enables us to compute the surface energy σ given by the excess of free energy per unit area of the interface. It is obtained by subtracting the bulk free energy per unit surface area of an homogeneous system as

$$\sigma = \int_{-\infty}^{\infty}\left[\frac{1}{2}b\left(\frac{dm}{dx}\right)^2 + f(m) - f(m_0)\right]dx = b\int_{-\infty}^{\infty}\left(\frac{dm}{dx}\right)^2 dx. \tag{2.41}$$

In the dynamic case, following up with the 1-d mechanical analogy, we can analyse the problem as follows. Due to the applied field, the two maxima of the effective potential are of different height. Thus, the friction term must be such that the particle leaves the highest maximum, which corresponds to a metastable minimum of the free energy, and stops on the lower one, that corresponds to the equilibrium minimum. Indeed, this solution exists only for a given value of v/Γ, which determines the velocity v of the moving interface. Taking into account, as in the steady state case, that for $x \to \pm\infty$, $m(x) \to \pm m_0$, integration of Eq. 2.40 yields,

$$\int_{-\infty}^{\infty}\frac{d}{dx}\left[\frac{1}{2}b\left(\frac{dm_d}{dx}\right)^2 - f + hm_d\right]dx = -\frac{v}{\Gamma_{nc}}\int_{-\infty}^{\infty}\left(\frac{dm_d}{dx}\right)^2 dx, \tag{2.42}$$

which indicates that the loss of (free) energy associated with the interface movement is dissipated by the effective friction.[1] Therefore,

$$[f(m_0) - hm_0] - [f(-m_0) + hm_0] = 2m_0h \cong \frac{v\sigma}{\Gamma_{nc}b}, \qquad (2.43)$$

where we have assumed that $\sigma/b \cong \int_{-\infty}^{\infty}(dm_d/dx)^2dx$. Finally, it is obtained that the interface moves with a velocity which is proportional to the applied field

$$v = \frac{2\Gamma_{nc}bm_0}{\sigma}h. \qquad (2.44)$$

Example 2: Planar interface in a binary alloy

It is interesting to compare previous results with those corresponding to a planar interface in the case of a system with conserved order parameter. For the sake of concreteness, we will consider the case of a binary alloy *A*-*B*. The order parameter is the concentration $c(\boldsymbol{r}, t)$ defined as the difference of mol fractions per unit volume of A and B atoms. The conservation of the order parameter supposes that the average concentration of the alloy is constant. This system can be described by the following free energy functional

$$F[c(\boldsymbol{r}), T] = \int d^3r \left\{ f[T, c(\boldsymbol{r})] + \frac{1}{2}b[\nabla c(\boldsymbol{r})]^2 \right\}, \qquad (2.45)$$

where the local free energy density is a symmetric function, $f[T, c(\boldsymbol{r})] = \frac{1}{2}a_2(T)c^2(\boldsymbol{r}) + \frac{1}{4}a_4c^4(\boldsymbol{r})$. The thermodynamic force that drives the flux of atoms in the dynamic process, which gives rise to phase separation, is the gradient of the difference of chemical potentials of A and B atoms. This difference will be denoted as $\mu(\boldsymbol{r})$ and is given by,

$$\mu(\boldsymbol{r}) = \frac{\delta F}{\delta c(\boldsymbol{r})}. \qquad (2.46)$$

Due to the conservation of the order parameter, the dynamics of this system must be accounted for by model B. Therefore, neglecting the noise term as in the previous example, the equation of motion of the order parameter will be

$$\frac{\partial c(\boldsymbol{r}, t)}{\partial t} = \Gamma_c \nabla^2 \left[\frac{\partial f}{\partial c} - b\nabla^2 c \right], \qquad (2.47)$$

where the parameter Γ_c has been assumed constant. This equation is often known as the Cahn-Hilliard equation since it was first proposed by these authors to describe phase separation in alloys [58, 59].

---

[1] Note that the right-hand side of this equation can be interpreted as the work dissipated by the friction force acting on the particle of mass m.

When μ is constant, $\partial c/\partial t = 0$, which means that no flux of atoms in the system can occur. Therefore, a stationary solution of the Cahn-Hilliard equation must satisfy

$$\frac{\partial f}{\partial c} - b\nabla^2 c - \mu = 0, \tag{2.48}$$

which is equivalent to

$$\frac{\delta}{\delta c(\boldsymbol{r})}\left\{F[c(\boldsymbol{r}), T] - \mu \int c(\boldsymbol{r})\right\} d^3 r = 0. \tag{2.49}$$

This last equation can be understood as a minimization of the functional F with the constraint that global composition is constant. From this point of view, μ can be viewed as a Lagrange multiplier. It is clear that this equation is equivalent to Eq. 2.38 corresponding to the non-conserved case, and thus determines the stationary states of the system. Actually, the difference of chemical potential of the two species in the present conserved case plays the role of the applied magnetic field in the non-conserved case.

To be more explicit, let us consider two phases, α and β, of different composition, in equilibrium with each other separated by a planar interface. We may consider that the α-phase is an A-rich phase, while the β-phase is a B-rich phase. In equilibrium no matter flux will exist across the interface, and thus $\mu = \mu_s$ must be a constant. Therefore, according to Eq. 2.49, this requires that $f(c) - \mu_s c$ must have the same value in both α and β phases. This condition is in fact the well-known common tangent equilibrium condition for phases with different composition [66].

Interface motion will require a flux of atoms in the system induced by a gradient of composition that supposes the existence of a gradient of chemical potential in the system. In the region close to the interface, the full Cahn-Hilliard equation will be needed to describe the process. Anyway, outside this region, the problem can be described conveniently within a simpler diffusion approximation, which means that, due to the conservation constraint, matter has to be transported by diffusion from a far away region for the interface to move. We will analyse this approximation in detail later on when considering the growth of spherical droplets.

2.3 Nucleation

We have seen that metastable states are local minima of the free energy. The decay to the deepest stable minimum is initiated by means of a nucleation mechanism, which supposes the appearance of germs or nuclei of the new phase. Whether the order parameter is a conserved quantity or not, this process is made possible by local aggregation via diffusion or by local ordering of constituents, respectively. Therefore, nucleation is associated with localized, high amplitude fluctuations. Once a critical nucleus is formed, its size increases in time in a growth process that ends when a new homogeneous or phase separated phase is formed.

Classical theory of homogeneous nucleation is based on the simple thermodynamic point of view that nucleation of a nucleus of the stable phase supposes a gain of the bulk free energy, that compensates for the excess of the surface free energy associated with the creation of an interface, that separates the stable nucleus from the metastable initial phase [92]. An interesting example is the freezing of liquid water. Homogeneous nucleation of crystalline ice may occur only in very clean standing water, free of impurities (this is called pristine water), to avoid perturbations that can strongly influence the nucleation process. At normal pressure, pristine water can be supercooled tens of degrees below the equilibrium freezing point [61]. A maximum undercooling of about $-45°C$ can be reached. This is the so-called homogeneous nucleation temperature at which the liquid crystallizes almost instantly.

In the homogeneous nucleation case, since the surface energy per unit area is a positive quantity, the bulk free energy density between the two phases, Δg, will need to be negative for nucleation to be feasible. Therefore, if we assume that clusters or embryos of the new phase have a well-defined volume V and area A, the energy E_e that must be spent to form a droplet is given by

$$E_e = -|\Delta g|V + \sigma A. \tag{2.50}$$

If we consider that the droplet is spherical with radius R, then the energy E_e will depend on this radius that will represent a reasonable measure of the amplitude of thermal fluctuations needed to form the embryos. In this case,

$$E_e(R) = -\frac{4}{3}\pi|\Delta g|R^3 + 4\pi\sigma R^2. \tag{2.51}$$

The energy E_c as a function of R for given values of $|\Delta g|$ and σ is depicted in Figure 2.4.

Eventually, when one of these embryos reaches a large enough size, it will become a stable nucleus and will continuously grow to complete the

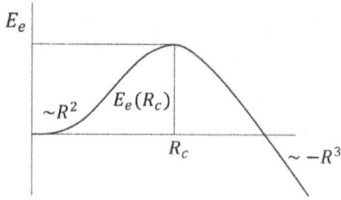

Figure 2.4 $E_e(R)$ as a function of R.

transition. The critical radius necessary to form a stable nucleus corresponds to the value $R = R_c$ that maximizes E_d. It is given by

$$R_c = \frac{2\sigma}{|\Delta g|}. \tag{2.52}$$

Note that in arbitrary dimension d, the critical radius will be given as $R_c = (d-1)\sigma/|\Delta g|$.

The bulk free energy term Δg is directly related to the degree of metastability. Thermodynamically, it can be estimated as

$$\Delta g = \Delta e - T\Delta s, \tag{2.53}$$

where Δe and Δs are enthalpy and entropy changes per unit volume between the stable and metastable states, respectively. In the simple case discussed in Section 2.1, the transition is driven by the field h at constant temperature and, due to the symmetry of the model $\Delta s = 0$, which leads to $|\Delta g| \simeq \Delta e = 2m_0 h$, where m_0 is a saturation magnetization corresponding to the state (T, h). Therefore, in this case,

$$R_c = \frac{\sigma}{m_0 h}, \tag{2.54}$$

and

$$E_e(R_c) = \frac{8}{3}\pi\frac{\sigma^2}{(hm_0)^2}. \tag{2.55}$$

In the preceding expression, the field h is a direct measure of the degree of metastability, since in the magnetic symmetric model considered the equilibrium transition occurs at $h = 0$.

In general, in an asymmetric system, if we suppose that the transition is temperature driven, nucleation will happen at a temperature $T < T_0$, where T_0 is the equilibrium temperature between the two phases. Then, thermodynamics enables us to express

$$\Delta e = \mathcal{L} + \int_{T_0}^{T} \Delta c \, dT, \qquad (2.56)$$

$$\Delta s = \frac{\mathcal{L}}{T_0} + \int_{T_0}^{T} \frac{\Delta c}{T} dT, \qquad (2.57)$$

where $\mathcal{L}$ is the transition latent heat, and Δc the difference of heat capacity between the two phases. Thus,

$$\Delta g = \frac{\mathcal{L} \Delta T}{T_0} + \int_{T_0}^{T} \Delta c \, dT - T \int_{T_0}^{T} \frac{\Delta c}{T} dT. \qquad (2.58)$$

If $\Delta c \simeq 0$, which is a reasonable approximation in many cases, $\Delta g \cong -\mathcal{L}\Delta T/T_0$. A less drastic approximation consists of assuming that Δc is temperature independent to a good approximation. Then,

$$\Delta g = \mathcal{L}\frac{\Delta T}{T_0} - \frac{1}{2}\Delta c \left[\frac{(\Delta T)^2}{T_0} - \frac{(\Delta T)^3}{T_0^2} + \ldots\right]. \qquad (2.59)$$

In the simplest case $\Delta c = 0$, the critical radius can be expressed in terms of the supercooling as

$$R_c \simeq \frac{2\sigma}{\Delta s_0 (T_0 - T)}, \qquad (2.60)$$

where $\Delta s_0 = \mathcal{L}/T_0$ is the entropy difference between the two phases in equilibrium at T_0. The corresponding energy barrier $E_c(R_c)$ is then given as

$$E_e(R_c) \simeq \frac{16}{3}\pi \frac{\sigma^3}{[\Delta s_0 (T_0 - T)]^2}. \qquad (2.61)$$

It is interesting to note that, in both field-driven and temperature-driven cases, the energy barrier to be overcome to form critical nuclei decreases as the degree of metastability increases while the critical radius decreases. In fact, in the metastable region, due to thermal fluctuations, small clusters (or embryos) of the ongoing phase will continuously form and decay toward the metastable matrix. From the thermodynamic theory of fluctuations [62, 63], the distribution of droplet sizes is expected to be proportional to $e^{-\beta W_{min}(R)}$, where W_{min} is the minimal work needed to form a cluster of size R. Thus, $W_{min}(R)$ must be identified with $E_e(R)$. Expanding $W_{min}(R)$ about $R = R_c$ leads to

$$W_{min}(R) = W_{min}(R_c) + \frac{1}{2}\left(\frac{\partial^2 W_{min}(R)}{\partial R^2}\right)_{R=R_c}(R - R_c)^2 + \ldots, \qquad (2.62)$$

where it has been taken into account that R_c corresponds to the maximum of $E_e(R)$. Thus, from Eqs. 2.51 and 2.52, it is obtained that

$$W_{min}(R) \simeq \frac{4\pi\sigma}{3} R_c^2 - 4\pi\sigma(R - R_c)^2, \qquad (2.63)$$

which leads to the gaussian distribution term,

$$\psi(R) \propto \psi_0(R_c) \exp\left\{\frac{4\pi(R - R_c)^2}{kT}\right\}, \qquad (2.64)$$

where $\psi_0(R_c) = e^{4\pi\sigma R_c^3/3kT}$. The probability density should include a prefactor that provides a time scale. It must be introduced by means of a kinetic theory and is essential to determine the nucleation rate. In general, the rate of formation of stable nuclei is obtained by considering the rate at which clusters grow and decay by adding or losing components. The idea was first introduced by Becker and Döring in the case of liquid nucleation from vapour. The model supposes that the growth or decay of a cluster occurs, respectively, by the absorption or evaporation of a molecule at a given time [44, 64]. Nevertheless, the estimation of the prefactor term is, in general, strongly dependent on specific features of the system considered, especially when dealing with nucleation in solids.

A complementary viewpoint is based on the idea that once a critical nucleus is formed it can grow and that its dynamics can be determined by either Eq. 2.33 or Eq. 2.34 depending on whether the order parameter is a conserved quantity or not. As an example, let us analyse the case of the uniaxial magnetic system discussed previously. For this system, the order parameter is non-conserved and, neglecting fluctuations, a kinetic equation similar to the one used for the analysis of a planar interface should be considered. Therefore, for a spherical nucleus, in d-dimensions, the equation of motion in spherical coordinates reads

$$\frac{\partial m}{\partial t} = \Gamma_{nc}\left\{b\frac{\partial^2 m}{\partial r^2} + b\frac{(d-1)}{r}\frac{\partial m}{\partial r} - \frac{df}{dm} + h\right\}, \qquad (2.65)$$

where the magnetic field favours the phase with positive magnetization, $+m_0$. Assuming a moving solution is of the type, $m(r,t) = m_d[r - R(t)]$, the preceding equation can be written as

$$-\frac{1}{\Gamma_{nc}}\frac{dR}{dt}\frac{dm_d}{dr} = b\frac{d^2m_d}{dr^2} + b\frac{d-1}{r}\frac{dm_d}{dr} - \frac{df}{dm_d} + h. \qquad (2.66)$$

This equation describes an interface of a certain width $\xi(T)$ in the vicinity of $r \simeq R$. We will assume that the interface is sharp, which means that $\xi \ll R$. This, indeed, will be a reasonable approximation if temperature is far below the critical temperature. Therefore, we may then assume that $r \simeq R$ because the variation of m_d will occur in a very narrow region, $R - \xi \leq r \leq R + \xi$. Then, in this approximation, the equation of motion can be written as follows:

$$b\frac{d^2 m_d}{dr^2} + \left[\frac{(d-1)b}{R} + \frac{v}{\Gamma_{nc}}\right]\frac{dm_d}{dr} - \frac{df}{dm_d} + h = 0. \qquad (2.67)$$

Note that this equation is equivalent to Eq. 2.40 corresponding to the problem of a planar interface with $\frac{(d-1)b}{R} + \frac{v}{\Gamma_{nc}}$ replaced by v. Therefore, the growth velocity will be given by

$$v = \frac{dR}{dt} = \frac{2bm_0 h}{\sigma} - \frac{\Gamma_{nc}(d-1)b}{R(t)}. \qquad (2.68)$$

Taking into account that radius of the critical nucleus is given by, $R_c = (d-1)\sigma/2m_0 h$, the velocity can be expressed as follows:

$$v = \Gamma_{nc}(d-1)b\left[\frac{1}{R_c} - \frac{1}{R(t)}\right]. \qquad (2.69)$$

This is an interesting equation that indicates that droplets with $R < R_c$ collapse while only those droplets with $R > R_c$ can grow. The case $R = R_c$ corresponds to an unstable stationary situation in which, $v = 0$. The equation also expresses that for $h = 0$, $R_c \to \infty$, so that in strict equilibrium, all droplets of finite size collapse. Therefore, in this model, a certain degree of metastability is required for the transition to start.

The study of the growth of critical nucleus is a relatively simple problem in a symmetric system with non-conserved order parameter, such as a uniaxial magnet. In asymmetric systems, the influence on the growth process of the release of latent heat may have a strong influence on the growth dynamics. In addition, in systems with conserved order parameter, the transport of matter by diffusion should also be considered [92]. From this point of view, it is interesting to illustrate this situation by considering the case of the diffusion controlled growth of a spherical droplet.

We will follow the notation introduced in Example 2 in Section 2.2.1 dealing with a planar interface in a binary alloy and consider a spherical droplet of the β-phase immersed in a metastable, supersaturated, α-phase matrix. Due to the interface curvature, the droplet can only be stationary if the chemical potential in the matrix differs from the equilibrium value μ_s corresponding to the planar interface stationary case. Inside the droplet, we can assume that the concentration is c_β, while the matrix has an excess of concentration, $c_\alpha + \delta c_\alpha$. The corresponding chemical potential in the matrix can then be assumed proportional to δc_α, that is,

$$\mu - \mu_s \equiv \delta\mu = \frac{\delta c_\alpha}{\chi_\alpha}, \tag{2.70}$$

where $\chi_\alpha = \partial c/\partial\mu$, which must be positive for stability reasons.

The radius of a stationary droplet must satisfy an equation analogous to the equation giving the critical nucleus. That is, it must be of the form $R = (d-1)\sigma/|\Delta g|$, where $|\Delta g|$ will be given by the jump discontinuity of $f - \mu c$ across the interface of the nucleus. It can be approximated as

$$\delta(f - \mu c) \cong \delta\mu\Delta c, \tag{2.71}$$

where $\Delta c = c_\beta - c_\alpha$. Therefore,

$$R = \frac{(d-1)\sigma}{\delta\mu\Delta c} = \frac{\chi_\alpha}{\delta c_\alpha}\frac{(d-1)\sigma}{\Delta c}. \tag{2.72}$$

Note that the critical radius may be obtained by just replacing δc_α by $\delta c_\infty \equiv c_\infty - c_\alpha$, where c_∞ is the concentration of the matrix far away from the interface.

We now consider the growth of a droplet of radius R. This requires flux of atoms associated with a concentration gradient in the region outside the interface. We will assume that the interface is sharp, which means that its width $\xi \ll R$. Since the full Cahn-Hilliard equation is only needed in the neighbourhood region of the interface, it can be assumed that the diffusion approximation is adequate to provide a reasonably good description of the problem. Then, within this approximation, the concentration at the interface R is fixed at $c_\alpha + \delta c_\alpha$, where δc_α is given by the value determined by Eq. 2.72. The situation is schematically illustrated in Figure 2.5.

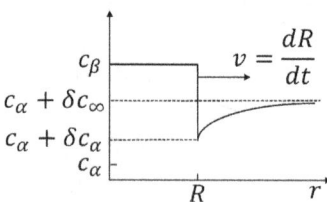

Figure 2.5 Concentration profile of a diffusion controlled growing droplet.

In analogy with the non-conserved case discussed above, we assume that near the moving interface, the concentration has the form $c(r,t) \cong c[r - R(t)]$. Therefore,

$$\frac{\partial c}{\partial t} = -\dot{R}\frac{\partial c}{\partial r}, \tag{2.73}$$

where $\dot{R} \equiv dR/dt$ is the interface velocity. We now take into account that outside the droplet, for $r > R$, the concentration can be described by a diffusion equation. In spherical coordinates, it has the form

$$\frac{\partial c}{\partial t} = D\left[\frac{d-1}{r}\frac{\partial c}{\partial r} + \frac{\partial^2 c}{\partial r^2}\right], \tag{2.74}$$

where D is a diffusion constant. We will suppose a quasistationary regime that is expected to occur when the diffusion time scale R^2/D is negligible compared with the growth time scale, $R/\dot{R}$. Then, for $d = 3$, the solution of the diffusion equation for $r \geq R$ will be of the form, $c(r) = c_\infty - C/r$, where it has been set that far away from the interface the concentration is c_∞. The constant C can be estimated imposing that $c(R) = c_\alpha + \delta c_\alpha$. Then, the following concentration profile results

$$c(r) = c_\infty - \frac{R}{r}(\delta c_\infty - \delta c_\alpha). \tag{2.75}$$

Conservation of the flux at the interface requires that

$$-\int_{R+\xi}^{R+\xi}\frac{\partial c}{\partial t}dr = \dot{R}\Delta c = D\left(\frac{\partial c}{\partial r}\right)_{R+\xi}, \tag{2.76}$$

where $(\partial c/\partial r)_{R+\xi} = (\delta c_\infty - \delta c_\alpha)/R$. Therefore, taking into account Eq. 2.72, the interface velocity is obtained as

$$\dot{R} = \frac{D}{R}\left[\frac{\delta c_\infty}{\Delta c} - \frac{2\sigma\chi_\alpha}{(\Delta c)^2}\frac{1}{R}\right]. \tag{2.77}$$

It is interesting to note that the quasistationary condition $R^2/D \ll \dot{R}/R$ is equivalent to $\delta c_\infty/\Delta c \ll 1$, which means, as expected, that the obtained equation provides a good description of the growing droplet problem when supersaturation is weak.

Introducing the critical radius, $R_c = \sigma\xi_\alpha/2\Delta c\delta c_\infty$, it is found that the interface velocity is given by

$$v = \dot{R} = \frac{\bar{D}}{R}\left[\frac{1}{R_c} - \frac{1}{R}\right], \tag{2.78}$$

where $\bar{D} = 2D\sigma\xi_\alpha/(\Delta c)^2$.[2] This equation is the analog of Eq. 2.69 giving the interface velocity of a growing droplet in a system with a non-conserved

---

[2] It must be noted that in the previous development, the growing droplet has been assumed to have spherical symmetry. Nevertheless, large spherical droplets (of radius about seven times the critical radius) growing by diffusion, as we have considered, are known to become unstable [65].

order parameter. It is interesting to remark that while in the non-conserved case $v \sim R^{-1}$, in the conserved case there is a term $\sim R^{-2}$ which is a consequence of diffusion.

2.3.1 Influence of long-range forces: elastic strain

When nucleation occurs in solids, volume or/and shape of a transformed droplet may be different from those in the original phase. This incompatibility results in a misfit between the droplet and the matrix that induces an elastic strain. The resulting elastic energy must then be added to the bulk free energy that provides the driving force for homogeneous nucleation and, consequently, can play a very relevant role in the nucleation of the new phase. This term is essentially long range and positive and has a significant influence on the interfacial term. Thus, this may strongly affect the nucleation barrier. In some cases, the elastic energy stored in the matrix can balance or even exceed the driving force provided by the stress-free bulk free energy difference between the two phases. This can be expressed as

$$\frac{G\epsilon_0}{|\Delta g|} \geq 0, \tag{2.79}$$

where ϵ_0 is a measure of the crystal lattice misfit, associated with either volume or shape change between the two phases, and G an effective elastic modulus that may depend on the nuclei shape. This elastic energy arises from a long-range effect that enables a sort of thermoelastic equilibrium which occurs at each temperature and applied field or stress. This happens, for instance, in some ferroelastic transitions and may give rise to the occurrence of interesting microstructures and to a remarkable thermodynamic behavior showing memory effects that we will discuss in Chapter 3.

2.3.2 Heterogeneous nucleation

Heterogenous nucleation may occur in the presence of special sites that are capable of lowering the energy barrier E_e to be overcome for the transition to start [66]. This happens commonly in solids in the neighbourhood of lattice defects that break translational symmetry such as grain boundaries or dislocations. The analysis of this situation depends significantly on specific features of the transition considered and lattice defects. Often, heterogeneous nucleation can be treated by simply modifying the homogeneous classical theory. As a simple example, we consider here the case of nucleation on dislocations [67] as a prototypical case that conveniently illustrates this problem.

A simple model consists of considering a cylinder of radius r of a linearly isotropic material with a dislocation in its centre. The elastic energy per unit length E_l of this cylinder shows a logarithmic dependence on r of the form [69]

$$\varepsilon_e(r) = b^2 G \ln \frac{r}{R_0}, \tag{2.80}$$

where b is the magnitude of the Burgers vector characterizing the dislocation and G is proportional to the shear modulus that, in fact, may depend on whether the dislocation is a screw or an edge one. R_0 is the effective core radius of the dislocation.

The basic argument consists of assuming that the entire region inside the cylinder of radius r transforms to the incoming new phase, which allows the dislocation energy to be released. By this mechanism, it is supposed that the dislocation favours the nucleation of the new phase that will replace its core region. The mechanism is considered incoherent since it supposes that a Burgers circuit around the dislocation in the matrix, surrounding the transformed cylindrical region, will still have a closure failure b.[3] In this situation, no elastic interaction between the strain field created by the nucleus of the new phase and the matrix must be considered. This is in contrast with the coherent case where the competition between the driving elastic field provided by the dislocation and the elastic field arising from the incompatibility misfit between both phases must be considered [68]. Thus, in the incoherent model the energy E'_e per unit length that must be spent to form an incoherent cylinder of radius r will be given by

$$E'_e(r) = -\pi r^2 |\Delta g| + 2\pi r \sigma - b^2 B \ln \frac{r}{R_0}. \tag{2.81}$$

The extreme values of this energy are solutions of the equation

$$|\Delta g| r_e^2 - \sigma r_e + \frac{b^2 B}{2\pi} = 0. \tag{2.82}$$

Solutions are controlled by the parameter $\alpha = b^2 B / \pi \sigma R_c$, where $R_c = \sigma / 2|\Delta g|$ is a characteristic radius corresponding to the homogeneous nucleation of a cylinder in the absence of dislocations ($b = 0$). They are given as $r_e = R_c[1 \pm \sqrt{(1 - \alpha)}]$. Two cases must be considered:

---

[3] This closure failure b permits defining the Burgers vector that characterizes the dislocation.

(*i*) $\alpha > 1$. The energy provided by the dislocation compensates the excess of surface energy and nucleation is barrierless, which means that the transition is controlled by growth kinetics.

(*ii*) $\alpha < 1$. In this case, a barrier still exists and Eq. 2.82 has two solutions

$$r_0 = R_c \left[1 - \sqrt{(1 - \alpha)} \right], \tag{2.83}$$

which is the radius of the minimum corresponding to the metastable initial state, and,

$$r_c = R_c \left[1 + \sqrt{(1 - \alpha)} \right], \tag{2.84}$$

which is the radius of the critical nucleus.

The energy $E_e'(r)$ scaled by the energy $\pi R_c^2 |\Delta g|$, which represents the energy barrier per unit length to be overcome for homogeneous nucleation of a cylinder of radius R_c, is plotted in Figure 2.6 as a function of r/R_c for $\alpha < 1$ and $\alpha > 1$.

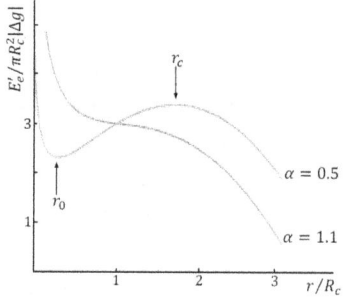

Figure 2.6 $E_e'(r)$ as a function of r. For $\alpha > 1$ the energy shows no maximum and thus nucleation process is barrierless. For $\alpha < 1$ the minimum at r_0 corresponds to the metastable phase and the maximum at r_c defines the critical nucleus.

The procedure discussed above is adequate to find the critical radius of the cylinder about the dislocation core and the corresponding radius in the metastable phase. However, it does not provide how this radius changes along the cylinder axis, which determines the shape of the critical nucleus. It is in fact possible to proceed in a better way by minimizing the total energy that must be spent to go from the metastable to the unstable region corresponding to the maximum of the barrier along the z-axis of the cylinder. This energy can be expressed in the functional form

$$\Delta E_e = \int_{-\infty}^{\infty} \Delta E_e'(r) dz, \tag{2.85}$$

where

$$\Delta E'_e(r, r') = E'_e(r, r') - E'_e(r_0)$$
$$= -\pi|\Delta g|(r^2 - r_0^2) - Bb^2 \ln \frac{r}{r_0} + 2\pi\sigma \left[r\sqrt{1+r'^2}\right] \quad (2.86)$$

with $r' = dr/dz$. The minimum energy function $r(z)$ that determines the shape of the critical nucleus (see Figure 2.7) will then be the solution of the corresponding Euler–Lagrange equation,

$$\frac{\partial \Delta E'_e(r, r')}{\partial r} - \frac{d}{dz}\frac{\partial \Delta E_e(r, r')}{\partial r'} = 0. \quad (2.87)$$

This is a non-linear differential equation that must be solved numerically.

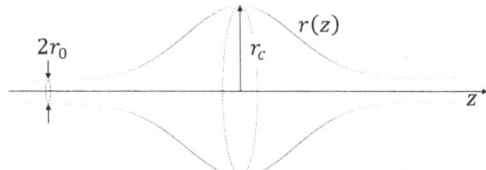

Figure 2.7 Schematic representation of the shape of an incoherent critical nucleus formed along the axis of a dislocation core. The critical radius and the radius of the metastable phase are indicated.

2.4 Spinodal decomposition

Spinodal decomposition is the mechanism by which a phase transition occurs in a system described by a conserved order parameter when it is brought into an unstable state.[4] Therefore, in contrast with the nucleation mechanism, spinodal decomposition is a barrierless mechanism that occurs globally in the system induced by low amplitude, long wavelength fluctuations. In particular, this is the relevant mechanism that gives rise to phase separation that may occur, for instance, in binary alloys rapidly driven or quenched into an unstable state within the two-phase coexistence region.

We can suppose, as in Example 2 of Section 2.2.1, a binary A-B alloy of average concentration c_0 that is quenched into an unstable state at a temperature T. The alloy will decompose into an A-rich, α-phase and a B-rich, β-phase that, in equilibrium, have concentrations c_α and c_β, respectively. Because the order parameter is conserved, this decomposition process

---

[4] The corresponding process taking place in a system with non-conserved order parameter that decays from an unstable state is usually denoted as an ordering process.

will be controlled by diffusion. Initially, we will assume that the system is in a state with homogeneous concentration c_0. We now focus on the early stages of the decomposition and we expand the term $\partial f / \partial c$ about $c = c_0$ in Eq. 2.47, that is,

$$\frac{\partial f}{\partial c} = \left(\frac{\partial^2 f}{\partial c^2}\right)_{c_0} (c - c_0) + \frac{1}{2}\left(\frac{\partial^3 f}{\partial c^3}\right)_{c_0} (c - c_0)^2 + ..., \tag{2.88}$$

where it has been taken into account that in the initial homogeneous state, $(\partial f / \partial c)_{c_0} = 0$. With the aim of focusing on the fluctuations of the concentration field, we introduce the Fourier modes of the deviation field, $u(\mathbf{r}, t) = c(\mathbf{r}, t) - c_0$, which are given as

$$u(\mathbf{r}, t) = \frac{1}{V} \sum_{\mathbf{k}} \hat{u}(\mathbf{k}, t) \exp(i\mathbf{k} \cdot \mathbf{r}), \tag{2.89}$$

where V is the system volume, and the sum must be carried out over all wave vectors $\mathbf{k}$ of the first Brillouin zone. The usual convention $\hat{u}(\mathbf{k} = 0) = 0$ is assumed. In the early stages of evolution, stability analysis can be performed with the linearized version of the Cahn-Hilliard equation, which is obtained by keeping only the first term in the expansion of Eq. 2.88. In Fourier representation this linearized equation takes the form,

$$\frac{\partial \hat{u}(\mathbf{k}, t)}{\partial t} = \omega(k)\hat{u}(\mathbf{k}, t), \tag{2.90}$$

where the amplification factor, $\omega(k)$, is given by

$$\omega(k) = -\Gamma_c k^2 \left[bk^2 + \left(\frac{\partial^2 f}{\partial c^2}\right)_{c_0} \right]. \tag{2.91}$$

According to this equation, when $(\partial^2 f / \partial c^2)_{c_0} \geq 0$, which indicates that the initial homogeneous state corresponds to a metastable minimum of the free energy [note that $(\partial^2 f / \partial c^2)_{c_0} = 0$ defines the spinodal line], $\omega \leq 0$ for all k. Therefore, in this case, all fluctuations decrease with time. However, when $(\partial^2 f / \partial c^2)_0 < 0$, which indicates that the initial state is unstable, fluctuations with wavenumber in the range $0 < k < k_0 = [-(\partial^2 f / \partial c^2)_{c_0}/b]^{1/2}$ can grow. The wavenember dependence of ω is depicted in Figure 2.8 for positive, zero and negatives values of $(\partial f / \partial c)_{c_0}$.

The most unstable wave length $\lambda_{max} = 2\pi/k_{max} = \sqrt{8}\pi/k_0$ corresponds to the wave number $k_{max} = k_0/\sqrt{2}$ for which the amplification factor ω is maximum. Actually, the existence of a most unstable wavelength $\lambda_{max} > 0$ means that random fluctuations will decompose into a pattern with a

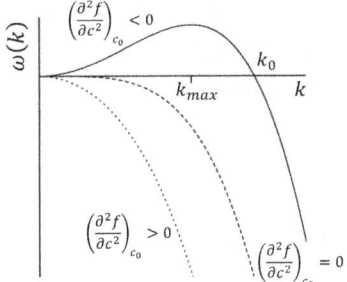

Figure 2.8 Amplification factor ω as a function of the wave number k for positive (metastable state), zero (marginally unstable state) and negative (unstable state) values of the second derivative of the free energy about the initial homogeneous state. Only in this last case fluctuations with wave number in the range $0 < k < k_0$ can grow.

characteristic length scale. This behaviour is schematically illustrated in Figure 2.9 that shows the expected time evolution of the local concentration.

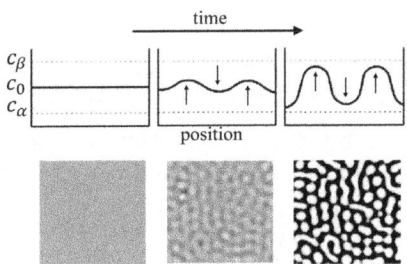

Figure 2.9 Schematic representation of the time evolution of the local concentration in a binary alloy that undergoes a phase separation by a spinodal decomposition mechanism. Top figure schematically illustrates the decomposition process in a one-dimensional case. The bottom figure shows snapshots of the microstructure evolution obtained numerically in a two-dimensional model.

It is interesting to note that in a system with non-conserved order parameter $\Gamma_c k^2$ should be replaced by Γ_{nc}. In this case, for an unstable initial state, which would correspond to $(\partial^2 f / \partial c^2)_{c_0} < 0$, ω is maximum for $k = 0$, which corresponds to a wave length $\lambda \to \infty$. This means that the whole system may decay toward the equilibrium state through a pure relaxation process without the constraint associated with the need for diffusion imposed by the conservation.

The linear theory developed above is reasonably adequate to account for the very early stages of spinodal decomposition. Shortly after the system develops the spinodal instabilities, the approximation is expected to break down. In this context, it is important to note that the sharp spinodal line that establishes whether phase separation occurs through an activated nucleation process or through a spinodal process is, in fact, not a well-defined physical line. At most, it gives only a rough idea of where the crossover from smooth to barrier controlled activated process must occur. To better understand this point consider that still in the metastable region the barrier is smaller than kT. Then, due to thermal fluctuations, the barrier should not play any relevant role. On the unstable side, the existence of a small barrier can even be effective in initiating the phase separation process since this would avoid initiating the process by establishing the unstable, very slow long wave length modes. Therefore, obtaining realistic solutions for longer times requires taking into account non-linear terms in the Cahn-Hilliard equation and one shall also need to explicitly consider thermal fluctuations that should be included through a noise term as taken into account in the general model-B equation. Indeed, this is only possible by using numerical methods.

The time evolution of the complex pattern produced by spinodal decomposition as shown in Figure 2.9 is well described by the structure factor, which is the spatial Fourier transform of the two-point, equal time correlation function. It is defined as

$$\hat{S}(\boldsymbol{k}, t) = \langle |\hat{u}(\boldsymbol{k}, t)|^2 \rangle = \int \langle u(\boldsymbol{r}, t) u(\boldsymbol{r}', t) \rangle e^{i\boldsymbol{k} \cdot (\boldsymbol{r} - \boldsymbol{r}')} d^3 r d^3 r'. \qquad (2.92)$$

This quantity is especially interesting since it is accessible through x-ray or neutron elastic scattering experiments.

In the linear approximation discussed above, the structure function is proportional to $\exp[2\omega(k)t]$ and shows a peak at the characteristic inverse length k_{max} that exponentially grows with time. Instead, wavelength fluctuations with $k > k_0$ are exponentially damped. For all times, it vanishes at $k = 0$, which simply expresses the conservation condition that imposes that $\int u(\mathbf{r}) d^3 r = 0$. This behaviour represents an adequate description for very small times. At longer times, the equation of motion of the structure factor that takes into account the effect of noise, at the lowest order non-linear contribution has the form

$$\frac{\partial \hat{S}(\boldsymbol{k}, t)}{\partial t} = -2\Gamma_c k^2 \left[bk^2 + \left(\frac{\partial^2 f}{\partial c^2} \right)_{c_0} \hat{S}(\boldsymbol{k}, t) + \frac{1}{6} \left(\frac{\partial^4 f}{\partial c^4} \right)_{c_0} \hat{S}_4(\boldsymbol{k}, t) \right]$$
$$+ 2kT\Gamma_c k^2, \qquad (2.93)$$

where $\hat{S}_4(\boldsymbol{k}, t)$ is the Fourier transform of the four-point correlation function. Numerical solution of this equation requires a reasonable closure scheme that permits establishing a relation between $\hat{S}_4(\boldsymbol{k}, t)$ and $\hat{S}(\boldsymbol{k}, t)$. A scheme has been proposed in Ref. [70], based on the assumption that the composition probability distribution can be expressed as the sum of two gaussian functions.

Figure 2.10 shows the time evolution of structure factor measured during the spinodal decomposition of an $Al_{0.62}Zn_{0.38}$ alloy quenched from high temperature to a temperature below the critical phase separation temperature. In the figure, the continuous line is a numerical solution of the previous Eq. 2.93 based on the closure scheme reported in Ref. [70].

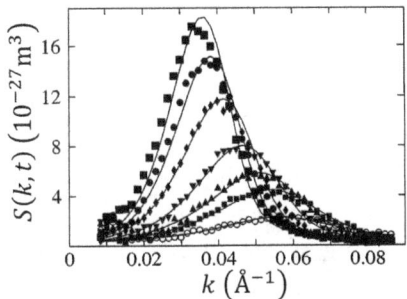

Figure 2.10 Structure factor measured during spinodal decomposition in $Al_{0.62}Zn_{0.38}$ as a function of time. The prequench scattering pattern has been subtracted. Symbols from empty circles to large solid squares correspond to measurements performed at $t = 0, 0.1, 0.2, 0.4, 0.8, 1.2$, and 1.6 s. Adapted from Ref. [71].

2.5 Coarsening and domain growth

We are now interested in discussing the late stages of the ordering process that occurs when a system, with both conserved and non-conserved order parameters, in an initial homogeneous phase, is quenched into a two-phase region [72]. The homogeneous state evolves and, at long times, it is expected that a complex interconnected pattern comprising domains of the two phases develops and coarsens. This late stage regime is, to a large extent, independent of the initial mechanism, nucleation or spinodal decomposition, that triggers the process. In this regime, domains compete with the aim of minimizing the excess energy, which essentially locates on the interfaces that separate the domains. A typical example is a uniaxial magnet quenched in the absence of an applied magnetic field from a temperature above its critical temperature, T_c, to a temperature $T < T_c$. Immediately after the quench, at the temperature T, the system is in an unstable disordered state corresponding to the initial temperature, before the quench. Domains of the two equilibrium phases with positive and negative magnetization, $\pm m_0$, can grow and a network of domains of these two phases develops. Other systems where a similar phenomenon may transpire are binary alloys that undergo an order-disorder transition. When such a system is quenched from above its critical order-disorder temperature, to a temperature T below this temperature, antiphase domains grow in a way similar to the one described in the case of a magnetic system since the order parameter is also non-conserved. This is in spite of the fact that the alloy concentration is actually conserved. Nevertheless, this conservation condition can be satisfied without a long-range diffusion process. Indeed, this is in contrast with a binary alloy that undergoes phase separation. In this case we know that the growth is controlled by (long-range) diffusion which is expected to have a relevant influence in the coarsening regime.

Let us consider more explicitly the case of a uniaxial magnet. At early times after the quench, we can consider that the system consists of spherical droplets that grow according to Eq. 2.69. At longer times, domains should not be spherical and a complex two-phase topology with domains inside domains is expected to occur. In this coarsening stage, due to the complexity of the microstructure, we may adopt a scaling point of view and consider that Eq. 2.69 can still be used to analyse the behaviour of the typical length scale that characterizes the coarseness of the pattern that is expected to control the dynamic behaviour of the system. Therefore, in this scaling sense, we assume that the motion of an arbitrary piece of radius of curvature R is described by Eq. 2.69. For $h = 0$, $R_c \to \infty$ and thus $v < 0$ and all convex domains shrink. At a time t the minimum initial size, $\bar{R}$, of domains that have not shrunk to zero can be obtained by integration of Eq. 2.69 with $h = 0$. This leads to

$$R^2(t) - \bar{R}^2 = -2\Gamma_c(d-1)bt. \qquad (2.94)$$

Therefore, when $R(t) = 0$, $\bar{R} \propto \sqrt{t}$. According to this equation, domains with characteristic length larger than $\bar{R}$ shrink most slowly, while shrinkage accelerates as R decreases. This means that the only domains that remain at time t are those with size of the order of $\bar{R}$, which is the characteristic scale in the coarsening regime. Therefore, the dynamic behaviour of ordering at late stages will be described by the scaling equation

$$\bar{R} \sim t^{1/2}, \qquad (2.95)$$

which is usually known as the Allen-Cahn law [73]. The latter already confirmed the behaviour predicted by this law in experiments of antiphase domain coarsening in the Fe-Al alloys.

In the case of a system with conserved order parameter, the problem shows some analogies to the non-conserved case but Eq. 2.78 must be considered instead of 2.69. In this case, as time goes on, it is expected that supersaturation decreases due to the flux of atoms required for the interfaces to move. Consequently, it is expected that R_c increases, which should give rise to a regime of domain competition. In this late stage regime, droplets with $R \leq R_c$ should occur that decrease in size. Therefore, atoms must diffuse from these small droplets to larger droplets with $R > R_c$. Hence, in this regime the average size of the droplets must be identified with R_c that determines the characteristic length scale of the coarsening problem. This implies that $R \sim R_c$ which means that both terms on the right-hand side of Eq. 2.78 are of the same order. Thus, this equation can be written in the form, $R^2 \dot{R} \cong$ const. Integration then leads to the scaling equation

$$\bar{R} \sim t^{1/3}, \tag{2.96}$$

where $\bar{R}$ is the characteristic length scale. Similarly to the conserved case, the late stage dynamics is described by a power-law behaviour, but with an exponent $1/3$ instead of $1/2$. This dynamic scaling law of coarsening is usually known as the Lifshitz-Slyozov-Wagner law [74, 75].

The derivation of both Allen-Cahn and Lifshitz-Slyozov-Wagner laws is based on the scaling hypothesis that assumes that, at late times, there exists a simple characteristic length, $\bar{R}(t)$, such that the complex domain structure is, in a statistical sense, independent of time when lengths are scaled by $\bar{R}$. The situation has a number of similarities with the equilibrium critical phenomena. In the coarsening problem $\bar{R}$ is, in fact, larger than all microscopic lengths in the system and follows a power-law dependence, $\bar{R} \sim t^x$, on time. This algebraic behaviour reflects indeed the existence of scale invariance. This supposes that the system may appear self-similar under proper rescaling of both length and time. This behaviour is illustrated in Figure 2.11. The possibility of universality would imply that only few values of the exponent x are possible, each one defining a universality class. Then, different systems, exhibiting a wide variety of growth processes, will be classified into these few universality classes, which should depend only on very general features such as conservation laws, ground-state degeneracy, and number of components of the order parameter, among few others. It seems, however, that the existence of a conservation law or not for the order parameter is the essential feature that determines the late stages coarsening scaling law.

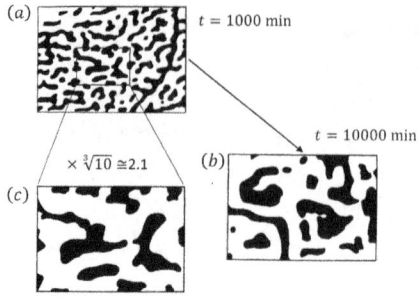

Figure 2.11 Domain structure formed during late stages of a phase separation process in a Fe-24.7 at.% Al alloy quenched from 630°C and annealed at 570°C during (a) $t = 1\,000$ min and (b) $t = 10\,000$ min. (c) corresponds to a blow-up of the region limited by a rectangle in (a) by a linear factor $\sim \sqrt[3]{10}$. After the rescaling consistent with the growth law, $\bar{R} \sim t^{1/3}$, (b) and (c) appear self-similar. Schematic plot adapted from experimental results reported in Ref. [77].

The scaling hypothesis can be expressed more conveniently in terms of the two-point, equal time correlation function, $C(\mathbf{r}, t) = \langle \phi(\mathbf{r}_0 + \mathbf{r}, t)\phi(\mathbf{r}_0, t)\rangle$, and its Fourier transform, the structure function $\hat{S}(\mathbf{k}, t)$. In the preceding expression, ϕ is the order parameter, that can be a conserved or non-conserved quantity, and $\langle \cdot \rangle$ denotes average over initial conditions. The existence of a single characteristic length, $\bar{R}$, supposes that these two functions must satisfy that

$$C(\boldsymbol{r}, t) = \mathcal{C}(r/\bar{R}), \qquad (2.97)$$

$$\hat{S}(\boldsymbol{k}, t) = \bar{R}^d \mathcal{S}(k\bar{R}). \qquad (2.98)$$

It is expected that $\bar{R} \sim k_{max}^{-1}$, where k_{max} is the wave number of the peak maximum of the structure function. Actually, other estimations of the characteristic wave number have been proposed. For instance, it has been also determined from the width of the structure function as

$$\bar{k}^2 = \frac{\sum_{\boldsymbol{k}} k^2 \hat{S}(\boldsymbol{k}, t)}{\sum_{\boldsymbol{k}} \hat{S}(\boldsymbol{k}, t)}. \qquad (2.99)$$

The scaling predicted by Eq. 2.98 has been experimentally demonstrated in systems undergoing phase separation as well as systems undergoing an order–disorder transition. In Figure 2.12, we show an example of the scaling of the structure peak during the late stages of the ordering process in a Ni_3Mn alloy quenched from above its critical order–disorder temperature. This is a face-centred cubic alloy that exhibits a fourfold degenerate ground state. Therefore, four different domains can grow during the ordering process, which are separated by two different kinds of antiphase boundaries, flat and curved interfaces [78, 79]. Thus, only the latter have the driving force to grow. The characteristic wave number $\bar{k}$ was estimated according to Eq. 2.99, and has been shown to follow a power-law dynamics characterized by an exponent that, at the late stages, is consistent with the Allen-Cahn exponent $x = 1/2$ expected for systems with a non-conserved order parameter. It is interesting to mention that this regime is preceded by an intermediate stage characterized by an exponent $1/4$, as shown in Figure 2.12 (c). In the limit of zero temperature, this is the expected singular Allen-Cahn behavior in systems with pinning effects associated with the existence of flat interfaces [80]. At finite temperature a hierarchical movement of interfaces occurs such that curved interfaces have to wait for flat interfaces to disappear before the expected late stage Allen-Cahn mechanism becomes dominant. This is an intermediate regime where the $t^{1/4}$ behavior can be observed.

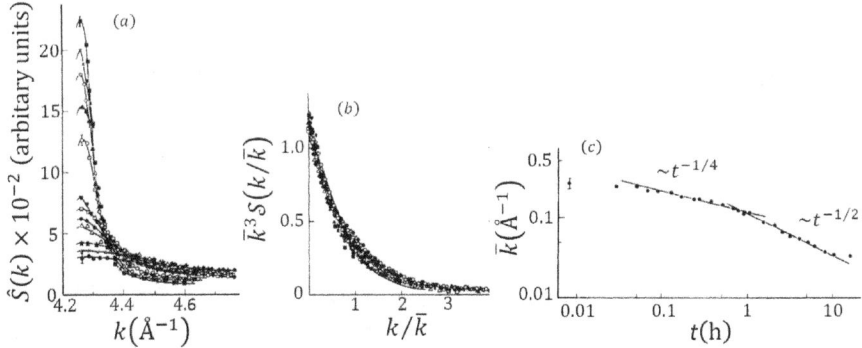

Figure 2.12 Structure factor of a Ni_3Mn alloy as a function of the annealing times t measured at $T = 470\,°C$ below the order–disorder temperature $T_c = 510\,°C$ after quenching from $600\,°C$. (a) (211) superlattice diffraction peak at $t = 0.01$ h (solid circles), 0.03, 0.1, 0.25, 0.5, 0.75, 1, 3, 6, 12, 20 and 34 h (solid squares). (b) Scaling function obtained according to Eq. 2.98. (c) Time dependence of the characteristic wavenumber $\bar{k}$ as a function of time in log-log scale. Exponents $x = 0.24 \pm 0.2$ and $x = 0.46 \pm 0.3$ have been fitted in the intermediate and late stage regions, respectively. The figure has been adapted from Ref. [76].

In general, at a temperature T below the critical temperature T_c of the system, the scaling hypothesis should apply in the range $r \gg \xi$ and $\bar{R} \gg \xi$, where ξ is the equilibrium correlation length. Therefore, taking into account that, in mean-field, we have found that the correlation length is proportional to the inverse of the square of the equilibrium order parameter (Eq. 1.121,

Chapter 1), ϕ_0, we expect that $\mathcal{C}(r/\bar{R}) \propto \phi_0^2$, where ϕ_0 should contain all the temperature dependence. As a matter of fact, it is expected that any remaining temperature dependence should be accounted for in the characteristic scale $\bar{R}$. This simple argument shows that the late stage exponent x should not depend on temperature.

Exercises

2.1 Consider an Ising system consisting of N spins $S_i = \pm 1$. The ferro-magnetic exchange coupling parameter of each spin with its z nearest neighbours is $J > 0$ and it is assumed that the system is in contact with a heat bath at temperature T. Suppose that $w_j(S_j)$ is the probability that the spin at site j flips from S_j to $-S_j$. If $p(\{S_i\}, t)$ is the probability of finding the system in the configuration $\{S_i\}$ at time t, a simplified version of the master equation 2.12 is:

$$\frac{d}{dt}p(\{S_i\}, t) = -\sum_{j=1}^{N} w_j(S_j)p(S_1, ..., S_j, ...S_N, t)$$

$$+ \sum_{j=1}^{N} w_j(-S_j)p(S_1, ..., -S_j, ...S_N, t).$$

(a) Imposing the principle of detailed balance, show that in equilibrium,

$$\frac{w_j(S_j)}{w_j(-S_j)} = \frac{\exp[-\beta h_j S_j]}{\exp[\beta h_j S_j]} = \frac{1 - S_j \tanh(\beta h_j)}{1 + S_j \tanh(\beta h_j)},$$

where $h_j = h + J\sum_{\langle ij \rangle_{nn}} S_i$ is the local field acting on spin S_j.

(b) Choosing for $w_j(S_j)$ the expression, $w_j(S_j) = \frac{1}{2\alpha}(1 - S_j \tanh \beta h_j)$, where α is a constant with units of inverse of time, and defining the expectation value of the spin j at time t as, $\langle S_j \rangle \equiv \sum_{\{S_i\}} S_j p(\{S_i\}, t)$, where the sum extends over 2^N configurations, show that the dynamic equation of this expectation value is:

$$\alpha\frac{d}{dt}\langle S_j \rangle = -\langle S_j \rangle + \langle \tanh \beta h_j \rangle.$$

(c) Show that within the mean-field approximation, in the absence of an applied field:

$$\alpha\frac{d}{dt}m = -m + \tanh \beta z J m,$$

where m is the mean value of the spin. Note that the steady state solution of this equation is, as expected, the mean-field solution of the ferromagnetic Ising model.

(d) Suppose that at $t = 0$ the system is in the fully ordered state, $m = 1$. Find $m(t)$ for $T < T_c$, $T = T_c$ and $T > T_c$, where $T_c = zJ/k$ is the critical temperature.

2.2   Consider a material that undergoes a first-order transition between two
      solid phases with different symmetry. The change can be described as
      a shape deformation that occurs through a mechanism that involves a
      shear strain along a given crystallographic plane plus a strain normal
      to this plane. The elastic coherency strain between the two phases
      is minimized if the product phase forms as thin layers lying in the
      shear plane. Then, to a good approximation, the morphology of the
      resulting product phase can be described by an oblate spheroid with
      semi-axes, r, r and c, with $r \gg c$. The coherency strain energy per unit
      of transformed volume is $E_{elas} \simeq Ac/r$ where A is a shear modulus.

  (a) Suppose that γ is the interface energy per unit area of a nucleus.
      Find the expression for the size and shape parameters of a coherent
      critical nucleus of the new phase.

  (b) Find an expression of the activation barrier for the formation of a
      coherent critical nucleus of the new phase.

  (c) Supposing that the chemical driving force at the transition is $\Delta g =$
      -170 MJ/m^3, $\gamma = 150$ mJ/m^2 and $A = 2.410^3$ MJ/m^3, comment
      on the likelihood of homogeneous nucleation of the product phase.

2.3   Consider an $A_x B_{1-x}$ alloy, which is quenched from high temperature
      inside the unstable two-phase region at a temperature T. Just after
      the quench, the alloy is in an homogeneous disordered state of average
      concentration $c_0 = 2x - 1$ and undergoes a spinodal decomposition into
      A-rich and B-rich phases that have unit cells of different volumes and,
      thus, interact elastically, which produces internal strains. At the early
      stages of decomposition, it can be assumed that the system remains
      coherent, which means that the elastic energy associated with the vol-
      ume difference varies continuously across interfaces. Assume a planar
      concentration fluctuation of the form

$$c(z) = c_0 + \delta c \cos kz,$$

      where k is the wave vector and δc the amplitude of the fluctuation.
      Within linear isotropic elasticity, elastic strain components are, $\epsilon_{xx} =$
      $\epsilon_{yy} = -\frac{1-\nu}{2\nu}\epsilon_{zz} = -\alpha_c \delta c \cos kz$ and $\epsilon_{ij} = 0$ for $i \neq j$. The corresponding
      components of the stress are $\sigma_{xx} = \sigma_{yy} = -\alpha_c \delta c \frac{E}{1-\nu} \cos kz$, $\sigma_{zz} = \sigma_{ij} =$
      0, with $i \neq j$. In the preceding expressions, E is the Young's modulus,
      ν the Poisson ratio, and α_c a constant parameter. Then, the elastic
      contribution to the free energy can be obtained as

$$F_{elas} = \frac{1}{2} \int_\Omega d^3r \sum_{i,j} \sigma_{ij}\epsilon_{ij}.$$

Show that:

(a) In this case, the amplification factor, $\omega(k)$, is given as:

$$\omega(k) = -\Gamma_c k^2 \left[bk^2 + 2\alpha_c^2 Y + \left(\frac{\partial^2 f}{\partial c^2}\right)_{c_0} \right], \tag{2.100}$$

where $Y = \frac{E}{1-\nu}$.

(b) The coherency strain introduces an additional barrier to the spinodal line, which is now determined by the condition:

$$\left(\frac{\partial^2 f}{\partial c^2}\right)_{c_0} + 2\alpha_c Y = 0. \tag{2.101}$$

2.4 Experimentally, the progress of spinodal decomposition can be monitored by isothermally measuring at given temperatures the amount of decomposition as a function of time. Results are often represented as plots called *time-temperature-transformation* diagrams or, simply TTT diagrams.

Starting from the amplification factor obtained in the preceding exercise, derive an expression for the temperature at which the rate of spinodal decomposition is maximum. Show that this leads to a TTT curve that has a 'C' shape, and find the temperature of the nose of these C-curves. Assume that the free energy density is the same as that in Exercise 1.8 which can be expressed in terms of c as:

$$f(c,T) = \frac{1}{8}z\epsilon(1 - c^2) + \frac{kT}{2}\left[(1 + c)\ln\frac{1+c}{2} + (1 - c)\ln\frac{1-c}{2}\right].$$

Part II
Applications

3
Ferroic materials

Ferroic materials constitute a family of materials characterized by a property, denoted as ferroic, which can occur in two or more orientation states that can be switched from one to another by means of an external field thermodynamically conjugated to the ferroic property [81]. This property occurs due to the spontaneous ordering of some microscopic *entities* that takes place through a phase transition from a high-temperature para-phase to the low-temperature ferroic phase. The differently oriented states are called domains and are regions where the ferroic property is homogeneous. These domains are separated by domain walls and are formed due to the symmetry breaking that characterizes each ferroic transition. In all these materials, the field-induced switch from an oriented state to another, usually occurs through a hysteresis cycle as illustrated in Figure 3.1.

The first known example of this class of materials are ferromagnetic materials where ferromagnetism, which is the corresponding ferroic property, is associated with the spontaneous ordering of magnetic moments. Ferroelectricity is an analogous phenomenon related to the spontaneous ordering of electric dipolar moments. La Rochelle salt ($KNaC_4H_4O_6 \cdot 4H_2O$) was the first discovered material of this class in the early twentieth century when it was shown that in the presence of an electric field, the electric polarization exhibited a striking behaviour depending on whether the field was increasing or decreasing. At present, we know that this is a characteristic feature of all ferroic materials and is a consequence of hysteresis effects associated with the field-induced ferroic property switching.

More recently, ferroelasticity, which is related to the spontaneous deformation of a material, was also recognized as a ferroic order since the macroscopic behaviour of the deformation as a function of stress is similar to that of the magnetization versus magnetic field or the polarization versus

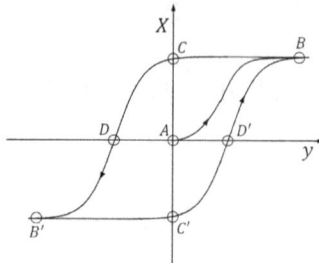

Figure 3.1 Schematic ferroic hysteresis curve, which shows the dependence of the ferroic property X as a function of the conjugated field y, for increasing and decreasing fields. X can be magnetization, polarization or strain for ferromagnetic, ferroelectric and ferroelastic materials for increasing and decreasing the corresponding conjugated field y. Then, y is magnetic field, electric field and stress, respectively. The state A is the multi-domain state reached when the phase transition from the para- to the ferroic phase occurs at zero applied field. In this state usually the average value of the ferroic property is zero. The path $A - B$ represents the first ferroization path that eventually brings the material to a saturation state B. When the field is reversed, X shows a finite remanence at C corresponding to a zero value of the field. X vanishes at the state D, which is reached at the coercive value of the field. States B', C' and D' are symmetric states to B, C and D with opposite signs of X and y. The ferroic remanence and coercive field are often used to characterize the hysteresis loop.

electric field.[1] Nevertheless, the corresponding microscopic entities that get ordered in the ferroic phase[2] are not well defined in the case of ferroelastics as are magnetic moments and electric dipolar moments in the case of ferromagnetism and ferroelectricity, respectively. In any case, the fact that the macroscopic ferroic property can be manipulated by the application of the field that is thermodynamically conjugated to this property is the relevant feature that provides to this class of materials interesting functionalities with immense technological potential. Actually, the term *ferroic* was introduced by Kêitsiro Aizu [81] with the aim of unifying a single-family ferromagnetic, ferroelectric and ferroelastic materials given the strong analogies that give rise to qualitatively similar macroscopic behaviours.

It is well known that ferromagnetic materials are of great importance in modern technology. They are essential for efficient energy applications such as power generation, conversion and transformation [83]. They are

---

[1] The term ferroelastic and the corresponding symmetry classification was provided by Aizu in the late sixties [82].

[2] These entities should likely be related to unit cell-level strains.

important as well, as sensors, actuators and memory grids among many others. The importance of ferroelectrics and especially ferroelastics is probably less tangible. In any case, ferroelectrics have been used for quite long time as actuators, piezoelectric transducers and pyroelectric detectors. The relevance of these materials rose considerably with the possibility of integrating ferroelectic thin films in semiconductor chips. At present, they are the basis of random access memories and high-frequency capacitors [84]. Ferroelastics are interesting as sensors and actuators. Especially interesting is the family of shape memory materials, which are ferroelastic alloys displaying very large recoverable deformations that can be as large as 10% or even more. This class of alloys shows shape memory properties. Specifically, the shape memory effect refers to the fact that when these materials are highly deformed in the low-temperature ferroelastic (usually denoted in this context martensitic) phase, they are able to recover their original shape by the reverse transition upon heating. In the high-temperature paraelastic phase (often denoted in this context austenite), the same systems display another unique property called superelasticity that consists of the possibility of recovering, upon loading and unloading, a very large strain associated with the stress-induced transformation [85].

Ferromagnetic, ferroelectric and ferroelastic materials are often assumed to constitute the group of primary ferroic materials. Nevertheless, other classes of ferroic materials associated with a property, different from magnetization, polarization or strain that can be switched by means of a conjugated field cannot be discarded. With this idea in mind, there has been some interest in designing materials, which display novel types of ferroic order. Among them, ferrotoroidicity associated with the ordering of toroidal moments defined as the moment of a loop of magnetic moments has already been demonstrated. Ferrotoroidic materials are, indeed, the most studied class of these new materials that have been included in the family of ferroic materials. In fact, nowadays, it is broadly accepted that a material can be denoted as ferroic if it pertains to the class of materials that show ferromagnetism, ferroelectricity, ferroelasticity and ferrotoroidicity.[3]

An interesting feature regarding the macroscopic behaviour and, in general, the functional capacities of ferroic materials, is the fact that it strongly depends on the size, distribution and morphology of the domains that form at the phase transition. The number of domains is determined by the ground

---

[3] It is interesting to note that the etymology of the name ferroic has the origin in the Latin name of iron, ferrum, which is one the first materials known to show ferromagnetism. Most of the ferroelectric, ferroelastic and even ferromagnetic materials do not have iron in them, but the terms have persisted in the literature due to the (hysteresis) analogy with ferromagnetism.

state degeneracy that reflects the symmetry breakdown that takes place at the ferroic phase transition. These regions across which a uniform order is maintained, appear mainly as a consequence of compatibility constraints[4] but the actual domain pattern is also influenced by history-dependent effects, including kinetic effects. Actually, the size and shape of domains are strongly sensitive to extrinsic factors such as grain size in polycrystals or annealing procedures that can influence the amount and distribution of crystallographic defects. In fact, the complexity of all these factors may even hide the relevance of intrinsic parameters in the macroscopic behaviour of ferroic materials [86], which is established to a large extent by compatibility conditions. All these effects have a crucial influence on the hysteresis pathway under an applied field, which is dominated by domain boundary motion. Kinetic effects and the influence of disorder on ferroic hysteresis loops will not be considered in this chapter and will be discussed in Chapter 8.

3.1 Phase transitions in ferroic materials

The ferroic property is the adequate order parameter that reflects the symmetry properties of the materials and, thus, the symmetry breaking that takes place at the phase transition at which ferroicity emerges. The different classes of ferroic materials are characterized by a given vectorial or tensorial nature and symmetry properties of this order parameter. The order parameters of ferromagnets, ferroelectrics and ferrotoroidics have a vector nature,[5] while it is a rank-2 tensor in the case of ferroelastics.

3.1.1 Vectorial ferroics

In vectorial ferroics, the order parameter is a consequence of the ordering of microscopic moments which, within a classical picture, are well defined on the basis of Maxwell equations of electromagnetism. Therefore, consider distributions of charges, $\rho(\boldsymbol{r})$, and currents, $\boldsymbol{j}(\boldsymbol{r})$, localized in a given region of space of volume Ω. These distributions create electric and magnetic fields

---

[4] For instance, the magnetostatic self-energy in ferromagnets or the electrostatic self-energy in ferroelectrics is minimized when the system is subdivided into domains. Similarly, in ferroelastic materials, excess of elastic energy associated with the matching conditions of the original and deformed lattices is minimized by domain formation.

[5] Polar vectors (or true vectors) and axial vectors (or pseudovectors) are included in this group. Geometrically, under reflection symmetry, the parallel component to the mirror plane of polar vectors is conserved and the perpendicular component changes sign, while it is the perpendicular component which is conserved and the parallel one which changes sign in the case of axial vectors.

that, according to Maxwell equations in the presence of matter, within the continuum dipolar approximation, must satisfy

$$\nabla \times \boldsymbol{H} - \frac{\partial \boldsymbol{D}}{\partial t} = \boldsymbol{j}, \tag{3.1}$$

$$\nabla \cdot \boldsymbol{D} = \rho, \tag{3.2}$$

$$\nabla \times \boldsymbol{E} + \frac{\partial \boldsymbol{B}}{\partial t} = 0, \tag{3.3}$$

$$\nabla \cdot \boldsymbol{B} = 0, \tag{3.4}$$

where $\boldsymbol{E}$ and $\boldsymbol{B}$ are the electric field and magnetic induction, respectively, and the magnetic field, $\boldsymbol{H}$, and the electric displacement, $\boldsymbol{D}$, are defined as:

$$\boldsymbol{D} = \varepsilon_0 \boldsymbol{E} + \boldsymbol{P}, \tag{3.5}$$

$$\boldsymbol{H} = \frac{1}{\mu_0} \boldsymbol{B} - \boldsymbol{M}, \tag{3.6}$$

with $\boldsymbol{P}$ and $\boldsymbol{M}$ being the polarization and the magnetization of the medium and ε_0 and μ_0 are the electric permittivity and the magnetic permeability of free space, respectively. Polarization and magnetization are introduced as volume densities of the electric, $\boldsymbol{p}$, and magnetic, $\boldsymbol{m}$, moments, which are defined from a multipole expansion far from charge and current distributions of the electric scalar potential, φ, and the magnetic vector potential, $\boldsymbol{A}$, at the dipolar order, respectively [87]. In static conditions,[6] these potentials are defined from equations 3.1 and 3.4 as, $\boldsymbol{E} = -\nabla \times \varphi$ and $\boldsymbol{B} = \nabla \times \boldsymbol{A}$, respectively. Thus, electric and magnetic moments are given as

$$\boldsymbol{p} = \int_\Omega \boldsymbol{r}\, \rho(\boldsymbol{r}) d^3 r, \tag{3.7}$$

$$\boldsymbol{m} = \frac{1}{2} \int_\Omega \boldsymbol{r} \times \boldsymbol{j}(\boldsymbol{r}) d^3 r, \tag{3.8}$$

where the integral is over the volume Ω of the charge and current distributions.

Magnetization is the order parameter in the case of ferromagnetic materials and is associated with the ordering of magnetic dipole moments, which are usually determined by the spin and orbital motion of electrons in the material. Both, magnetic moment and spin are axial vectors that, from a semi-classical approach, can be viewed as current loops that change their sense of rotation upon time reversal. Therefore, magnetization is the axial vector property that emerges when time-reversal symmetry is broken.

---

[6] Under these conditions, ρ is time independent and, from the continuity equation expressing charge conservation, $\frac{\partial \rho}{\partial t} + \nabla \cdot \boldsymbol{j} = 0$, it follows that $\nabla \cdot \boldsymbol{j} = 0$.

It must be taken into account that besides the widely known parallel alignment of magnetic moments in ferromagnets, more complex types of magnetic order, which may also show macroscopic magnetization, can exist. Examples are, ferrimagnets, which are materials with a partially compensated alignments of magnetic moments or certain types of non-collinear magnets, including uncompensated helical, sinusoidal or cycloidal arrangements of magnetic moments. All these systems show spontaneous magnetization that can be switched under a magnetic field and, thus, are considered to belong to the class of ferroic materials.

In ferroelectric materials, the order parameter is the polarization that is related to the collectively aligned microscopic dipole moments. This electric dipole is associated with a non-centrosymmetric charge distribution that may be characterized by an electric dipole moment, which is related to a displacement and is thus a polar vector. Consequently, polarization is the polar vector property that emerges when space-inversion symmetry is broken in the system. Therefore, the property of ferroelectricity occurs in materials characterized by a non-centrosymmetric point group (see Appendix).

Similar to the case of magnetism, besides the strict alignment of electric moments, other types of order can exist with partially compensated alignment of electric moments that may show spontaneous polarization.

The dipolar approximation is not sufficient for some peculiar electric and magnetic moment configurations [88]. Suppose, for instance the case of a spiral-like configuration of electric or magnetic moments. For this configuration, the polarization or the magnetization occurs along the z-axis of the spiral, while the projection on the xy-plane vanishes since it has a circular configuration. This circular-like configuration can be understood as being originated from a toroidal configuration of electric dipoles or loop currents in the magnetic case. It is clear that the polarization, $\boldsymbol{P}$, or the magnetization, $\boldsymbol{M}$, alone does not provide sufficient information about the ordering of the corresponding moments. In fact, in this case an arbitrary amount of a divergence-free polarization or magnetization distribution can be added to $\boldsymbol{P}$ or $\boldsymbol{M}$ without affecting the external field created by the distribution [89]. This divergence-free term can be written as the curl of some vector that characterizes the circular-like configuration of moments in the xy-plane. In the electric case, this term has the form $\nabla \times \boldsymbol{T}^E$ where $\boldsymbol{T}^E$ is the electric toroidization, while in the magnetic case, it is given as $\nabla \times \boldsymbol{T}^M$, with $\boldsymbol{T}^M$ being the magnetic toroidization. It is interesting to note that no broken symmetry is associated with the emergence of electric toroidization while the existence of magnetic toroidization supposes that both, time-reversal and spatial inversion symmetries are broken. Therefore, magnetic toroidization should be represented by an axiopolar (or time-odd polar) vector.

When electric and magnetic toroidizations are taken into account, P must be replaced by $P + \nabla \times T^E$ and M must be replaced by $M + \nabla \times T^M$. Therefore, Maxwell equations take the same formal expressions as the preceding ones after the following redefinitions of the electric displacement and magnetic field

$$D \to \widehat{D} \equiv \varepsilon_0 E + P + \nabla \times T^E = \varepsilon_0 E + \widehat{P}, \qquad (3.9)$$

$$H \to \widehat{H} \equiv \frac{1}{\mu_0} B - M - \nabla \times T^M = \frac{1}{\mu_0} B - \widehat{M}. \qquad (3.10)$$

Similarly to polarization and magnetizartion, toroidization, both electric and magnetic, may be understood as volume densities of the corresponding toroidal moments that characterize a distribution of charges and current with specific configurations of electric and magnetic moments. The toroidal moments appear in the second-order term of the multipole expansion of the electric scalar and magnetic vector potentials. For instance, in the magnetic case this term can be expressed as the sum of magnetic quadrupolar and toroidal terms. This last term enables to introduce the magnetic toroidal moment, t, which is given as:

$$t = \frac{1}{4} \int_\Omega r \times [r \times j(r)] \, d^3 r. \qquad (3.11)$$

Defining $m \equiv \frac{1}{2}[r \times j(r)]$ as the distribution of magnetic moments, the toroidal moment can be expressed as the moment of the magnetic moment, that is,

$$t = \frac{1}{2} \int_\Omega r \times m \, d^3 r. \qquad (3.12)$$

Ferrotoroidicity is associated with the spontaneous ordering of microscopic magnetic toroidal moments. The appropriate order parameter for this class of ferroic materials is chosen as the cross-product of magnetization and polarization. This product defines the toroidization that is expected to emerge when both space-inversion and time-reversal symmetries are broken. In principle, in addition to magnetic toroidal moments, electric toroidal moments may also be envisaged. However, the symmetries associated with this moment are trivial since this property does not change sign under a spatial inversion or time reversal. Vectorial ferroic orders with the corresponding symmetry properties are summarized in Figure 3.2.

The ideas previously discussed can be generalized by considering systems with spiral-like configurations of toroidal moments, either electric or magnetic, which arise from given distributions of electric or magnetic moments. This kind of double-vortex configurations should be characterized

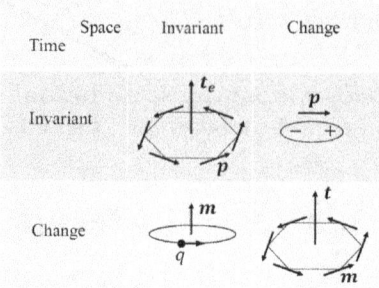

Figure 3.2 Vectorial ferroic orders and corresponding symmetry properties. m is the magnetic moment associated with a loop current and p is the electric dipolar moment associated with a dipolar charge distribution. The magnetic toroidal moment t is defined as the moment of the magnetic moment and the electric toroidal moment, t_e, as the moment of the electric dipolar moment.

by divergence-free vectors expressed as the curl of higher-order moments. These moments are denoted as hypertoroidal moments [90]. Indeed, this can be generalized further to any order by defining higher-order hypertoroidal moments. For instance, in the magnetic case, at nth-order, M in the Maxwell equations should be replaced by $\widehat{M}_n$ given by

$$\widehat{M}_n = M + \nabla \times T^M_{(1)} + \nabla \times \nabla \times T^M_{(2)} + ... + \nabla \times ... \times \nabla \times T^M_{(n)}, \quad (3.13)$$

where $T^M_{(1)}$ is the magnetic toroidal moment T^M.

At all orders, hypertoroidization vectors can be considered as volume densities of a corresponding hypertoroidal moment. For instance, the second-order hypertoroidal moment should be defined as the moment of the toroidal moment. This is schematically illustrated in Figure 3.3.

3.1.2 Tensor ferroics

In ferroelastics, the order parameter is the strain tensor, which is a rank-2 tensor.[7] In these materials, ferroicity emerges when rotational symmetry is broken (see Appendix for more details). In most of these materials, the point group of the low-temperature phase is a subgroup of the point group of the high-temperature phase. Examples of lattice symmetries with point groups that satisfy a group–subgroup relationship and can thus be involved in ferroelastic phase transitions are shown in Figure 3.4. Therefore, for these

---

[7] Strictly, this is a rank-2 polar tensor. It must be noted that there are physical properties described by a rank-2 axial tensor such as magnetogyration, which is an optical activity induced by a magnetic field that occurs, for instance, in some spin-half antiferromagnets.

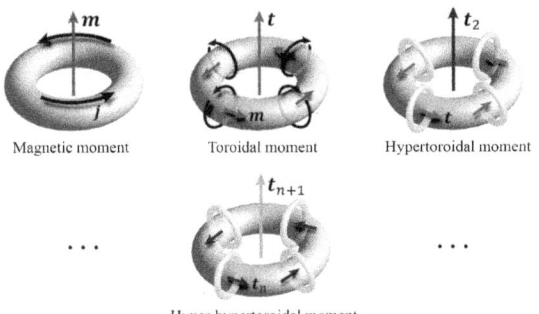

Figure 3.3 Schematic representation of the magnetic moment, toroidal moment and the generation of successive higher-order hypertoroidal moments.

transitions to occur only *small* collective displacements of individual atoms from their original equilibrium positions are required, where in this context, small means that displacements represent a small fraction of the lattice parameter. For this reason, this class of transitions are diffusionless and are often denoted as displacive. For these transitions, the ratio between the number of symmetry elements of the high- and low-temperature phases determines the number of domains that may occur at the phase transition. For instance, in the case of a ferroelastic transition from a cubic to a tetragonal phase the number of domains is 3, while it is 2 for a transition from a tetragonal to an orthorhombic phase. This is illustrated in Figure 3.5.

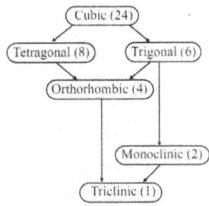

Figure 3.4 Relationship between lattice symmetries belonging to different point groups. A group at the lower level is a subgroup of those at the upper level. In each case, the number of symmetry elements is indicated.

The order parameter in this class of systems is the strain tensor that describes the deformation at the transition. Consider first a continuum medium and let $\boldsymbol{u}(\boldsymbol{r})$ be the displacement field that determines the medium

Ferroic materials

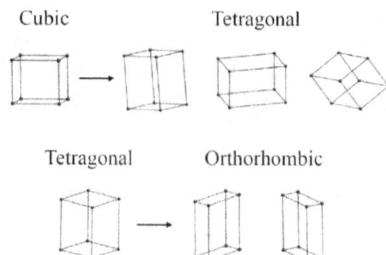

Figure 3.5 Two schematic illustrations of cubic to tetragonal and tetragonal to orthorhombic transitions showing the 3 tetragonal domains that can occur in the first case and the 2 orthorhombic domains that can occur in the second case. Adapted from Ref. [91].

deformation. The components of the distortion tensor, $\mathbf{D}$, are defined from the displacement field as

$$D_{\mu\nu} = \frac{\partial u_\mu(\boldsymbol{r})}{\partial r_\nu}, \qquad (3.14)$$

where the indices μ and ν run over x, y, z. The tensor $\mathbf{D}$ can be decomposed as the sum of a symmetrized tensor, $\boldsymbol{\epsilon}$, and an antisymmetrized tensor, $\boldsymbol{\omega}$, such that

$$\boldsymbol{\epsilon}(\boldsymbol{r}) = \frac{1}{2}[\mathbf{D} + \mathbf{D}^T], \qquad (3.15)$$

$$\boldsymbol{\omega}(\boldsymbol{r}) = \frac{1}{2}[\mathbf{D} - \mathbf{D}^T], \qquad (3.16)$$

where the superscript T denotes transpose. The components $\epsilon_{\mu\nu}$ of the symmetrized tensor $\boldsymbol{\epsilon}$ define the linear strain tensor, while $\boldsymbol{\omega}$ is a rotation tensor.

In the case of ferroelastics, the strain must be introduced at the level of a unit cell. Therefore, strains must be, in principle, defined from discrete differences of lattice points of the distorted and original lattices. Consider that $\boldsymbol{r}_i$ are the lattice points of the original lattice and $\boldsymbol{R}_i = \boldsymbol{r}_i + \boldsymbol{u}(\boldsymbol{r}_i)$ those of the deformed lattice. For nearest neighbour points in the lattice, $\boldsymbol{r}_i$ and $\boldsymbol{r}_j$, the components of the discrete displacement difference vectors are

$$\Delta_\mu u_\nu(\boldsymbol{r}_i) = u_\mu(\boldsymbol{r}_j) - u_\nu(\boldsymbol{r}_i). \qquad (3.17)$$

Then, the unit-cell symmetric strain tensor components, $E_{\mu\nu}$, can be defined as the changes of atom separations in given directions, which may be written as

$$E_{\mu\nu} \equiv \frac{1}{2}[\Delta_\mu u_\nu(\boldsymbol{r}_i) + \Delta_\nu u_\mu(\boldsymbol{r}_i) + \Delta_\mu \boldsymbol{u}(\boldsymbol{r}_i) \cdot \Delta_\nu \boldsymbol{u}(\boldsymbol{r}_i)]. \qquad (3.18)$$

The last term represents the geometric non-linearity, which can be relevant when deformations are large. Therefore, this definition of the strain tensor is adequate for large deformations in discrete systems. However, the continuum approximation is often assumed, and the non-linear strain tensor is defined in terms of the distortion tensor as

$$\mathbf{E}(\boldsymbol{r}) = \frac{1}{2}[\mathbf{D} + \mathbf{D}^T + \mathbf{D}^T\mathbf{D}], \tag{3.19}$$

which is denoted as the Lagrangian-strain tensor.

Consider, as an example, the case of $3d$ cubic lattices. The Lagrangian strain tensor has six components, E_{ij} $(i, j = x, y, z)$. It is convenient then to introduce physical strain components e_μ with $\mu = 1, 2, ..., 6$ defined as normalized linear combinations of the E_{ij}, which describe compressional, e_1, deviatoric, e_2, e_3 and shear e_4, e_5, e_6 symmetrized distortions. These strains transform as one-, two- and three-dimensional irreducible representations of the cubic point group, respectively (see Appendix). They are defined as

$$e_1 = \frac{1}{\sqrt{3}}(E_{xx} + E_{yy} + E_{zz}), \tag{3.20}$$

$$e_2 = \frac{1}{\sqrt{2}}(E_{xx} - E_{yy}), \quad E_3 = \frac{1}{\sqrt{6}}(E_{xx} + E_{yy} - 2E_{zz}), \tag{3.21}$$

$$e_4 = 2E_{yz}, \quad e_5 = 2E_{zx}, \quad e_6 = 2E_{xy}. \tag{3.22}$$

For the cubic lattice, with X, Y, Z transforming as $3d$ cartesian coordinates, e_1 transforms as $X^2 + Y^2 + Z^2$, e_2 as $X^2 - Y^2$, e_3 as $X^2 + Y^2 - 2Z^2$, and the shears e_4, e_5 and e_6 as YZ, ZX and XY, respectively.

Any distortion of the cubic lattice can be quantified by appropriate combinations of these physical strains. For instance, a tetragonal (or orthorhombic) distortion can be determined with the deviatoric strains, e_2 and e_3. Therefore, these strains must be chosen as the primary two-component order parameter to describe a cubic-tetragonal ferroelastic transition. On the other hand, in the case of a transition from a cubic lattice to a trigonal lattice, the three-component order parameter, e_4, e_5, e_6, must be considered as the primary order parameter.

Note that, in general, in d-dimensions the displacement field has d independent components at any point $\boldsymbol{r}$, whereas the symmetric strain tensor nominally has $d(d + 1)/2$ independent components. This means that the physical strains are not all independent variables. As strain components are composed of gradients of the displacement field, symmetrized distortions cannot vary arbitrarily and therefore their components must satisfy some constraints. If lattice integrity is preserved, which means that no defects such as dislocations are considered, these constraints should be derived from the condition that expresses that if the unit cell is deformed, then the neighbouring and further-off cells must also deform in a decreasing way such that all fit together in a smoothly *compatible* manner as illustrated in Figure 3.6. Therefore, the deformation gradient, $\mathbf{F} = \mathbb{I} + \mathbf{D}$, where $\mathbb{I}$ is the identity

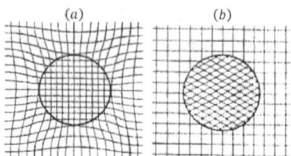

Figure 3.6 Illustration of the compatibility condition. (*a*) Rectangular deformation of a square lattice that preserves lattice integrity. The unit-cell undergoes a rectangular deformation so that neighbouring and further-off cells must also smoothly deform without breaking the lattice. (*b*) Oblique and square regions that coexist without compatibility. The lattice is broken in this case in the region of the interface. Adapted from Ref. [92].

tensor, must be an irrotational tensor. Thus, the following general compatibility condition must be satisfied:

$$\nabla \times \mathbf{D} = 0. \tag{3.23}$$

For small deformations, when non-linear terms can be neglected in the definition of strains, this condition leads to the Saint-Vénant compatibility condition that can be expressed as

$$\nabla \times [\nabla \times \epsilon]^T = 0. \tag{3.24}$$

To better understand this result, consider the case of a 2*d* lattice. In this simpler situation, Eq. 3.23 reduces to the following two identities:

$$\frac{\partial^2 u_x}{\partial y \partial x} = \frac{\partial^2 u_x}{\partial x \partial y}, \tag{3.25}$$

$$\frac{\partial^2 u_y}{\partial y \partial x} = \frac{\partial^2 u_y}{\partial x \partial y}. \tag{3.26}$$

Introducing the linear strain tensor and rotation, $\omega = \frac{1}{2}\left(\frac{\partial u_x}{\partial y} - \frac{\partial u_y}{\partial x}\right)$, the preceding compatibility equations can be written in the form

$$\frac{\partial(\epsilon_{xy} + \omega)}{\partial x} = \frac{\partial \epsilon_{xx}}{\partial y}, \tag{3.27}$$

$$\frac{\partial(\epsilon_{xy} - \omega)}{\partial y} = \frac{\partial \epsilon_{yy}}{\partial x}. \tag{3.28}$$

The rotation ω can be eliminated by differentiating with respect to x the first equation, and with respect to y the second equation. Subtraction of the two obtained equations leads to

$$\frac{\partial^2 \epsilon_{xx}}{\partial y^2} + \frac{\partial^2 \epsilon_{yy}}{\partial x^2} - 2\frac{\partial^2 \epsilon_{xy}}{\partial x \partial x} = 0, \tag{3.29}$$

which is the usual form of the Saint-Vénant compatibility condition 3.24 in $2d$. There is a second equation, which involves the rotation that often is not taken into account since, as we will see later, the free energy is independent of rotations. In $2d$ this equation is obtained by differentiating Eqs. 3.27 and 3.28 with respect to x and y, respectively. It takes the form,

$$\nabla^2\omega + \left(\frac{\partial^2}{\partial x^2} - \frac{\partial^2}{\partial y^2}\right)\epsilon_{xy} = \frac{\partial^2(\epsilon_{xx} - \epsilon_{yy})}{\partial x \partial y}. \tag{3.30}$$

It is worth pointing out that a general compatibility condition can be obtained in terms of the non-linear strains that reduces to the Saint-Vénant compatibility form in the limit of small deformations even in the case of arbitrarily large rotations [93]. This is interesting since the linear strain tensor is not rotationally invariant and, consequently, this approximation is unable to predict microstructures that involve large rotations.

It is important to stress that the preceding results apply to dislocation-free ferroelastics. It is also interesting to point out that in materials with dislocations, the *plastic* part of the distortion tensor introduces a displacement mismatch that can be removed by elastic relaxation. The elastic strains are then no longer compatible in the sense of the compatibility conditions discussed above, which results in an incompatibility tensor that is proportional to the gradient of the dislocation density tensor [94].

In some cases, the change of symmetry must be described by a combination of a pure homogeneous deformation together with a shuffle displacement, which describes small movements of atoms within the unit cell. As a result, pure shuffles must not result in a shape change of the unit cell. Therefore, shuffles can be viewed as phonon modes of the high-temperature phase that become frozen in the low-temperature phase. Shuffles are needed to account for the change of symmetry in many martensitic transitions, which is a class of ferroelastic transitions that is often observed in metals and alloys that show low-temperature supermodulated structures related to shuffle modes.

It is interesting to indicate that the symmetry classification given in Figure 3.2 for vectorial ferroics might, in principle, be conceivably generalized to include *tensor ferroics*. Taking into account rank-2 tensor ferroics only, the figure should be enlarged with two entries in the left column that must be strain (rank-2 polar tensor) and magnetogyration (rank-2 axial tensor) but, at present, the tensor analogs of polarization and toroidization in the right column that would complete the table have not been properly identified.

Beyond rank-2 tensor ferroics, higher-order ferroics can be considered. In most of them the higher tensorial order is in fact a consequence of the combination of elasticity and other ferroic degrees of freedom. This is the case of ferroelastoelectric systems. Pure elastic higher-order ferroics are ferrobielastic systems which are materials where domains are distinguished by different orientation of elastic compliances. An example of a material showing this class of ferroic transition is $FeSiF_6 \cdot 6H_2O$ [95].

Finally, it is interesting to stress that not all lattice symmetry point groups show a group–subgroup relationship. For instance, this is the case of transitions from cubic to hexagonal structures. This class of transitions is also classified as ferroelastic but can only occur through a reconstruction of the lattice, which may entail large atom displacements. They are often denoted as reconstructive and, to describe them, site symmetry and paths that connect high- and low- temperature phases must be taken into account. Then, the description requires the identification of primary and secondary order parameters corresponding to these paths. The general treatment of this class of transitions is much more involved than that of displacive transitions. A brief introduction can be found in Appendix and it is discussed in detail in Ref. [96].

3.2 Ferromagnetic materials

Two main classes of ferromagnetic materials are usually considered. Materials where magnetism arises from magnetic moments localized on the atoms and those where it is due to itinerant electrons. The localized magnetism picture is based on the Heisenberg exchange model, which is essentially adequate to deal with insulators. It works also well in most 4f elements (such as lanthanides) and their compounds due to the fact that 4f electrons are tightly bound to the atom core[8]. On the other hand, metals and alloys are good examples of itinerant magnetism. In this case, unpaired electrons, responsible for the magnetic moment, are no longer localized and accommodated in energy levels belonging exclusively to a given magnetic atom. In this class of materials, magnetization is a consequence of an exchange splitting associated with a significant difference in the population of spin-up and spin-down

---

[8] Valence-fluctuating rare-earth based materials are the exception. In this case, the exchange interaction is due to an indirect mechanism through conduction electrons.

electrons in the minimum-energy configuration that occurs spontaneously without the presence of an external field. The most prominent examples of itinerant-electron systems are metallic materials based on transition elements, where 3d electrons are responsible for the magnetic properties. In this class of materials, the occurrence of ferromagnetism can be explained within a band or collective electron theory, which can predict why Fe or Co are ferromagnetic while the higher transition metals, Cu or Zn, are not [97].

It must be pointed out that localized and itinerant magnetism represent limiting situations that in any case represent good approximations for certain materials. However, it must be indicated that neither of these approaches is strictly adequate to address many magnetic materials that share features of both models. The different origin of magnetism in localized and itinerant materials is reflected by a number of specific features. For instance, while the magnetic moment per atom must originate from an integer number of electrons and must be the same in the paramagnetic and ferromagnetic phases in localized magnetism, this should not be the case in some itinerant materials. In spite of that, the temperature dependence of the spontaneous magnetization shows similar behaviour in both cases since the mechanism producing ferromagnetism is ultimately an exchange mechanism that can be accounted for reasonably well within a mean-field approach in both cases. Therefore, in what follows, we will describe the behaviour of ferromagnetic materials within the same framework regardless of whether magnetism is localized or itinerant.

3.2.1 Energy contributions and modelling

The exchange energy is the essential contribution to the free energy that explains the existence of a phase transition from a paramagnetic to a ferromagnetic phase. Nevertheless, the properties of ferromagnetic materials and especially their behaviour under an externally applied magnetic field are strongly influenced by magnetostatic and anisotropy energies that are not taken into account by the Heisenberg model. In what follows, the relevance of these terms is discussed.

Exchange energy

In the case of materials with localized magnetism, the exchange contribution to the energy can be accounted for by the Heisenberg model (Eq. 1.46). Strictly, this model is only adequate when all the contributions to the atomic magnetic moment arise from electron spins and their orbital contribution vanishes. Nevertheless, the model is commonly used even in cases where this condition does not apply. Therefore, it is assumed that the magnetic moment is associated with the total angular momentum $\boldsymbol{J} = \mathbf{L} + \mathbf{S}$ and is given by $\boldsymbol{\mu} = g\mu_B\boldsymbol{J}$, where g is the Landé factor.[9] In the presence of an external field, $\boldsymbol{B}$, the energy of the magnetic dipole is $-\boldsymbol{\mu} \cdot \boldsymbol{B} = g\mu_B\mathcal{J}_z B$, where $\mathcal{J}_z$ is the z-component of $\boldsymbol{J}$, that can take the $2\mathcal{J}+1$ values, $-\mathcal{J}, \mathcal{J}+1, ..., \mathcal{J}-1, \mathcal{J}$.

In this case, the addition of angular and spin magnetic moments leads to the relation $g\mathcal{J} = L + 2S$ [98]. Therefore, combining this equation with the equation $\boldsymbol{J} = \mathbf{L} + \mathbf{S}$, leads to $\mathbf{S} = (g-1)\boldsymbol{J}$. Assuming exchange interactions between nearest neighbour dipole moments and the presence of an external field, the Heisenberg model can be then expressed as

$$\mathcal{H}_H = -2J(g-1)^2 \sum_{\langle i,j \rangle_{nn}} \boldsymbol{J}_i \cdot \boldsymbol{J}_j - g\mu_B \boldsymbol{B} \cdot \sum_i \boldsymbol{J}_i, \qquad (3.31)$$

where J is the exchange parameter. Then, similarly to the case of the Ising model discussed in Chapter 1, a mean-field solution of the Heisenberg model can be obtained by replacing the actual dipole exchange interaction by an effective field proportional to the magnetization. Hence, the mean-field hamiltonian is given as $\mathcal{H}_{H_{MF}} = -\boldsymbol{\mu} \cdot \boldsymbol{B}_{eff}$, where the effective field is given by

$$\boldsymbol{B}_{eff} = \frac{2zJ(g-1)^2\langle\boldsymbol{\mu}\rangle}{g^2\mu_B^2} + \boldsymbol{B} = \lambda\boldsymbol{M} + \boldsymbol{B}, \qquad (3.32)$$

where λ is the molecular field parameter, z is the lattice coordination number and $\boldsymbol{M}$ the magnetization. Finally, the obtained solution for the magnetization is the so-called Weiss solution, which is given by

$$M(T, B) = M_s\mathcal{B}_{\mathcal{J}}(x), \qquad (3.33)$$

where $M_s = n\mu$, with $n = N/V$ being the density of magnetic dipoles and $\mathcal{B}_{\mathcal{J}}(x)$ the Brillouin function

$$\mathcal{B}_{\mathcal{J}}(x) = \frac{2\mathcal{J}+1}{2\mathcal{J}} \coth\left(\frac{2\mathcal{J}+1}{2\mathcal{J}}x\right) - \frac{1}{2\mathcal{J}} \coth\left(\frac{1}{2\mathcal{J}}x\right). \qquad (3.34)$$

In the preceding expression $x = \frac{\mu}{kT}(B + \lambda M)$. It is interesting to note that an Ising-type solution is recovered for $\mathcal{J} = 1/2$ when the Brillouin function $\mathcal{B}_{1/2}(x)$ reduces to the hyperbolic tangent function. In the absence of an applied magnetic field, the model predicts a critical temperature given by

$$T_c = \frac{2zJ(g-1)^2}{3k_B}\mathcal{J}(\mathcal{J}+1). \qquad (3.35)$$

---

[9] The Landé factor or spectroscopic splitting factor is given by
$g = 1 + \frac{\mathcal{J}(\mathcal{J}+1)+S(S+1)+L(L+1)}{2\mathcal{J}(\mathcal{J}+1)}$.

The dependence of the magnetization obtained from Eq. 3.33 is shown in Figure 3.7. Results show that taking $\mathcal{J} \simeq S = 1/2$ model predictions agree quite well with experimental data for Fe, Ni and Co, which are typical examples of itinerant magnetic materials. Actually, in the case of Ni with 10 valence electrons per atom, a calculation based on the rigid band theory[10] gives that 9.46 electrons are in 3d bands and 0.54 in 4s bands so that the exchange splitting is negligible for 4s electrons but very significant for 3d electrons [99]. Therefore, the theory predicts a saturation magnetization of $0.54n\mu_B$ instead of $0.5n\mu_B$, which would correspond to the $\mathcal{J} = 1/2$ case.

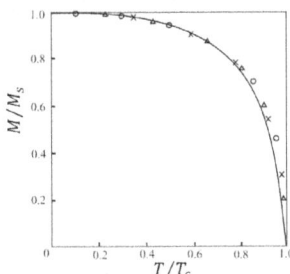

Figure 3.7 The continuous line gives the reduced magnetization, M/M_s, vs. reduced temperature, T/T_c for $\mathcal{J} \simeq S = 1/2$ in the absence of an applied external field. T_c is the critical temperature of the Weiss model that is given by $T_c = 12J/kT$ in the case of an *fcc* material with $z = 12$. Symbols are data for *fcc* iron (circles), nickel (triangles) and cobalt (crosses). Adapted from Ref. [100].

The free energy can be obtained by first rewriting the equation of state 3.33 in the form

$$\frac{g\mu_B \mathcal{J}}{k_B T}(B + \lambda M) = \mathcal{B}_{\mathcal{J}}^{-1}\left(\frac{M}{ng\mu_B \mathcal{J}}\right). \tag{3.36}$$

Then, Taylor expansion of the inverse Brillouin function leads to

$$\frac{B}{M} = \frac{1}{C}(T - T_c) + \frac{A_2}{C^2}TM^2 + \frac{A_3}{C^3}TM^4 + ..., \tag{3.37}$$

where $C = n\mathcal{J}(\mathcal{J}+1)g^2\mu_B^2/3k_B$ is the Curie constant, and the coefficients, A_2, A_3, ..., are the following functions of $\mathcal{J}$:

$$A_2 = \frac{2\mathcal{J}^2 + 2\mathcal{J} + 1}{10k_B \mathcal{J}(\mathcal{J}+1)n}, \tag{3.38}$$

$$A_3 = \frac{44(2\mathcal{J}+1)^2(\mathcal{J}^2 + \mathcal{J} + 1) - (2\mathcal{J}+1)^2}{2800k_B^2 \mathcal{J}^2(\mathcal{J}+1)^2 n^2}. \tag{3.39}$$

---

[10] The rigid band approximation assumes that 3d and 4s bands do not change markedly across the first transition element series and thus any differences in electronic structure must be associated with changes in the Fermi energy.

Integration of Eq. 3.37 over M gives the magnetic free energy,

$$F = F_m = F_0 + \frac{1}{2}C^{-1}(T-T_c)M^2 + \frac{1}{4}A_2C^{-2}TM^4 + \frac{1}{6}A_3C^{-3}TM^6 + ... - BM,$$
(3.40)

where F_0 is a term that only depends on temperature.

The preceding free energy is the Landau-type free energy corresponding to the Weiss model of ferromagnetism that includes the contributions associated with the exchange and Zeeman energies.

The magnetostatic energy is a direct consequence of Maxwell equations in the presence of matter. In a magnetized body subjected to a magnetic field, the total magnetic induction, $\boldsymbol{B} = \mu_0(\boldsymbol{H} + \boldsymbol{M})$, must satisfy the Maxwell equation 3.4 that leads to $\nabla \cdot \boldsymbol{H} = -\nabla \cdot \boldsymbol{M}$. The magnetic field inside the material can then be expressed as the sum, $\boldsymbol{H} = \boldsymbol{H}_0 + \boldsymbol{H}_{ms}$, with $\nabla \cdot \boldsymbol{H}_0 = 0$. Therefore, $\nabla \cdot \boldsymbol{H}_{ms} = -\nabla \cdot \boldsymbol{M}$, where the magnetostatic field satisfies $\nabla \times \boldsymbol{H}_{ms} = 0$. Inside the sample, this field is oriented antiparallel to the magnetization and is often denoted as the demagnetizing field, $\boldsymbol{H}_d$. Due to its irrotational character, the magnetostatic field can be assumed as created from bulk, ρ_m, and surface, σ_m, density distributions of magnetic charges formally given by [87]

$$\rho_m = -\nabla \cdot \boldsymbol{M}, \tag{3.41}$$

$$\sigma_m = \boldsymbol{n} \cdot \boldsymbol{M}, \tag{3.42}$$

where $\boldsymbol{n}$ is the vector normal to the boundary surface of the body. From this viewpoint, any inhomogeneity in the magnetization acts like a magnetic charge density and the magnetization discontinuity at interfaces or boundaries as surface magnetic charge density. Therefore, the magnetostatic field can be associated with a scalar potential, U, that inside the body satisfies

$$\nabla^2 U_{in} = \rho_m, \tag{3.43}$$

while, outside the body, since $\boldsymbol{M} = 0$

$$\nabla^2 U_{out} = 0. \tag{3.44}$$

Hence, at the surface of a magnetic material, the following boundary conditions must apply:

$$U_{in} = U_{out}, \tag{3.45}$$

$$\frac{\partial U_{in}}{\partial n} - \frac{\partial U_{out}}{\partial n} = \boldsymbol{M} \cdot \boldsymbol{n}. \tag{3.46}$$

When the preceding differential equations are integrated taking into account these boundary conditions, the magnetostatic field can be obtained. It is found that, in the absence of an external field, the magnetostatic energy contribution is given by

$$E_{ms} = \frac{1}{2}\mu_0 \int_\Omega \boldsymbol{M} \cdot \boldsymbol{H}_d \, d^3r, \qquad (3.47)$$

where Ω is the volume of the magnetic material. The field created by the magnetized body in the surroundings of the material, $\boldsymbol{H}_s$, is denoted as the stray field. Therefore, the total magnetostatic energy should include the energy associated with this field. Minimization of this energy supposes avoiding the existence of a net distribution of magnetic charges on the surface of the material. This is accomplished by creating an heterogeneous multi-domain state in which domains of homogeneous magnetization cancel the total created magnetic field [101]. The magnetostatic energy given by Eq. 3.47 has a long-range character and, consequently, its effect depends on the shape of the material considered. Often, for this reason, the magnetostatic effect is associated with a shape anisotropy.

An interesting example is the case of a uniformly magnetized sample of ellipsoidal shape, the demagnetizing field can be expressed as, $\boldsymbol{H}_d = -\mathbb{N}\boldsymbol{M}$, where $\mathbb{N}$ is a symmetric trace-1 tensor [102]. Thus, the corresponding magnetostatic energy is proportional to $\int [N_x M_x^2 + N_y M_y^2 + N_z M_z^2] d^3r$, where $N_x + N_y + N_z = 1$, and the coefficients depend on size dimensions. Actually, this expression is often used as a first step to make an approximate estimation of the magnetostatic energy in systems of more complicated shape.

In general, the magnetic anisotropy is a term that describes the energetically favourable orientations of magnetic moments in a crystal. This defines the easy axes and easy planes that in turn define directions along which a lower applied magnetic field is required to reach the saturation magnetization [103]. Besides the macroscopic shape anisotropy associated with the magnetostatic energy, anisotropy also has a microscopic contribution, which is associated with the symmetry of the local environment of magnetic atoms in the underlying crystal lattice.

The symmetry of the local environment determines which are the electronic orbitals with lower energy due to the crystal field splitting. Therefore, this magnetocrystalline anisotropy is an intrinsic property of the material, which is a consequence of the spin–orbit interaction of the electrons.[11] Indeed, this contribution is not restricted to ferromagnetic materials but

---

[11] The spin–orbit effect is a relativistic interaction of the spin of a particle with its motion inside a potential. It can be described as an effective magnetic field, which is seen by the spin of an electron in the rest frame due to the orbital motion.

instead can be relevant in any magnetic material, including materials with compensated magnetism such as antiferromagnetic materials. The magnetocrystalline energy density, e_{mc}, can be expressed in terms of a series expansion at different orders of the magnetization that has the general form,

$$e_{mc} = K_0 + \sum_i K_i^{(1)} M_i + \sum_{i,j} K_{i,j}^{(2)} M_i M_j + \sum_{i,j,k} K_{i,j,k}^{(3)} M_i M_j M_k + ..., \quad (3.48)$$

where M_i are cartesian components of the magnetization and the parameters $K^{(n)}$ denote the components of anisotropy rank-n tensors. For instance, in the case of uniaxial magnetocrystalline anisotropy, which usually occurs in hexagonal, tetragonal and trigonal lattices, expressing the magnetization components in spherical coordinates as $M_x = M_0 \cos\theta \cos\phi$, $M_y = M_0 \cos\theta \sin\phi$, $M_z = M_0 \sin\theta$, to the lowest order, the magnetocrystalline energy can be expressed as

$$E_{mc} = \int_\Omega K^{(2)} M_0^2 \sin^2\theta \, d^3 r, \quad (3.49)$$

where θ is the angle between the magnetization and the z-axis. $K^{(2)}$ is the second-order anisotropy constant and terms of higher order have been neglected. M_0 is the magnitude of the magnetization, which may be a function of temperature. The preferential alignment of magnetic moments is along the z-axis if $K^{(2)} > 0$, while it is perpendicular to the z-axis for $K^{(2)} < 0$. In the first case the z-axis is the easy axis, while in the second it is a hard axis and the material is said to display easy-plane anisotropy. Indeed, when higher-order terms are taken into account, a more complex angular dependence of the anisotropy energy may emerge.

Sometimes, the magnetostrictive energy associated with the coupling of magnetism and lattice is also included within the contributions to the energy of magnetic materials. This contribution may affect the anisotropy properties of the material but, in any case, it is often weak and has little effect in most ferromagnetic materials. In the next chapter, on multiferroic materials, we will consider in detail the case of materials where due to a strong interplay between magnetism and structure, this term is essential to understand their magnetic properties.

Phase field micromagnetic models

To understand magnetic domain configurations in ferromagnetic materials of given shapes, the competition between the three contributions associated with exchange, magnetostatics and anisotropy energies, must be taken into consideration. The magnetostatic term prefers the establishment of multidomain states that ensure the closure of the magnetic flux. Nevertheless, due to the anisotropy effect, magnetization cannot rotate continuously and only domains with magnetization along given directions can occur. Additionally, the energy excess associated with the boundary separating domains, which originated from the exchange energy, must be taken into account. This boundary is also called a domain wall and, usually, it is classified according to the angle between the magnetization in the two domains. The most common types are Bloch walls in which the magnetization rotates in a plane parallel to the plane of the wall. However, other less common situations can occur. For instance, in the case of Néel walls, the magnetization rotates in a plane perpendicular to the plane of the wall.

In general, all these effects must be considered in order to determine the domain configuration that minimizes energy and to model magnetization processes in real materials. Usually, this is undertaken by means of phase field or micromagnetic models. These models are formulated within the continuum approximation and are especially suited to describe the behaviour of ferromagnetic materials.

The total magnetic free energy can thus be expressed as the sum of the exchange, Zeeman, magnetostatic and magnetocrystalline contributions

$$F_M = F_{ex} + F_z + F_{ms} + F_{mc}. \qquad (3.50)$$

The exchange contribution is included through a local Landau term that accounts for the phase transition from the paramagnetic to the ferromagnetic phase.[12] In the ferromagnetic phase, this term is responsible for the preferentially ferromagnetic collinear alignment of neighbouring magnetic moments. This effect is essential to take into account the excess of energy associated with the existence of domains in the ferromagnetic phase. This term is included within a classical approach and is captured by the usual Ginzburg gradient term proportional to $\int_\Omega (\nabla M)^2 d^3r$, where the integral extends over the volume Ω of the material.

As an example, it is insightful to consider the equilibrium configuration of magnetic domains of ferromagnetic materials exhibiting uniaxial anisotropy. Apart from constant terms, the magnetic free energy can be written in the form

$$F_M(\{\boldsymbol{m}\}) = \int_\Omega \left[K(\nabla \boldsymbol{m})^2 + e_{mc}(\boldsymbol{m}; \boldsymbol{n}) - \mu_0 M_s (\frac{1}{2}\boldsymbol{H}_{ms} + \boldsymbol{H}_0) \cdot \boldsymbol{m} \right] d^3r, \qquad (3.51)$$

where it is supposed that magnetization saturation, M_s, is reached inside magnetic domains and, then, $\boldsymbol{m} = \boldsymbol{M}/M_s$ is a unit vector that determines

---

[12] Micromagnetic models dealing only with the behaviour of magnetization within the ferromagnetic phase do not include this term, which is supposed to be constant.

the magnetization direction. Here n is also a unit vector that defines the direction of the easy axis. The problem of obtaining the domain configuration requires minimization of the preceding functional, which must be accomplished with the constraint $|m| = 1$. The most general variation, δm, of the reduced magnetization compatible with this constraint is, $\delta m = m \times \delta\theta$, where θ is a vector that describes a small arbitrary rotation. Then, it is obtained that extremum solutions of F_M for any arbitrary variation θ must satisfy the equations,

$$m \times H_{eff} = 0, \tag{3.52}$$

$$m \times \frac{\partial m}{\partial n} = 0, \tag{3.53}$$

where H_{eff} is an effective field that, in addition to the external field, takes into account the effect of demagnetization and anisotropy. It is given by

$$H_{eff} = \frac{2K}{\mu_0 M_s} \nabla^2 m - \frac{1}{\mu_0 M_s} \frac{\partial e_{mc}}{\partial m} + H_0. \tag{3.54}$$

Equation 3.52 indicates that the torque induced by the effective field on magnetization vanishes at equilibrium. On the other hand, Eq. 3.53 represents the boundary condition that must be satisfied by the magnetization on the surface of the considered body. Note that the information required to solve Eqs. 3.52 and 3.53 are the parameters K^{13} and M_s, the form of the anisotropy density function as well as the information about the shape of the body and the applied external field.

It is interesting to briefly discuss how a material, which is initially in a non-equilibrium state, approaches the equilibrium configuration. It is important to remind that the magnetization vector has a non-conserved direction and a conserved magnitude. It is then expected that the rate of change of the magnetization direction is proportional to $\delta F_M/\delta m$, which is the corresponding driving force. This leads to a dynamic equation of the form

$$\frac{\partial m}{\partial t} = \gamma m \times H_{eff}. \tag{3.55}$$

This equation describes a purely gyroscopic dynamics of the magnetic moment. Hence, if the magnetization is not initially in equilibrium, this term predicts that it will precess around the field but will never approach equilibrium. Therefore, a dissipation term must be included that permits the system to relax to the equilibrium configuration. We assume a relaxation equation of the form, $H_\perp - \alpha \partial m/\partial t = 0$, where $\alpha > 0$. In this equation,

---

[13] This parameter is assumed constant but, in a more general treatment, it could be supposed as position dependent.

only the component of H_{eff} perpendicular to m must be considered since the constraint $|m| = 1$ requires that no changes of m parallel to itself can occur. Note that, $\partial m/\partial t$ is, indeed, perpendicular to m due to the constraint. Therefore, the preceding relaxation equation can be written in the equivalent form[14]

$$m \times \left(H_{eff} - \alpha \frac{\partial m}{\partial t} \right) = 0. \tag{3.56}$$

In conclusion, the general equation that takes into account the precession of the magnetic moment and dissipation can be written in the form

$$\frac{\partial m}{\partial t} = \gamma m \times \left(H_{eff} - \alpha \frac{\partial m}{\partial t} \right). \tag{3.57}$$

This equation is often known as Gilbert's dynamic equation. It is interesting to remark that this equation is consistent with the equilibrium condition given by Eq. 3.52 since in equilibrium $\partial m/\partial t = 0$.

Other magnetic orders

Many magnetic materials, below a given temperature, show magnetic order different from the ferromagnetic order. These orders can be classified within the categories of antiferromagnetic and ferrimagnetic orders. In the former case, all magnetic moments are assumed to be identical, while they are not so in the latter case. Antiferromagnetic order can show collinear and non-collinear compensated ordering of magnetic moments characterized by zero magnetization in the absence of an applied magnetic field. Ferrimagnetic orders are, from a general point of view, intermediate situations between ferromagnetism and strictly compensated antiferromagnetism. Therefore, these magnetic structures still show magnetization.

In relation to a functional point of view, while the enormous importance of ferromagnetism is clear, the practical exploitation of materials with compensated arrangement of magnetic moments is much less evident. Notwithstanding, antiferromagnetic configurations may be quite interesting since they are very robust against external perturbations and they usually exhibit ultrafast magnetic response. However, there is, in practice, no conjugated field[15] that allows manipulation of the antiferromagnetic order, which means that these materials are usually assumed not to belong to the family of ferroic materials. We will see in the next section that the exceptions are those

---

[14] It is taken into account that a generic equation of the form $a = 0$ is equivalent to $\nabla \times a = 0$.
[15] Indeed, a staggered field couples to the antiferromagnetic order, but such a field is not experimentally feasible.

compensated magnetic configurations that are characterized by a toroidal moment of large enough magnitude.

In general, the emergence of compensated ordered configurations of magnetic moments should be discussed within the framework of the Heisenberg model. In particular, antiferromagnetic collinear structures can be accounted for in terms of sublattices as discussed in Chapter 1 for the simple case of the Ising model where only *up* and *down* nearest neighbour spins in lattices with no frustration effects are considered, which corresponds to a strong uniaxial anisotropy case. The sublattice method can be easily generalized to deal with compensated magnetic configurations where more than two sublattices are needed to properly establish the magnetic ground state of the system [104]. To address more complex non-collinear situations, the angle between neighbouring pairs of magnetic moments must be considered. This is, for instance, the case of helicoidal magnetic structures, which are observed in a number of Mn-based compounds such as the MnO_2 oxide [105]. These helicoidal structures may form in materials with easy-plane anisotropy and may be understood as constituted of planes with magnetic moments ferromagnetically aligned which rotate in the direction of the ordered moments along the z-axis, which is the axis of the helix.

This class of systems can be modelled within a good approximation assuming that planes are perfectly ordered with a saturation magnetization $\boldsymbol{M}_s$. Then, the material can be described as a linear chain by means of the following Heisenberg-type hamiltonian operator

$$\mathcal{H} = -J_1 \sum_{i=1}^{N_p} \boldsymbol{M}_{s_i} \cdot \boldsymbol{M}_{s_{i+1}} - J_2 \sum_{i=1}^{N_p} \boldsymbol{M}_{s_i} \cdot \boldsymbol{M}_{s_{i+2}}, \tag{3.58}$$

where the subscript i corresponds to ferromagnetically ordered planes perpendicular to the z-axis, N_p is the number of planes, and J_1 and J_2 are effective exchange parameters that account for the nearest and the next-nearest neighbour plane interactions, respectively.

Within a classical approach, it can be assumed that plane magnetization can continuously rotate along the z-axis. In this situation, the energy competition between neighbouring planes is the crucial feature that permits establishing a helical ordering. This can be shown by writing the hamiltonian in the form

$$\mathcal{H} = -J_1 M_s^2 \sum_{i=1}^{N_p} \cos\theta_{i,i+1} - J_2 M_s^2 \sum_{i=1}^{N_p} \cos\theta_{i,i+2}, \tag{3.59}$$

where $\theta_{i,j}$ is the angle formed by the magnetization vectors of planes i and j. Since it is assumed that all magnetic moments are identical, the magnetization of all planes must have the same orientation with respect to the magnetization of their neighbours. Consequently, $\theta_{i,i+2} = 2\theta_{i,i+1} = 2\theta$ for all values of i. Therefore, the ground-state energy can be written as

$$E(\theta) = -2N_pM_s^2(J_1\cos\theta + J_2\cos 2\theta). \qquad (3.60)$$

Minimization of this energy with respect to θ leads to $J_1\cos\theta + 4J_2\sin\theta$ $\cos\theta = 0$, which is the condition to be satisfied by the angle between consecutive planes in the ground state. One possible solution corresponds to $\sin\theta = 0$, which is accomplished for $\theta = 0$ and $\theta = \pi$. The ground-state energy, E_{gs}, is then

$$E_{gs} = -N_pM_s^2J_2\left(\pm\frac{J_1}{J_2}+1\right), \qquad (3.61)$$

where the signs $+$ and $-$ correspond to the ferromagnetic and antiferromagnetic collinear solutions, respectively. There is another solution with $\sin\theta \neq 0$. It is given by, $\cos\theta = -J_1/4J_2$. In this case, the ground-state energy is given by

$$E_{gs} = -N_pM_s^2J_2\left(\frac{J_1^2}{8J_2^2}-1\right). \qquad (3.62)$$

This solution corresponds to a non-collinear helicoidal configuration that has lower energy than the ferromagnetic and antiferromagnetic configurations for $J_2 < 0$ and $|J_1| < 4|J_2|$. Note that the helicoidal order corresponds to a compensated configuration, that, as discussed in Section 3.1.1, could be characterized by magnetic toroidization. The phase diagram of the model in the space J_1-J_2 is shown in Figure 3.8.

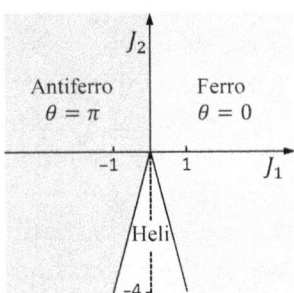

Figure 3.8 Phase diagram in the J_1-J_2 space of the model described by hamiltonian 3.58. The regions of stability of ferromagnetic, antiferromagnetic and helimagnetic ground-state configurations are shown.

3.3 Ferroelectric materials

Ferroelectrics is a class of vectorial ferroic materials which shows many similarities with the ferromagnetic materials, both at macroscopic and mesoscopic scales. However, microscopic features at the origin of ferroelectricity and ferromagnetism are quite different. This is a consequence of the fact that while the electric moment and thus ferroelectricity, are related to an asymmetry in charge, magnetic moment and ferromagnetism arise from an up-down asymmetry in the electronic spin. In both classes of materials, the occurrence of spontaneous polarization and magnetization is a consequence of the interaction of these microscopic electric and magnetic moments. However, the physical origin of this interaction is different in both classes of materials. It is the exchange interaction in the case of magnetism while it is a dipole–dipole interaction in ferroelectrics. It is interesting to note that in some cases, competing effects overcome this dipole–dipole interaction and electric moments may order in a configuration with no polarization. These structures are classified as antiferroelectric structures.

It is important to take into account that the phase transition to the ferroelectric state can be classified as either of displacive or order-disorder type. In the former case, the high-temperature phase is a non-polar centrosymmetric phase. The transition can then be understood in terms of an ion displacement that results in an asymmetrical shift in the equilibrium ion positions and hence leads to a permanent dipole moment in the ferroelectric phase. This is the case of many oxides that possess a perovskite structure, which is shown in Figure 3.9. In the order–disorder limit, there is a dipole moment in each unit cell, but at high-temperatures, the moments are pointing in random directions. Upon lowering the temperature and going through the phase transition, the dipoles order and polarization emerges. These two classes of ferroelectric transitions can be viewed as limiting situations since, often, the transition exhibits the elements of both types. For instance, barium titanate ($BaTiO_3$) is a typical example of displacive ferroelectrics where the ionic displacements that induce the electric moment are associated with the relative position of the titanium ion within the oxygen octahedral cage. However, in lead titanate ($PbTiO_3$), although the structure is rather similar to barium titanate, the driving force for ferroelectricity is more complex with interactions between lead and oxygen ions also playing an important role.

In displacive ferroelectrics such as ideal perovskites, the high-temperature phase is symmetric due to a balance between short-range repulsions between electron clouds of adjacent ions that favour a non-polar symmetry,

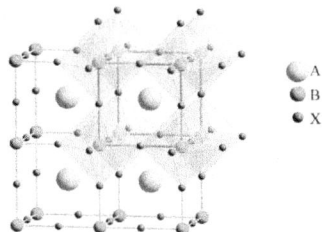

Figure 3.9 Basic ABX$_3$ perovskite structure. Adapted from Ref. [106].

and bonding effects that tend to stabilize the necessary distortion for polarization to occur. The change of chemical bonding can be understood as pseudo- or second-order Jahn-Teller mechanism [107] which stabilizes the distortion that causes the loss of centrosymmetry and gives rise to electric moments and thus to polarization.[16]

Similarly to the ferromagnetic case, domains also form at ferroelectric transitions with a configuration aimed at cancelling the polarization, either for displacive or order–disorder transitions. This is mainly a consequence of the minimization of electrostatic energy in a polarized material, which must take into account interface and anisotropy effects. The problem is formally comparable to that of the magnetostatic energy in a magnetized body discussed in the preceding section. Actually, in the absence of net charge, taking into account Eqs. 3.2 and 3.5, in a polarized material $\varepsilon_0 \nabla \cdot \boldsymbol{E} = -\nabla \cdot \boldsymbol{P}$ and an electrostatic field can be assumed to be created by distributions of bulk, ρ_e, and surface, σ_e, density distributions of charges given by

$$\rho_e = -\nabla \cdot \boldsymbol{P}, \tag{3.63}$$

$$\sigma_e = \boldsymbol{n} \cdot \boldsymbol{P}, \tag{3.64}$$

where $\boldsymbol{n}$ is the vector normal to the boundary surface of the body. The field created inside the body is denoted as the depolarizing field, $\boldsymbol{E}_d$, and the associated electrostatic energy is given by

$$E_{es} = \frac{1}{2} \int_\Omega \boldsymbol{P} \cdot \boldsymbol{E}_d \, d^3 r. \tag{3.65}$$

It is important to note that when elastic effects may play an important role, as can happen in displacive ferroelectrics, the problem of the formation of domains is more complex as ferroelectric and ferroelastic domains may form

---

[16] This Jahn-Teller mechanism requires d-elements with (empty) d^0-shells. Other mechanisms are also known to give rise to ferroelectricity. In some cases, ferroelectricity is magnetically induced and this will be discussed in Chapter 4.

simultaneously at the transition. In this case, a configuration with ferroelectric domains inside ferroelastic domains may occur. This situation will be discussed in the case of materials with large magnetostructural coupling in Chapter 4.

Phase field models for ferroelectrics are usually formulated in a way quite parallel to that discussed for ferromagnetic materials. Therefore, the free energy functional includes a Ginzburg–Landau term that incorporates the symmetries of the material considered and a gradient term accounting for the excess of energy associated with inhomogeneous polarization states, a term related with the electrostatic energy and the coupling with an externally applied field, equivalent to the Zeeman term in magnetism. Anisotropy of the polarization is, often, directly included in the local part of the Ginzburg–Landau[17] free energy density contribution, which is responsible for the phase transition from the para- to the ferroelectric phase.

In contrast to magnetism, the polarization is not allowed to precess in ferroelectrics and the approach to equilibrium is usually described by a pure relaxational dynamics equation.

3.4 Ferrotoroidic materials

A number of materials have been identified that show ferrotoroidal order that is expected to occur due to the coupling between individual toroidal moments. These toroidal moments are either a consequence of a vortex-like configuration of magnetic moments in a magnetic solid or, rather, they must be associated with a moment configuration of a single molecule, which is assumed to carry the toroidal moment. The former class of materials is expected to be antiferromagnets with compensated vortex-like toroidal alignment of magnetic moments of the unit cell that may spontaneously lead to the emergence of a macroscopic toroidal moment, which must be considered the primary order parameter of the transition to a ferrotoroidic phase. However, in these materials, the toroidal response is, in general, very weak and the experimental corroboration of the existence of a ferrotoroidal order had been elusive for many years. This changed in 2007 when ferro-toroidal order was acknowledged to occur in the $LiCoPO_4$ compound [108]. In this material, magnetic vortex patterns have been identified and, below a certain temperature ferrotoroidic domains have been detected by means of optical second harmonic generation [108].[18] Moreover, it has been shown

---

[17] Often denoted in the context of ferroelectrics as the Landau–Devonshire free energy.
[18] In this technique, electromagnetic light field $E(\omega)$ of given frequency is incident on a crystal and induces a polarization at double the field frequency, which acts as a wave source. Since

that the toroidal order can be switched under a conjugated toroidal field with associated hysteresis [109]. This is a crucial point, which must be understood as it is the hallmark confirming the existence of the ferrotoroidal state.

The occurrence of ferrotoroidal order has been suggested recently to be possible in a number of molecular solids. In this case, toroidal moments are carried by single molecules that constitute the solid. These molecules show a complex structure with unique polygonal-shaped topology of magnetic moments that provide the toroidal moment. An example is the molecular solid comprising $[Dy_3(\mu_3\text{-}OH)_2L_3Cl(H_2O)_5]Cl_3$ molecules (L is the anion of ortho-vanillin, see [110]), where magnetic moments localized at Dy atoms order in a triangle-shaped configuration with cancellation of the net magnetic moment. The structure of these Dy_3 clusters is depicted in Figure 3.10. Therefore, this class of molecular solids is an interesting candidate to display a ferrotoroidal transition and ferrotoroidicity associated with single-molecule toroidic moments [111].

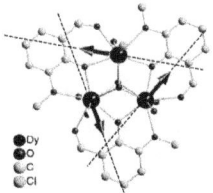

Figure 3.10 Ground-state structure of the molecule of a trinuclear Dy_3 cluster with the local magnetic moments located at Dy atoms oriented along local anisotropy axes indicated by dashed lines. For the sake of simplicity, hydrogen atoms, the chloride counter-anions and the solvent molecules of crystallization have been omitted. The existence of a toroidal moment is a consequence of the compensated triangle-shaped ordering of the magnetic moments. Adapted from Ref. [112].

A third category of ferrotoroidic materials can be considered. These are artificial ferrotoroidics, which are metamaterials created as artificial crystals, which consist of mesoscopic entities characterized by a magneto-toroidal moment. This interesting possibility has been demonstrated in Ref. [113] by growing a crystal formed of mesoscopic planar nanomagnets with a magneto-toroidal-ordered ground state.

Regardless of the class of materials that display ferrotoroidal order, the important question is, which is the field conjugated to the order parameter

the symmetry affects the corresponding susceptibility, the second harmonic generation light from domains with opposite order has a phase shift of 180°, which allows the detection of ferrotoroidic domains.

that enables switching the ferrotoroidal order? In reality, the existence of this field is the essential point that allows classifying materials with ordered toroidal moments as ferroic materials. To answer this question, consider a distribution of magnetic moments characterized by a moment density $\hat{\boldsymbol{M}} = \boldsymbol{M} + \nabla \times \boldsymbol{T}^M$. In the presence of a magnetic field induction $\boldsymbol{B}$, the energy density of the distribution is $-\boldsymbol{B} \cdot \hat{\boldsymbol{M}}$. Hence, the coupling energy of the external field with magnetic toroidization is of the form

$$
\begin{aligned}
E &= -\int_\Omega \boldsymbol{B} \cdot (\nabla \times \boldsymbol{T}^M) \, dr^3 \\
&= -\int_\Omega \nabla \cdot (\boldsymbol{T}^M \times \boldsymbol{B}) \, d^3 r - \int_\Omega (\nabla \times \boldsymbol{B}) \cdot \boldsymbol{T}^M \, d^3 r. \quad (3.66)
\end{aligned}
$$

Taking into account the Gauss theorem, the first integral on the right-hand side of the preceding equation vanishes and, thus, the energy of a toroidal distribution is given by $E = -\int_\Omega (\nabla \times \boldsymbol{B}) \cdot \boldsymbol{T}^M \, d^3 r$. This result may be interpreted in the sense that $\nabla \times \boldsymbol{B}$ is the conjugated field of the magnetic toroidization. Nevertheless, controlling the magnetic toroidization by means of such a field appears quite unfeasible since this would require the action of coherent loop currents of a very small size, comparable to the characteristic size of the unit cell of the material considered [114].

An alternative conjugated field can be foreseen by taking into account that materials with toroidal moment must display intrinsic magnetoelectric response.[19] Indeed, in these materials, an applied magnetic field $\boldsymbol{B}$ breaks the inversion symmetry and thus induces polarization. On its turn, an applied electric field $\boldsymbol{E}$ breaks time-reversal symmetry and induces magnetization. Therefore, we expect that these materials respond to applied electric and magnetic fields according to

$$
\boldsymbol{P} = \chi_e \boldsymbol{E} + \alpha \boldsymbol{B}, \quad (3.67)
$$

$$
\boldsymbol{M} = \alpha \boldsymbol{E} + \chi_m \boldsymbol{B}, \quad (3.68)
$$

where χ_e and χ_m are, respectively, electric and magnetic susceptibility tensors, and α is the magnetoelectric tensor. From a thermodynamic point of view, the polarization $\boldsymbol{P}$ and magnetization $\boldsymbol{M}$ should be derived from the free energy $\mathcal{F}$ as, $-\partial \mathcal{F}/\partial \boldsymbol{E}$ and $-\partial \mathcal{F}/\partial \boldsymbol{B}$, respectively. Hence, it is expected that the free energy contains a magnetoelectric term of the type $\mathcal{F}_{me} = -\boldsymbol{E}\alpha\boldsymbol{B}$ (or, expressed in coordinates, $-\alpha_{ij} E_i B_j$). The decomposition

---

[19] These cross-response multiferroic effects will be discussed in detail in Chapter 4.

of this term into pseudoscalar, vector and symmetric traceless terms allows to express $\mathcal{F}_{me}$ in the form [115],

$$\mathcal{F}_{me} \sim -\boldsymbol{E} \cdot \boldsymbol{B} - \boldsymbol{T}' \cdot (\boldsymbol{E} \times \boldsymbol{B}) - \boldsymbol{E}(\mathbf{Q} + \mathbf{Q}^T)\boldsymbol{B}, \qquad (3.69)$$

where the last term is the symmetric traceless term where the superscript T denotes transpose and $\boldsymbol{T}'$ is a vector with the same symmetry properties as the toroidal moment and magnetic toroidization. Identifying this vector with the magnetic toroidization $\boldsymbol{T}^M$ supposes that its conjugated field is precisely, $\boldsymbol{G} = \boldsymbol{E} \times \boldsymbol{B}$. It is interesting to note that this identification is not purely formal but rather it is in agreement with experiments that show that the toroidal moment can be controlled by simultaneous application of perpendicular electric and magnetic fields [116].

The term $\boldsymbol{G} \cdot \boldsymbol{T}^M$ indicates that ferrotoroidal materials are expected to show asymmetric magnetoelectric response. This is an interesting result that suggests a method to indirectly infer the existence of ferrotoroidal order, based on the observation that components of the magnetoelectric tensor must satisfy that $\alpha_{ij}^T \neq \alpha_{ji}^T$. This asymmetric behaviour has been reported to occur in some compounds such as the $Ni_3B_7O_{13}Br$ boracite [117] or the $Ga_{2-x}Fe_xO_3$ [118] and Cr_2O_3 oxides [119], which have been suggested to be good candidates to display ferrotoroidal order.

According to the symmetry properties of vectors $\boldsymbol{P}$, $\boldsymbol{M}$ and $\boldsymbol{T}^M$, a term such as $\boldsymbol{T}^M \cdot (\boldsymbol{P} \times \boldsymbol{M})$ is a symmetry invariant term permitted in a Landau free energy expansion. Taking this fact into account, a simple Landau expansion for ferrotoroidic materials has been proposed in Ref. [120]. The free energy function is of the form

$$F = \frac{1}{2}\chi_e^{-1}P^2 - \boldsymbol{P} \cdot \boldsymbol{E} + \frac{1}{2}\chi_m^{-1}M^2 - \boldsymbol{M} \cdot \boldsymbol{B}$$
$$+ \frac{1}{2}A_0(T - T_c^0)(T^M)^2 + \frac{1}{4}C(T^M)^4 + \kappa\boldsymbol{T}^M \cdot (\boldsymbol{P} \times \boldsymbol{M}). \quad (3.70)$$

In this free energy function, the external electric and magnetic fields directly couple to their corresponding conjugated properties, polarization and magnetization, and the effect of these fields on magnetic toroidization is taken into account by the last term that couples the magnetic toroidization to polarization and magnetization [121]. From this free energy, the equilibrium values of polarization and magnetization can be obtained by minimization. The result is

$$\boldsymbol{P} = \chi_e\boldsymbol{E} - \kappa\chi_e(\boldsymbol{M} \times \boldsymbol{T}^M), \qquad (3.71)$$
$$\boldsymbol{M} = \chi_m\boldsymbol{B} - \kappa\chi_m(\boldsymbol{T}^M \times \boldsymbol{P}). \qquad (3.72)$$

Substitution, for instance, of the preceding first equation in the second one leads, to the leading order in $\boldsymbol{T}^M$, to

$$\boldsymbol{P} = \chi_e \boldsymbol{E} - \kappa \chi_e \chi_m (\boldsymbol{B} \times \boldsymbol{T}^M), \qquad (3.73)$$

where the last term describes the magnetoelectric effect, which is proportional to the magnetic toroidization. Thus, the magnetoelectric tensor is given as $\boldsymbol{\alpha} = -\kappa \chi_e \chi_m (\mathbb{I} \times \boldsymbol{T}^M)$, where $\mathbb{I}$ is the identity tensor. Therefore, the components of this tensor can be expressed as, $\alpha_{ij} = \partial P_i / \partial B_j = -\kappa \chi_e \chi_m \sum_k \epsilon_{ijk} T_k$, where ϵ_{ijk} is the Levi-Civita symbol.[20] This expression shows the antisymmetric character of the magnetoelectric tensor.

An effective free energy can now be obtained by replacing the equilibrium values of $\boldsymbol{P}$ and $\boldsymbol{M}$ given by Eqs. 3.71 and 3.72, respectively, in the free energy Eq. 3.70. For the sake of simplicity, it is convenient to consider that electric and magnetic fields are, respectively, applied along x and y directions that, assuming isotropy, lead to $\boldsymbol{P} = (P, 0, 0)$, $\boldsymbol{M} = (0, M, 0)$ and $\boldsymbol{T}^M = (0, 0, T^M)$. Then, the obtained free energy has the following form

$$\begin{aligned} F_{eff} = F_0 + \frac{1}{2} A_0 (T - T_c)(T^M)^2 \\ + \frac{1}{3} \beta(G)(T^M)^3 + \frac{1}{4} C(T^M)^4 - G' T^M, \end{aligned} \qquad (3.74)$$

where, $F_0 = -(\chi_p E^2 + \chi_m B^2)/2$, $T_c = T_c^0 + \kappa^2 \chi_p \chi_m (\chi_p E^2 + \chi_m B^2)$, $\beta(G) = -3\kappa^3 \chi_p^2 \chi_m^2 G$ and $G' = -\kappa \chi_p \chi_m G$. It must be noted that G' is proportional to $G = EB$, which is the field conjugated to the toroidal moment. Therefore, the obtained free energy corresponds to the free energy of a toroidal material subjected to a toroidal field G. It must also be noted that the coefficient of the third-order term is proportional to this toroidal field. Hence, in the absence of an applied toroidal field, this free energy describes a continuous transition from a paratoroidic high-temperature phase to a ferrotoroidal low-temperature phase that should occur at the critical temperature T_c, while in the presence of the field, the transition becomes first order. It must be pointed out that this prediction has not been experimentally confirmed yet.

3.5 Ferroelastic materials

Ferroelastic phase transitions can be understood as a problem of elasticity concerning the occurrence of a product material with a given

---

[20] The Levi-Civita symbol is defined so that $\epsilon_{ijk} = 0$ if any two indices are the same, it is $\epsilon_{ijk} = 1$ if i, j, k is an even permutation of 1,2,3 and it is $\epsilon_{ijk} = -1$ if i, j, k is an odd permutation of 1,2,3. The components of the cross-product of two vectors, $\boldsymbol{a}$ and $\boldsymbol{b}$, can be expressed as, $(\boldsymbol{a} \times \boldsymbol{b})_i = \sum_{j,k} \epsilon_{ijk} a_j b_k$.

crystallographic symmetry embedded in a matrix of a parent phase of different crystallographic symmetry. Historically this problem is related to the problem of determining the elastic field of an inclusion in an infinite elastic body that dates back to the seminal work of Eshelby [122]. In the case of martensitic transitions, a habit plane geometrical solution between product and parent phases has been proposed long time ago in real [123] and reciprocal space [124] formulations, and still is a subject of interest. From a general point of view, the solution of the inclusion problem is a result of strain accommodation and energy minimization. With this idea in mind, a number of approaches to deal with the formation of microstructure at a ferroelastic transition have been proposed, all based on the minimization of a free energy functional that takes into account the elastic inclusion energy and lattice coherency. The difference between the various approaches mainly arises from the way coherency is introduced. Among the different formulations, those proposed first by Khachaturyan [125] and Ball and James [126] are worth mentioning. The former is a phase field approach in which the total strain, which is supposed linear, is the sum of elastic and transition strain contributions, and the free energy is assumed harmonic with respect to the elastic strain. This elastic energy is then effectively eliminated in terms of a phase field variable that labels ferroelastic domains or variants.[21] Therefore, in this approach, minimization is performed in terms of the phase field variable. Instead, the latter approach is based on the minimization of an energy potential with respect to the total strain, and compatibility at any interface is introduced via the general continuum elasticity Hadamard jump condition [127].

The approach discussed here is also a strain-based approach formulated in terms of a Ginzburg–Landau functional that can be adapted to any change of symmetry taking place at the ferroelastic transition considered. Coherency across interfaces is then taken into account through the general compatibility condition given by Eq. 3.23. This formulation has the advantage of taking naturally into account the required invariance of the free energy under global uniform rotations. Since, in this Landau context, strain is the natural order parameter, the total physical strain is separated into order parameter and non-order parameter symmetry adapted strains. Therefore, the free energy functional is expressed as

$$F = \int_\Omega \left(f_L + f_{NOP} + f_G\right) d^3r, \tag{3.75}$$

---

[21] In the context of ferroelastic and, especially, martensitic transitions, ferroelastic/martensitic domains are often denoted as variants.

where f_L is a Landau free energy density associated with the order parameter strains, f_{NOP}, the non-order parameters density contribution, and f_G the gradient or Ginzburg term. It has been already discussed in Section 3.1.2 that in dimension d there are $d(d+1)/2$ strains from which, the number of order parameter strains, N_{OP}, depends on the symmetry reduction taking place at the ferroelastic transition. For instance, it has been indicated that in $3d$, e_2, e_3 are adequate to describe a tetragonal distortion of a cubic lattice. Therefore, for a cubic to tetragonal ferroelastic transition, as reported for the Fe-Pd or In-Tl intermetallic compounds, e_2 and e_3 will define the order parameter, while the four remaining strain components will be non-order parameters. In general, the free energy contribution associated with non-order parameters only keeps harmonic terms. On the other hand, the term that accounts for the energy of an externally applied stress, σ, is included in the Landau term and is expressed as the tensorial product of stress and strain tensors. In any case, the relevant terms of this product are those that couple the stress and the order parameter components of the strain tensor, since these terms are the ones that may enact a symmetry breaking of the high-temperature phase that favour some of the variants that can grow at the transition over the others.

Assuming linear strains, compatibility can be imposed by minimizing the free energy functional given by Eq. 3.75 under the Saint-Vénant compatibility constraint given by Eq. 3.24. Then, for an infinite system, this enables to express the non-OP strain components in terms of order parameter strains, which results in effective repulsive forces that arise from forcing strain compatibility at interfaces. This naturally leads to long-range elastic interactions. From this point of view, the Saint-Vénant condition must be considered as a non-defect compatibility condition in a sense analogous to the zero-divergence condition of the magnetic field that supposes a non-monopole condition.

Example: 2d square-to-rectangle and square-to-oblique ferroelastic transitions

For a better understanding of the problem, it is convenient to consider in detail the case of the ferroelastic transitions in a $2d$ square system. In this case, a deviatoric e_2 deformation breaks the $\pi/2$ rotational invariance of the square phase, which gives rise to a rectangular phase. Therefore, we can describe a square-to-rectangle transition with the following free energy density (the derivation of this free energy function is discussed in detail in Appendix)

$$f_{sr} = \frac{1}{2}A_2 e_2^2 + \frac{1}{4}B_2 e_2^4 + \frac{1}{6}C_2 e_2^6 + \frac{1}{2}A_1 e_1^2 + \frac{1}{2}A_3 e_3^2 + \frac{1}{2}\kappa_2|\nabla e_2|^2, \quad (3.76)$$

where the coefficient A_2 of the harmonic term is related to twice the elastic constant $C' = [C_{11} - C_{12}]/2$, and is assumed to linearly depend on temperature as, $A_2 = a_2(T - T_c)$ where T_c is the limit of stability of the high-temperature phase. The remaining parameters are supposed to be positive constants except B_2 that must be negative to ensure the first-order character of the transition. The non-order parameter contribution is related to the strain components e_1 and e_3, and, as it is usual, is assumed harmonic. The corresponding coefficients, $A_1 = C_{11} + C_{12}$, related to the bulk modulus, and $A_4 = 2C_{44}$, related to the shear modulus, are taken as independent of temperature.

In the particular case of a $2d$ system, the Sain-Vénant condition is given by Eq. 3.29. For the system considered here, it can be expressed in the form

$$\left(\frac{\partial^2}{\partial x^2} + \frac{\partial^2}{\partial y^2}\right)e_1 - \left(\frac{\partial^2}{\partial x^2} - \frac{\partial^2}{\partial y^2}\right)e_2 = \sqrt{8}\frac{\partial^2 e_3}{\partial x \partial y}. \tag{3.77}$$

Then, minimization of free energy functional $\int f_{sr} d^3r$ under the Saint-Vénant constraint via Lagrange multipliers allows expressing the non-order parameters contribution in terms of the order parameter e_2 as

$$f_{NOP}(e_2) = \int e_2(\boldsymbol{r})U(\boldsymbol{r} - \boldsymbol{r}')e_2(\boldsymbol{r}')\, d^2r. \tag{3.78}$$

The preceding equation indicates that due to the Saint-Vénant compatibility condition, the elastic-free energy contribution associated with e_1 and e_3 gives rise to non-local effects described by the kernel $U(\boldsymbol{r} - \boldsymbol{r}')$ that relates the deformations in any two volume elements located at $\boldsymbol{r}$ and $\boldsymbol{r}'$. To better understand the characteristics of these long-range effects, it is convenient to express the kernel in reciprocal space. It has the form

$$\hat{U}(\boldsymbol{k}) = \frac{1}{2}A_1\frac{(k_x^2 - k_y^2)^2}{k^4 + 8\frac{A_1}{A_3}(k_x k_y)^2}, \tag{3.79}$$

and is represented in Figure 3.11. From this Figure and the preceding expression, it is clear that the elastic energy associated with e_1 and e_3 is minimized for $k_x = \pm k_y$. Therefore, it means that this term would favour microstructures constituted of twin-related variants with boundaries along the diagonals of the high-temperature square lattice. Note that in real space, this interaction should asymptotically vary as $\cos(\theta - \theta')/r^2$. The decay going as $\sim 1/r^2$ is a consequence of the $2d$ formulation.

At this point, it is important to note that to properly account for the habit plane defining interfaces between high- and low-symmetry phases, the periodicity of twin-related variants must be determined such that the elastic

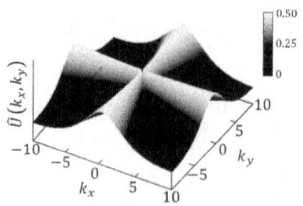

Figure 3.11 Long-range elastic kernel $\hat{U}$ corresponding to the non-order parameter free energy density in reciprocal k-space for $A_1 = 1$ and $A_3 = 2$. Significant valleys occur along the diagonal directions, which determine preferred directions for interfaces between transformed regions. The singularity at $k = 0$ is related to the fact that the uniform state is not subjected to the compatibility constraint. Hence, it represents the ground state of the system since in this configuration the gradient term vanishes.

energy of the elastic field created in the high-symmetry phase is minimized. This problem has been analyzed in detail for the square-to-rectangle transition in Ref. [128]. Imposing continuity and compatibility of the total strain determined by e_1, e_3 and the transition strain, e_2, across the habit plane, results in the fact that the wave length, λ, of the lamellar configuration of twin-related groups of variants of a ferroelastic region of size $L \times L_1$ embedded in parent phase matrix as shown in Figure 3.12, must satisfy the scaling relation, $\lambda \sim \sqrt{L}$. In fact, this result can be understood by simply determining the periodicity λ that minimizes the sum of the competing elastic energy, E_{el}, caused by the misfit across the habit plane and the surface energy, E_s, associated with interface energy of the lamellar low-temperature phase. That is, taking into account that $E_{el} \sim L\lambda e_2^2$, and $E_s \sim L^2/\lambda$, minimization of $E(\lambda) = E_{el} + E_s$ with respect to λ leads immediately to the scaling, $\lambda \sim \sqrt{L}$. It is interesting to note that using similar arguments, the same scaling has been obtained in the case of a striped ferromagnetic phase when the competition between exchange, associated interface term and anisotropy energies is considered [129].

The model discussed so far can be generalized to account for both square-to-rectangle and square-to-oblique ferroelastic transitions. For that purpose, it must be taken into account that the symmetry-adapted strain component e_3 describes a transformation of a square phase into a centred rectangle phase so that a combination of both e_2 and e_3 strains may result in an oblique phase. These $2d$ models are often considered to mimic the behaviour of cubic-to-tetragonal and cubic-to-monoclinic ferroelastic transitions in bulk materials.

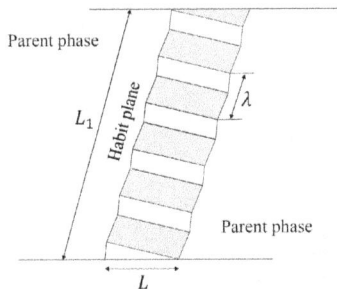

Figure 3.12 Schematic representation of a ferroelastic lamellar phase constituted of groups of twin-related variants with periodicity λ. Ferroelastic and parent phases are separated by a habit plane.

Now, the starting point is a general Landau free energy density that includes symmetry allowed terms not only in e_2 but also in e_3 and their coupling. Hence, f_L is now split into three terms, $f_L = f_1 + f_2 + f_{23}$, given as

$$f_1 = \frac{1}{2}A_2 e_2^2 + \frac{1}{4}B_2 e_2^4 + \frac{1}{6}C_2 e_2^6, \tag{3.80}$$

$$f_2 = \frac{1}{2}A_3 e_3^2 + \frac{1}{4}B_3 e_3^4 + \frac{1}{6}C_3 e_3^6, \tag{3.81}$$

$$f_{23} = \frac{1}{2}A_{23} e_2^2 e_3^2 + \frac{1}{6}C_{23} e_2^4 e_3^2 + \frac{1}{6}C_{32} e_2^2 e_3^4. \tag{3.82}$$

The coefficients A_2 and A_3 of the harmonic terms, related to the elastic constants $2C' = C_{11} - C_{12}$ and $2C_{44}$, respectively, are now both assumed to depend on temperature as, $A_2 = a_2(T - T_c)$ and $A_3 = a_3(T - T_c)$, with the same stability limit T_c, while the remaining parameters are supposed to be constants. B_2 and B_3 are taken negative to ensure the first-order character of the transition. Figure 3.13 depicts schematically the transition change corresponding to square-to-rectangle and square-to-oblique transitions with corresponding free energy contour maps corresponding to a temperature below the respective transition temperatures. In the first case, symmetry breaking leads to two variants, while in the second it leads to four variants. Phase diagrams obtained from minimization with respect to the order parameters of this free energy are also shown in Figure 3.13.

The non-order parameter contribution is, in this case, only related to the strain component e_1 and, as it is usual, is assumed harmonic. That is

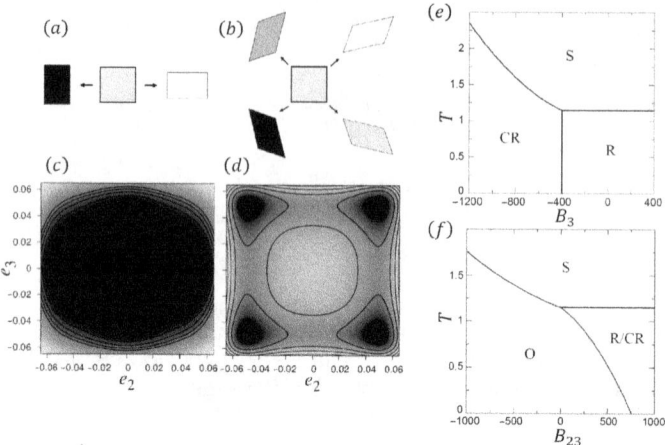

Figure 3.13 Schematics of a square-to-rectangle (*a*) and square-to-oblique (*b*) transitions. In the former case, two ferroelastic domains or variants can grow, while in the latter the number of variants is four. Contour maps of the Landau free energy in the e_2-e_3 space for square-to-rectangle (*c*) and square-to-oblique (*d*) cases are shown. Plots are obtained at $T = 0.8T_c$ for $B_{23} = 800$, $a_2 = a_3 = 1$, $B_2 = -400$, $C_2 = C_3 = 2 \times 10^5$ and $C_{23} = C_{32} = 0$. The parameter $B_3 = -200$ in (*c*), and $B_3 = -400$ in (*d*). With the same parameters, the phase diagrams T vs. B_3 and T vs. B_{23} are shown in (*e*) and (*f*), respectively. (S: square, R: rectangle, CR: centred rectangle, O: oblique).

$$f_{NOP} = \frac{1}{2}A_1 e_1^2, \tag{3.83}$$

with A_1 independent of temperature.

Finally, the gradient term takes now the following general form:

$$f_G = \frac{1}{2}\kappa_2|\nabla e_2|^2 + \frac{1}{2}\kappa_3|\nabla e_3|^2 + \frac{1}{2}\kappa_{23}\left(\frac{\partial e_2}{\partial x}\frac{\partial e_3}{\partial y} + \frac{\partial e_2}{\partial y}\frac{\partial e_3}{\partial x}\right). \tag{3.84}$$

The Saint-Vénant condition given by Eq. 3.77 allows to write e_1 as a function of e_2 and e_3 and leads to the following expression for the non-order parameter free energy density contribution

$$f_{NOP}(e_2, e_3) = \sum_{i,j=1}^{2} \int e_i(\boldsymbol{r})U_{ij}(\boldsymbol{r}-\boldsymbol{r'})e_j(\boldsymbol{r'})\, d^2r'. \tag{3.85}$$

In k-space the kernels $\hat{U}_{ij}$ are now given by

$$\hat{U}_{22}(\boldsymbol{k}) = \frac{1}{2} A_1 \frac{(k_x^2 - k_y^2)^2}{(k_x^2 + k_y^2)^2}, \tag{3.86}$$

$$\hat{U}_{23}(\boldsymbol{k}) = U_{32}(\boldsymbol{k}) = \frac{1}{2} A_1 \frac{\sqrt{8} k_x k_y (k_x^2 - k_y^2)^2}{(k_x^2 + k_y^2)^2}, \tag{3.87}$$

$$\hat{U}_{33}(\boldsymbol{k}) = \frac{1}{2} A_1 \frac{8 k_x k_y}{(k_x^2 + k_y^2)^2}, \tag{3.88}$$

which represent a generalization of the expression obtained above for the simpler square-to-rectangle model.

Dynamics

Dynamics plays a central role for a proper understanding of ferroelastic materials. Ferroelastic transitions are diffusionless, displacive transitions at which complex microstructures comprising multiple variants can develop, which gives rise to multiscale textures that occur by satisfying compatibility conditions. Therefore, the dynamic evolution of such class of materials from a non-equilibrium initial state is basically a strain-controlled process aimed at reducing the elastic strain energy and to reach a minimum at a stationary state.

In principle, it should be argued that since strain is a non-conserved quantity, a pure relaxational dynamics must be adequate to describe the dynamics of this class of materials. Nevertheless, strains are displacement gradients that should obey Newton's second law and, consequently, the influence of inertia effects should be important in ferroelastics. With this idea in mind, in a $1d$ ferroelastic, considering that displacement, u, and strain, ε, are related as, $e = \partial u / \partial x$, Lagrangian dynamics with Rayleigh dissipation leads to the following dynamic equation [130]

$$\rho_0 \frac{\partial^2 e}{\partial t^2} = \frac{\partial^2}{\partial x^2} \left(\frac{\delta F}{\delta e} + \frac{\delta R}{\delta \dot{e}} \right), \tag{3.89}$$

where $F(\varepsilon)$ is the free energy functional that describes the ferroelastic transition, and $R = \frac{1}{2} \gamma \dot{e}^2$ ($\dot{e} = \partial e / \partial t$) is the Rayleigh dissipation function. In the preceding equation, ρ_0 is related to the mass density and γ is a friction coefficient. These ideas have been generalized to address the dynamics of higher-dimensional ferroelastics [131]. It is then found that the dynamic equation for each order parameter strain, e_ℓ, has the more general form, $\rho_0 \partial^2 e_\ell / \partial t^2 = A_\ell \nabla^2 (\delta F_{com} / \delta e_\ell + \delta R_{com} / \delta \dot{e}_\ell)$, where now F_{com} and R_{com} are the effective free energy and Rayleigh dissipation functionals that result after eliminating non-order parameter strains using the Saint-Vénant compatibility constraint. In reciprocal space, the dynamics of an order parameter strain can be viewed as a set of non-linear coupled oscillators labeled by wavenumbers k, with competing short- and long-range couplings between k-modes, 'strain-mass' densities $\rho(k) \sim 1/k^d$ and damping proportional to $1/\rho(k)$. This supposes that large-k oscillators equilibrate first, giving rise to the formation of smaller-scale-oriented textures at earlier times. It is thus expected that after a quench from the high-temperature phase, this dynamics produces a sequential-scale elastic patterning, with hierarchical growth. At late times, inertial effects can be neglected, and the expected pure relaxational dynamics is recovered. An example of the evolving microstructure in a system undergoing a square-to-rectangle transition quenched from the high temperature to a temperature below the transition temperature is illustrated in Figure 3.14. Note that in numerical simulations, pure relaxational dynamics is commonly used when only the late-stage configurations are required.

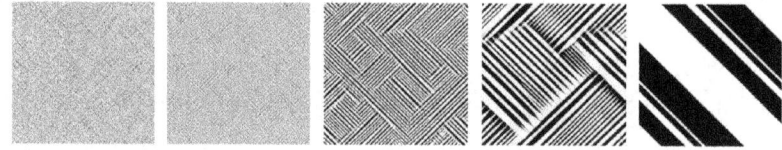

Figure 3.14 Evolution of the microstructure obtained numerically with inertial dynamics in a system undergoing a square-to-rectangle transition quenched from high temperature to a temperature below the transition temperature. Grey regions correspond to the high-temperature phase, while black and white regions correspond to ferroelastic variants with order parameter $e_2 > 0$ and $e_2 < 0$, respectively. From left to right the snapshots are obtained at 1.8 ps, 9.1 ps, 45.4 ps, 0.227 ns and 1.14 ns, after the quench. Adapted from Ref. [93].

3.5.1 Shape memory materials and shape memory effect

As already indicated, martensites and martensitic transitions are often the names used to denote ferroelastics and ferroelastic transitions in metals and alloys. An important subset of martensites are the so-called shape-memory materials [132]. These materials are characterized by a large transition strain and highly mobile twin boundaries, which confer them with relatively low transition hysteresis and make feasible the occurrence of a remarkable thermomechanical behaviour that is at the origin of the so-called shape memory properties. Some examples of materials belonging to this family include

Fe-Pd, Ni-Ti and Cu-based alloys such as the Cu-Zn-Al ternary alloy. Shape memory properties are related to the highly non-linear response to an applied stress that provides these materials the capacity of "remembering" an original shape after a drastic deformation process. Specifically, the shape memory effect refers to the fact that when these materials are deformed in the low-temperature martensitic phase, they are able to recover their original shape by the reverse transition upon heating. In the high-temperature phase, the same class of systems displays another unique property called superelasticity that consists of the possibility of recovering, upon loading and unloading, a large strain (in many cases > 10%) associated with the stress-induced transition. These two properties are schematically depicted in Figure 3.15. Indeed, these properties render this class of materials very attractive from a technological point of view. Actually, they may function as sensors as well as actuators and are promising intelligent materials since they have the ability to perceive *information* and to retain it as knowledge to be applied towards adaptive behaviour within a changing environment [133].

In these systems, memory can be considered as stored in the set of bonds between the atoms that constitute the material. Therefore, this requires that at the atomic level, displacements that take place at the transition should not be too large so that the bonds remain intact, which allows the atoms to return to their original positions upon heating or unloading. From this point of view, the system behaves similarly to an elastically deformed solid that will return to its original state when the load is removed. What makes memory effects different is not only the fact that the material can recover from very large deformations but, more importantly, the possibility of temporarily arresting a deformed configuration as is required in the shape memory effect. This is possible since the detwinned or partially detwinned deformed material is in a state closer to its ground state.

In shape memory materials, the large deformations require a cooperative movement of many atoms at the transition. Therefore, the diffusionless, displacive nature of the transition is an essential feature for shape memory properties to occur and can only be achieved when the symmetries of the high- and low-temperature phases show a group–subgroup relationship and, hence, no reconstruction is needed to form the product phase from the parent phase. The high mobility of interfaces is important in order to make feasible the shape memory behaviour with weak enough hysteresis effects.

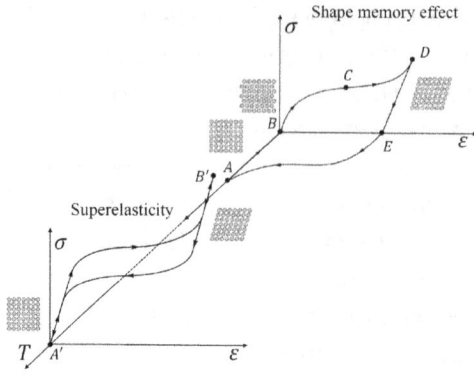

Figure 3.15 Schematic representation in a stress (σ)-strain (ε)-temperature (T) diagram of the shape memory effect and superelasticity. In the shape memory effect, the material is initially in the high-temperature parent phase, A, with a given shape. In the absence of applied stress, it is cooled down and it undergoes a transition to a twinned phase, B, that, on average, preserves the same shape it had in A. From B, it is severely deformed to C and D, which are partially detwinned and totally detwinned states, respectively. When the load is removed, in E, the material remains deformed with a shape different from the original one. Then, in the absence of stress, it is warmed up above the transition temperature and transforms back to the parent phase. In this transition, the deformation is removed and the material recovers its original shape. In the case of superelasticity, the material is initially in the parent phase, A'. Then, at a constant temperature, it is loaded and it transforms to a detwinned phase, B'. When the load is removed, the deformation is also removed and the material returns back to the original phase, A'.

Exercises

3.1 Consider an isotropic ferromagnetic material with localized magnetic moment located at the sites i of a lattice with z-nearest neighbours. Suppose that the magnetic moment $\boldsymbol{\mu} = g\mu_B \boldsymbol{S}$ arises from the total spin S, where μ_B is the Bohr magneton and g the Landé factor.

(a) If demagnetization field effects are not considered, show that within the Weiss molecular field approximation, which assumes that the magnetic moment exchange interaction can be approximated by an effective molecular field proportional to the magnetization, $\boldsymbol{B}_{eff} = \lambda \boldsymbol{M}$, where λ is the molecular field parameter and $\boldsymbol{M}$ the magnetization, the equation of state in the presence of a magnetic field $\boldsymbol{H}$ applied in the z-direction is given as

$$M = M_s \mathcal{B}_S(x), \tag{3.90}$$

which is of the form of Eq. 3.33, with $x = \frac{\mu}{kT}(B + \lambda M)$, $\mu = g\mu_B S$ and $M_s = n\mu$ is the saturation value of the magnetization. $n = N/V$ is the density of magnetic atoms.

(b) The parameter λ is given as:

$$\lambda = \frac{3k_B T_c}{ng^2\mu_B^2 S(S+1)}.$$

(c) Show that this solution is the same as the mean-field solution of the Heisenberg hamiltonian:

$$\mathcal{H}_H = -J \sum_{\langle ij \rangle_{nn}} \boldsymbol{S}_i \cdot \boldsymbol{S}_j - g\mu_B \boldsymbol{B} \cdot \sum_i \boldsymbol{S}_i.$$

Find the relation between λ and the exchange parameter J.

(d) In the case $S \to \infty$, show that this equation of state reduces to that of the classical Heisenberg model (see Exercise 1.3).

3.2 Find the mean-field critical exponents, α, β, γ and δ of the Heisenberg model.

3.3 The Bean-Rodbell model supposes that the exchange coupling depends isotropically on the distance between localized spins. Within the molecular field approximation, this effect can be taken into account assuming that the molecular field parameter λ_{BR} depends on volume as

$$\lambda_{BR} = \lambda(1 + \zeta\omega),$$

where λ is the molecular field parameter of the Weiss theory, $\omega = (v - v_0)/v_0$ the deformation of the unit cell volume and ζ a constant parameter that controls the strength of spin-volume coupling. The free energy density of the model is then assumed to be of the general form:

$$f_{BR} = f(M, T) + \frac{\omega}{2\kappa} - p\omega,$$

where $f(M, T)$ is the magnetic contribution to the free energy density, κ the compressibility and p the hydrostatic pressure. Therefore, the second and third terms on the right-hand side of the preceding equation are the elastic energy associated with volume deformation and the term coupling volume and pressure, respectively. Proceeding as in Section 3.2.1, the free energy density $f(T, M)$ can be expressed as a series expansion in powers of the magnetization.[22] Then, show that

(*a*) The magnetostriction is determined by the equation:

$$\omega = \frac{1}{2}\lambda\zeta\kappa M^2 - p\kappa. \tag{3.91}$$

(*b*) The model shows a tricritical point at $\zeta = \zeta_t$ given by:

$$\zeta_t^2 = \frac{3S^2 + 2S + 1}{5nk_B T_c \kappa S(S+1)},$$

so that the paramagnetic–ferromagnetic transition is continuous for $\zeta \leq \zeta_t$ and first order for $\zeta > \zeta_t$. In the preceding equation, T_c is the critical temperature of the Weiss model and $n = N/V_0$.

3.4   Consider a magnetic material with localized magnetic moments associated with a spin S, which sits on a *bcc*-lattice of parameter a, which interact via nearest and next-nearest neighbours with exchange parameters J_1 and J_2, respectively. Suppose that the lattice is divided into four interpenetrating *fcc*-sublattices of parameter $2a$, denoted as 1, 2, 3, and 4. An atom in sublattice 1, for instance, has four nearest neighbours on sublattice 2, the other four on sublattice 4 and the six next nearest neighbours on sublattice 3. Within the mean-field Weiss approximation,

(*a*) If λ_1 and λ_2 are nearest and next-nearest neighbour molecular field coefficients defined for the complete lattice, obtain the equations giving the effective fields acting on each sublattice, respectively.

---

[22] Note that the magnetic equation of state has the same form as that of the Weiss model of Exercise 3.1 after replacing λ by λ_{BR}.

(*b*) Find the expressions giving λ_1 and λ_2 in terms of J_1 and J_2.

(*c*) Show that at high temperature, in the paramagnetic phase, the susceptibility follows a Curie–Weiss law, $\chi = \frac{C}{T-\theta}$, with

$$\theta = \frac{S(S+1)}{3k}(8J_1 + 6J_2),$$

and C is the Curie constant.

(*d*) Show that the transition temperatures from the paramagnetic to the indicated ferromagnetic and antiferromagnetic ordered structures are:

1. Ferromagnetic structure: $M_1 = M_2 = M_3 = M_4$,

$$T_c = \frac{S(S+1)}{3k}(8J_1 + 6J_2).$$

2. Antiferromagnetic structure: $M_1 = -M_2 = M_3 = -M_4$,

$$T_{N_1} = \frac{S(S+1)}{3k}(-8J_1 + 6J_2).$$

3. Antiferromagnetic $M_1 = -M_3$, $M_2 = -M_4$,

$$T_{N_2} = -\frac{S(S+1)}{k}6J_2.$$

(*e*) Justify that for given values of J_1 and J_2, on cooling from high temperature, the system will transform from the paramagnetic phase to the magnetic structure with the largest transition temperature.

3.5 Consider an isotropic magnetic material with vectorial magnetization $\boldsymbol{M} = (M_x, M_y, M_z)$ that can point anywhere in space. In the magnetic state, the magnetization spontaneously chooses a direction in which to point. In the absence of an applied magnetic field, this direction is entirely arbitrary. Hence, the system must be rotationally invariant with respect to any angle of rotation. Therefore, the free energy must be invariant under the change $\boldsymbol{M} \to \mathbf{R}\boldsymbol{M}$, where $\mathbf{R}$ is a rotation matrix. Find a Landau model consistent with this symmetry.

3.6 Consider a ferrimagnet with two inequivalent sublattices, 1 and 2, such that atoms in sublattice 1 carry a spin S_1 while atoms in sublattice 2 carry a spin $S_2 \neq S_1$. Within the Weiss molecular field approximation, show that the effective field acting on sublattices 1 and 2 can be expressed as

$$B_{1,2}^{eff} = \mu_0 H + \lambda M_{1,2},$$

where H is an external applied field and $M_{1,2}$ are the magnetizations of sublattices 1 and 2, respectively.

(a) Show that the high-temperature susceptibility χ depends on temperature as

$$\chi(T) = \frac{1}{T^2 - \theta^2} \left[(C_1 + C_2)T + 2\lambda C_1 C_2 \right],$$

where C_1 and C_2 are the Curie constants of the two sublattices, respectively.

(b) Find the temperature θ and compare it with the critical temperature of the model.

(c) Show that if $C_1 = C_2 = C$, which corresponds to the case $S_1 = S_2$, the susceptibility reduces to that of an antiferromagnet if $\lambda < 0$ or to that of a ferromagnet if $\lambda > 0$.

3.7 Consider an infinite planar domain wall perpendicular to the z-axis. In the case of a Néel domain wall, the magnetization rotates about the y-axis, which is parallel to the wall, while in a Bloch domain wall, it rotates about the z-axis. These two types of walls can be described by functions $\theta(z)$ and $\phi(z)$ that provide the angles subtended by the magnetization with the z and y axes, respectively. Neglecting demagnetization effects, the corresponding energies per unit area of these wall can be approximated by a functional of the general form[23]

$$E = \int_{-\infty}^{\infty} \left[J \left(\frac{\partial^2 \varphi}{\partial z^2} \right) + K f(\varphi) \right] dz,$$

where J is an exchange constant, K an anisotropy constant and the angle φ stands for θ or ϕ for Néel or Bloch wall, respectively. In the former case, $f(\theta) = \sin^2 \theta$ and in the latter, $f(\phi) = \cos^2 \phi$.

Impose boundary conditions, $\theta(z \to -\infty) = 0$, $\theta(z \to \infty) = \pi$ and $\left(\frac{\partial \theta}{\partial z} \right) \to 0$ for $z \to \pm\infty$ in the Néel case and $\phi(z \to \pm\infty) = \pm\pi/2$ and $\left(\frac{\partial \phi}{\partial z} \right) \to 0$ for $z \to \pm\infty$ in the Bloch case.

---

[23] The demagnetization energy contribution is not included in this energy function. This contribution, which accounts for surface effects, should be included in a finite planar wall. For a Néel wall it is of the form $\varepsilon_d = \frac{1}{2}\mu_0 M_s N_y \sin^2 \theta$ while for a Bloch wall, $\varepsilon_d = \frac{1}{2}\mu_0 M_s (N_y \sin^2 \theta + N_x \cos^2 \phi)$, where M_s is the saturation magnetization and N_x and N_y are demagnetization factors.

(*a*) Show that the function $\varphi(z)$ that minimizes the surface energy must satisfy the equation

$$K f(\varphi) = J \left(\frac{\partial \varphi}{\partial z} \right)^2,$$

with φ being θ or ϕ in the Néel or Bloch case, respectively.

(*b*) Find, in both cases, the angle that describes the domain wall profile as a function of z.

3.8 Consider a simple cubic antiferroelectric crystal, in which the ions that belong to the sublattice a constituted of diagonal planes of the unit cell can be spontaneously polarized, with ions in neighbouring planes constituting sublattice b polarized in antiparallel direction. If P_a and P_b are the polarizations in sublattices a and b respectively, propose a Landau model adequate to describe the phase transition from a high-temperature paraelectric phase to a low-temperature antiferro-electric phase. Study the effect of an applied electric field. Can the para-antiferroelectric transition in the presence of an applied electric field be of first order?

3.9 Some materials display a ferroelastic transition from a high-temperature *bcc*-phase to a low-temperature phase that can be described as a long period modulated structure. An example is the $9R$ phase observed in Li and a number of alloys such as the β-brass Cu-Zn in which the unit cell consists of a stacking of close-packed planes with sequence $ABABCBCAC$, which is in contrast with the AB sequence in the *hcp* or the ABC sequence in the *fcc* structure. The transition path can be described in terms of homogeneous deviatoric symmetry-adapted strains e_2 and e_3 combined with a static transverse acoustic phonon displacement, $\frac{1}{3}\{110\}TA_2$. Strain e_2 shears the (110) planes along the $\pm[1\bar{1}0]$ direction and strain e_3 contracts the (100) planes along the [001] direction and stretches them in the $\pm[1\bar{1}0]$ direction. Both strains are associated with the elastic constant C'. The static phonon corresponds to a displacement $\boldsymbol{u}(\boldsymbol{r}) = A\hat{\boldsymbol{e}}_{[1\bar{1}0]} \sin(\boldsymbol{k} \cdot \boldsymbol{r} + \phi)$, where $\boldsymbol{k} = \frac{1}{3}(110)$ and $\hat{\boldsymbol{e}}_{[1\bar{1}0]}$ is a unit vector along the $[1\bar{1}0]$ direction. As usual, this distortion can be expressed as a two-component complex order parameter, $\psi = Ae^{i\phi}$. Then a Landau free energy function for the cubic to $9R$ transition can be proposed in terms of the order parameters e_2, e_3 and ψ. It is of the general form:

$$f(e_2, e_3, \psi, T) = f_{strain}(e_2, e_3, T) + f_{phonon}(\psi, T)$$
$$+ f_{strain-phonon}(e_2, e_3, \psi),$$

where $f_{strain}(e_2, e_3, T)$, $f_{phonon}(\psi, T)$ and $f_{strain-phonon}(e_2, e_3, \psi)$ are, respectively, the contributions associated with the homogeneous strain, the modulated distortion and the strain–phonon coupling.

(a) Propose an expression for $f_{strain}(e_2, e_3, T)$ that includes all the invariant terms up to the fourth order in e_2 and e_3. Assume that the expansion coefficients of second, third and fourth orders are, respectively, $C' = (C_{11} - C_{12})/2$, $C_{333}/3$ and $C_{2222}/4$.

(b) Suppose that $f_{phonon}(\psi, T)$ is expanded to eighth order in ψ as,

$$f_{phonon}(\psi, T) = \frac{1}{2}r_0|\psi|^2 + \frac{1}{4}u_0|\psi|^4$$
$$+ \frac{1}{6}v_0|\psi|^6 + \frac{1}{6}v_1[\psi^6 + (\psi^*)^6] + \frac{1}{8}w_0|\psi|^8,$$

and taking into account that the minimum order strain–phonon coupling is, $f_{strain-phonon}(e_2, e_3, \psi) = \kappa e_2[\psi^3 + (\psi^*)^3]$, show that in equilibrium, the order parameters satisfy the following relations:

$$e_3 = \frac{C_{333}}{C_{11} - C_{12}}e_2^2,$$
$$e_2 = -\frac{2\kappa}{C_{11} - C_{12}}A^3 \cos(3\phi),$$

and that this yields the following effective free energy density in terms of the phonon amplitude only

$$\tilde{f} = \frac{1}{2}r_0A^2 + \frac{1}{4}u_0A^4 + \frac{1}{6}v_{eff}A^6 + \frac{1}{8}w_0A^8,$$

where $v_{eff} = v_0 + 2v_1 - 12\frac{\kappa^2}{C_{11}-C_{12}}$.

(c) Justify why $f_{phonon}(\psi, T)$ must be expanded to the eighth order in ψ and find the condition for the *bcc* $\to$ *9R* transition to be first order.

4

Coupling of ferroic degrees of freedom: multiferroics

Schmid coined the term *multiferroics* to refer to a class of materials that exhibits simultaneously more than one primary ferroic property in a single phase [134]. Originally this class included the combination of ferromagnetism, ferroelectricity and ferroelasticity but, at present, ferrotoroidicity is also considered among the four primary ferroic orders. Moreover, the definition is often expanded to include non-primary order parameters such as antiferromagnetism or ferrimagnetism and composite materials constituted of parts of individual ferroics. In all these materials, some coupling between degrees of freedom associated with the different ferroic orders may occur. The interesting situation occurs when this coupling is strong, which supposes that a ferroic property can be manipulated by a non-conjugated field. For instance, in the presence of a strong enough magnetostructural coupling magnetization can be switched by an external stress and strain can be induced by the application of a magnetic field. Similarly, in the presence of electrostructural coupling, application of stress can be used to switch polarization while electric field can be used to induce strain. From this point of view, the most interesting class of multiferroics is that which displays strong magnetoelectric coupling, which in turn implies that magnetization can be switched by electric field and polarization by magnetic field.[1] The general situation defining magnetostructural, electrostructural and magnetoelectric coupling is schematized in Figure 4.1. It must be taken into account that non-multiferroic materials can display either magnetostructural, electrostructural, or magnetoelectric coupling, which are rather common effects that occur in a more widespread class of materials [135, 136]. For instance, piezoelectric materials do not need to be either ferroelectric or ferroelastic. In that case, polarization occurs when the material is subjected to an

---

[1] Since the existence of magnetostructural and electrostructural coupling is considered quite obvious, often the term multiferroic is simply used to designate magnetoelectric multiferroics.

applied electric field or to stress due to a linear electromechanical interplay between both mechanical and electrical degrees of freedom in materials
where no spontaneous breaking of spatial inversion symmetry occurs. The
situation is similar in magnetostrictive materials that mechanically respond
to an applied magnetic field. As has already been discussed in Chapter 3,
note as well that ferrotoroidic materials are intrinsically magnetoelectric but
not necessarily ferromagnetic or ferroelectric.

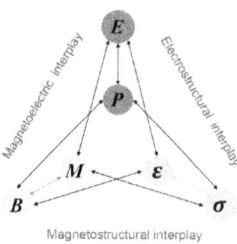

Figure 4.1 Schematic representation of the different types of coupling between magnetic, polar and structural orders. The corresponding ferroic order parameters can be induced directly by its conjugated field or through
a non-conjugated field.

From the point of view of basic physics, the main problem when dealing
with multiferroic materials is understanding which are the conditions and
constraints that determine the possibility of combining in a single phase two
or more ferroic properties. This issue is quite well understood in the case
of magnetostructural and electrostructural multiferroics, but it is a much
more challenging question in the case of magnetoelectric multiferroics since,
in many cases, the two kinds of order tend to exclude one another and
therefore the problem is quite non-trivial.

The interplay between different ferroic orders makes multiferroic materials
especially relevant in relation to applications. Magnetoelectric multiferroics
are certainly the most interesting class of multiferroic materials due to their
potential for controlling magnetism using an electric field or vice versa. The
former possibility is especially suitable since the generation of electric fields
supposes an energy cost much lower than the production of magnetic fields,
which require electric currents [137]. These capabilities suggest that this
class of materials has great potential related to a variety of promising new
technological applications such as magnetic memories electrically controlled
without currents, development of a new classes of four-state logic based on
up-down polarization and up-down magnetization or the design of magnetoelectric sensors.

Materials combining two or more ferroic properties are, in fact, known from quite long time ago. However, the interest in the study and development of new multiferroic materials has only increased at the beginning of the twenty-first century when the technological relevance of this class of materials was recognized. For instance, the well-known $BaTiO_3$ ferroelectric should, in fact, be classified as an electrostructural multiferroic. In fact, at ~ 393 K, this material undergoes a phase transition from a cubic to a non-centrosymmetric tetragonal structure and behaves both as a ferroelectric and a ferroelastic system [138]. However, this material has, for many years, been studied as either a ferroelectric or as a ferroelastic material but, instead, its multiferroic properties have not been exploited. Similarly, the single-element Dy, which is a ferromagnetic rare earth material, is known for quite some time to show magnetostructural properties typical of systems that simultaneously display ferromagnetism and ferroelasticity but, for many years, this striking behaviour was seen as an academic curiosity. In the case of magnetostructural multiferroics, the interest in this class of materials heightened after the discovery that the Heusler Ni_2MnGa intermetallic compound shows magnetic shape memory properties [140], which are related to the possibility of inducing large deformations by application of moderate magnetic fields. These properties are indeed a consequence of the multiferroic nature of this Heusler material. On the other hand, regarding magnetoelectric multiferroics, in the mid-1960s, a huge magnetoelectric effect was reported in the boracite $Ni_3B_7O_{13}I$ that allowed for hysteretic switching of a multiferroic state by either electric or magnetic field [143]. In fact, two of the currently most popular magnetoelectric multiferroics, $BiFeO_3$ [144] and the hexagonal manganites ($RMnO_3$, with R being Sc, Y, In, Dy-Lu) [145], were already identified at that time. In spite of that, only recently, the properties of these materials have been investigated intensively with the aim of exploiting their multiferroic nature.

4.1 Magnetoelectric multiferroics

Magnetoelectric multiferroics must combine magnetic and ferroelectric order in a single phase and, thus, both space inversion and time reversal symmetries must be broken. From the 122 Shubnikov colour point groups (see Appendix), only 13 simultaneously allow for the occurrence of spontaneous polarization and magnetization [141] which include a very restricted set of all possible structures and, thus, suggest that materials of this multiferroic family should be scarce. Beyond these symmetry considerations, electronic structure arguments also indicate contradicting requirements for ferroelectricity and ferromagnetism to develop in the same material [142]. On the one hand, ferroelectrics must be insulators since polarization introduces bound charges and, under the presence of free carriers, the charges will be screened and polarization will vanish. This means that ferroelectricity requires filled bands. On the other hand, magnetic materials tend to be metallic since ferromagnetism requires partially filled d or f bands of transition metal or rare-earth ions. Within this point of view, for instance, in perovskite oxide ferroelectrics, the second-order Jahn-Teller effect, that explains the off-centring mechanism and provides breaking of spatial inversion, can occur in transition metal ions with formally empty d-states. However, the existence of magnetism in this class of oxides requires localized d-electrons that provide the net spin, which permits the time-reversal symmetry breaking. Therefore, this supposes that there is an exclusion effect that is often denoted as the d^0 vs. d^n *problem* that limits the existence of magnetoelectric multiferroics. However, while this is a seemingly compelling argument, it strictly applies only to perovskite oxides. As a matter of fact, beyond this class of materials, several other compounds can display ferroelectricity with many of them also showing some kind of magnetic order.[2] This has led to the argument that, in fact, magnetoelectric multiferroics should be much more numerous than expected and, actually, this is corroborated by the fact that quite some materials have already been developed [146].

In general, magnetoelectric multiferroics are classified into two big groups that take into account the microscopic mechanism that allows for the coexistence of magnetic and ferroelectric order. These two families are commonly denoted as type-I and type-II multiferroics [147]. Type-I multiferroics consist of materials in which ferroelectricity and magnetism have different sources that, to a large extent, are independent of one another, although there is some coupling between the corresponding properties, but this coupling is weak. On the other hand, type-II multiferroics are materials in which magnetism causes ferroelectricity, which supposes that the coupling of magnetic and polar degrees of freedom is strong. Both families can be further classified into subclasses according to specific differences in the physical origin of ferroelectricity.

Type-I multiferroics

This class of magnetoelectric multiferroics are those that were first developed and remain as the more numerous ones. In general, they are good ferroelec-

---

[2] Note that while the microscopic origin of magnetism is essentially the same in all magnets and is related to the presence of localized electrons, mostly in partially filled d- or f-shells, there are a number of different microscopic mechanisms of ferroelectric order.

tric materials that show as well some kind of magnetic order. In general, the ferroelectric phase occurs at a temperature higher than the magnetic phase, with both transition temperatures typically well above room temperature. Depending on the mechanism giving rise to ferroelectricity, the following different subclasses can be considered.

Multiferroic perovskites. Probably the best known and among the most abundant ferroelectrics are perovskites. However, these ferroelectrics are in general not magnetic due to the d^0 *vs.* d^n problem discussed previously. Anyway, this is more of an empirical result rather than a physical law and, in fact, the existence of a d^n ion off-centre of the O_6 octahedra defining the perovskite structure cannot be excluded. Nevertheless, the most realistic strategy that can circumvent this problem is the development of mixed perovskites with d^0 and d^n ions. In this case, however, the strength of the magnetoelectric coupling is expected to be very weak.

Lone pairs. This mechanism is based on the spatial asymmetry created by the anisotropic distribution of unbonded valence electrons about a host ion. This occurs in some perovskite oxides with a non-magnetic ion that has two outer s-electrons that do not participate in the chemical bonding and are referred to as lone pairs. These pairs provide high polarizability, which means that they induce the displacement away from the centrosymmetric position in the oxygen surrounding these ions. Examples are the $BiFeO_3$, $BiMnO_3$, and probably the $PbVO_3$ oxides where the Bi^{3+} and Pb^{2+} ions have lone pairs of outer 6s electrons.[3] The existence of ferroelectricity in these compounds is then a consequence of the ordering of these pairs in a given direction. These materials can also display magnetic properties associated with a transition metal ion, which in $BiFeO_3$ and $BiMnO_3$ are iron and manganese ions, respectively. Therefore, in this class of perovskites, ferroelectricity and magnetism arise from distinct ions and thus can display multiferroicity.

Charge ordering. This mechanism of spatial inversion symmetry breaking may occur in compounds that contain transition metal ions with different valence and, more often, in materials with ions that carry different charge. Due to charge ordering, both sites and bonds can become inequivalent, which can lead to ferroelectricity [150]. Among other compounds, this mechanism has been shown to give rise to ferroelectricity in a number of magnanites and

---

[3] $BiFeO_3$ was the first discovered material showing large polarization and magnetization close to room temperature. However it was later found that the strong magnetization was in fact a non-intrinsic effect associated with the thin film nature of the studied samples.

nickelates [149] and also in the $LuFe_2O_4$ compound [151]. The interesting point is that ferroelectricity originating from this mechanism is not incompatible with the existence of magnetic order associated with a transition metal ion.

Geometric effects. In some materials such as the hexagonal manganite $YMnO_3$ [152], ferroelectricity simply originates from a buckling of the layered MnO_5 polyhedra, which is accompanied by displacements of the Y ions that occur due to a structural instability that yields a closer packing structure. As a consequence of this structural change, oxygen atoms move closer to the Y ions, which gives rise to the off-centring effect and thus to ferroelectricity. Indeed, this mechanism for ferroelectricity permits the coexistence of magnetism and ferroelectricity.

Type-II multiferroics

In this class of materials, ferroelectricity occurs due to the existence of a particular class of inhomogeneous magnetically ordered state that can induce the breaking of inversion symmetry. Therefore, it is the interplay of spins and charges that induces the non-centrosymmetry and drives the establishment of polar order. Thus, in these systems ferroelectricity is often denoted as improper. The three main mechanisms that can lead to this class of multiferroicity are described next.

Dzyaloshinskii-Moriya interaction. The Dzyaloshinskii-Moriya interaction is based on a mechanism in which a non-centrosymmetric environment induces an asymmetric magnetic interaction. In the present case, the effect is inverse and it is related to an acentric configuration of magnetic moments that induces a non-centrosymetric displacement of charges. This mechanism is schematically represented in Figure 4.2. This may occur, for instance, in materials showing a cycloidal spiral magnetic configuration. Then a polarization can be induced given by $P \sim r_{ij} \times (S_i \times S_j) \sim Q \times e$, where r_{ij} is the vector connecting neighbouring magnetic moments, Q the wave vector that describes the rotation and $e = S_i \times S_j$. This polarization is essentially a consequence of the optimization of the spin configuration, which arises from a spin–orbit interaction. The result is a one-to-one correlation between antiferromagnetic order and electric polarization. This class of magnetically induced ferroelectricity was reported for the first time in Cr_2BeO_4 [153] and later in other oxides such as $TbMnO_3$ [154], in which a giant magnetoelectric coupling was found. It is important to note that the existence of spiral magnetic order is related to some kind of competition between magnetic interactions[4] that, in some cases, may even lead to magnetic frustration effects. This is why most of this class of type-II multiferroics are found in magnetically frustrated systems.

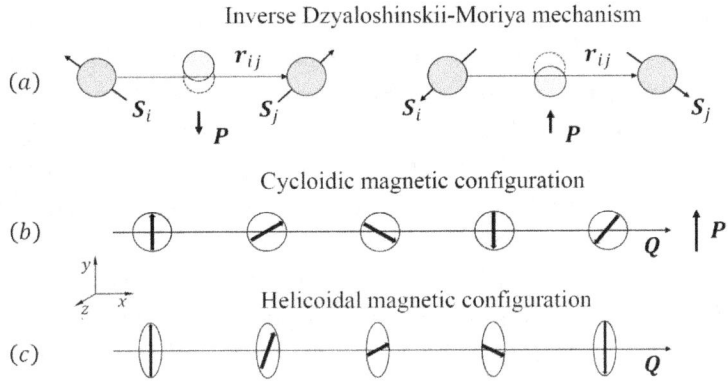

Figure 4.2 (*a*) Schematic illustration of magnetically induced ferroelectricity from inverse Dzyaloshinskii-Moriya interaction. (*b*) Cycloidic configuration of magnetic moments with the wave vector $Q = (Q_x, 0, 0)$. The magnetic moments rotate in the plane (x, y) and induce a non-zero polarization along the y-axis. (*c*) Helicoidal configuration. The magnetic moments rotate in a plane (y, z), perpendicular to Q. In this case, $e = S_i \times S_j$ and Q are parallel and, while the inversion symmetry is broken no polarization occurs, in general.

It is worth noting that this class of multiferroicity may arise in some domain walls. This, for instance, is the case of Néel domain walls in magnetic materials where the magnetization rotates in a plane containing the wave vector, which is perpendicular to the plane defining the boundary that separates domains with opposite magnetization. Therefore, a Néel domain wall can be considered as part of a cycloidal spiral and, thus, it is expected that in the domain wall region, electric polarization can emerge that is induced by an inverse Dzyaloshinskii-Moriya mechanism. With this idea, it has been suggested that materials with good multiferroic properties could be designed by controlling domain wall density.

---

[4] The competition giving rise to spiral configuration can be understood in a similar way as the one giving rise to helical order described in Chapter 3. However, it must be noted that no polarization is induced in systems with the helical magnetic order.

A similar situation occurs in systems with skyrmion textures[5] that also have physical ingredients which are common to those materials that give rise to type-II magnetoelectric multiferroics.

Exchange striction. This mechanism is a consequence of the fact that magnetic exchange varies with atomic positions. Therefore, it can occur in collinear magnetic structures without the involvement of spin–orbit effects and leads to an acentric displacement of charges, which happens in order to optimize the symmetric product $S_i \cdot S_j$ between neighbouring spins. For instance, in the case of a chain of ions that carry a net spin, it is expected that at high temperature the distance between pairs of neighbouring ions is always the same and thus the chain satisfies inversion symmetry with no polarization. If at low-temperature magnetic order of the type up-up-down-down ($\uparrow\uparrow\downarrow\downarrow$) is established, the distortion associated with ferro- and antiferromagnetic bonds is different and, consequently the inversion symmetry can be broken and polarization may appear such that, $P \propto R_{ij}(S_i \cdot S_j)$, where R_{ij} denotes the direction along which the magnetostriction occurs. This is illustrated in Figure 4.3. This mechanism of ferroelectricity has been found in Ca_3CoMnO_6, which consists of one-dimensional chains of alternating Co^{2+} and Mn^{4+} magnetic ions [155].

Exchange striction mechanism

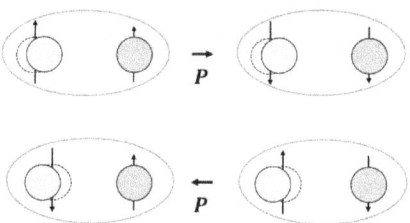

Figure 4.3 Schematic illustration of magnetically induced ferroelectricity from exchange striction mechanism.

Besides the preceding three main routes, other mechanisms that cause magnetically induced improper ferroelectricity are the spin-dependent $p - d$ hybridization and the electronic ferroelectricity that can occur in frustrated magnets. The former mechanism has been found in the $CuFeO_2$ delafossite and is a consequence of a spin-driven modulation of the chemical bonds

---

[5] Skyrmions are non-collinear spin textures which can be described as a given twist of magnetic moments with a non-coplanar pattern in which the directions of the spins can be mapped onto the surface of a sphere. These magnetic structures will be discussed in detail in Chapter 6.

between magnetic 3d orbitals and ligand 2p orbitals, which leads to a spontaneous polarization along the bond direction [156]. On the other hand, in a frustrated magnet, a spin-induced polarization may arise. For instance, in the prototypical geometrically frustrated triangular antiferromagnet, it can be shown that a polarization proportional to the correlation function $S_1 \cdot (S_2 + S_3) - 2S_1 \cdot S_2$ can be induced [157].

Composite magnetoelectric multiferroics

Discussion of multiferroic systems so far supposes that ferroelectricity and magnetism simultaneously coexist in a single-phase material. An alternative to such a situation, are hybrid or composite systems comprising ferroelectric and ferromagnetic materials. Usually, these composites are produced in a granular or layered form and are designed with the aim of improving the strength of the magnetoelectric coupling. In general, two types of coupling can be expected. Direct coupling based on electronic effects at the interfaces between both components and, perhaps more interesting, indirect coupling mediated by strain, which, contrary to the direct coupling, may induce a bulk effect. In this case, for instance, ferromagnetic magnetostrictive and ferroelectric piezoelectric materials are combined. Then, under an applied magnetic field, strain is induced in the magnetostrictive constituent, which is transferred via strain coupling between both constituents to the ferroelectric part, which responds with a voltage thanks to the piezoelectric effect. In this class of multiferroics, a very strong magnetoelectric coupling can be obtained and therefore these systems are very interesting from an applied point of view. The most interesting class of composite multiferroics is that of thin films layered heterostructures where the strain coupling is controlled by exploiting epitaxy effects. In reality, by selecting a convenient substrate, a wide range of tensile and compressive strains can be generated, which allows controlling the magnetoelectric coupling. For instance, it has been shown that ferroelectricity in $BaTiO_3$ crystals can be used to tune the metamagnetic transition temperature of epitaxially grown FeRh films and electrically drive a transition between antiferromagnetic and ferromagnetic order with only a few volts, just above room temperature [158]. One of the highest magnetoelectric couplings is found in a composite of terfenol-D ($Tb_xDy_{1-x}Fe_2$, with $x \sim 0.3$) and PZT ($Pb[Zr_xTi_{1-x}]O_3$).[6]

---

[6] Instead of terfenol-D, galfenol can be used which is a very high magnetostrictive gallium–iron alloy.

4.1.1 Free energy and magnetoelectric coupling

Any model aimed at describing magnetoelectric multiferroics must take polarization and magnetization as order parameters. Therefore, a Landau expansion of the free energy function for this class of materials cannot include terms of odd order in both order parameters since the free energy must be invariant under both time and space inversion operations. In an homogeneous system, the magnetoelectric coupling term of the lowest order that satisfies these symmetry conditions is a biquadratic term $\boldsymbol{P}^2\boldsymbol{M}^2$ [148]. This term is in fact a universal coupling term that must be considered in all magnetoelectric multiferroics. However, since it is a fourth-order term, of higher-order than the harmonic terms proportional to $\boldsymbol{P}^2$ and $\boldsymbol{M}^2$, its effect is expected to be indirect and weak. Actually, this term may cause temperature shifts of electric and/or magnetic transitions and lead to a renormalization of susceptibility values. However, it cannot promote a magnetically induced polarization or an electrically induced magnetization because the small energy gain, which is proportional to $-\boldsymbol{P}^2\boldsymbol{M}^2$, does not compensate for the cost of the polar or magnetic changes, which are proportional to $+\boldsymbol{P}^2$ or $+\boldsymbol{M}^2$, respectively. In any case, this biquadratic coupling term[7] is the essential term to consider for the magnetoelectric interplay in type-I multiferroics [149].

It is interesting to note that if magnetic ordering is inhomogeneous and, thus, magnetization varies over the crystal, symmetry also allows for a third-order term coupling linearly the polarization with the product of magnetization and magnetization gradient. In that case, even if the strength of such a coupling is weak, it can give rise to electric polarization, as soon as a magnetic order of a suitable type occurs. For cubic crystals, the magnetoelectric coupling contribution to the free energy, F_{me}, is then of the form [159]

$$F_{me} \propto \boldsymbol{P} \cdot [(\boldsymbol{M} \cdot \nabla)\boldsymbol{M} - \boldsymbol{M}(\nabla \cdot \boldsymbol{M})]. \tag{4.1}$$

Note that since the space derivative operation, ∇, breaks the space-inversion symmetry, this term will preserve both space-inversion and time-reversal symmetries in the free energy. Due to this term, minimization of the free energy leads to a magnetically induced electric polarization of the form $\boldsymbol{P} \propto [(\boldsymbol{M} \cdot \nabla)\boldsymbol{M} - \boldsymbol{M}(\nabla \cdot \boldsymbol{M})]$. Indeed, this contribution is only relevant when magnetization is not uniform and, therefore, it is important when dealing with type-II multiferroics. Note, for instance, that if a non-collinear cycloidic or helical magnetic configuration is considered, with magnetization components, $(M_x \cos(\boldsymbol{Q} \cdot \boldsymbol{r})x, M_y \cos(\boldsymbol{Q} \cdot \boldsymbol{r})y, M_z z)$, where $\boldsymbol{Q}$ is the

---

[7] In a material with ferroelectric-antiferromagnetic couplig characterized by two sublattices, this term is expected to be of the form $\boldsymbol{P}^2\boldsymbol{M}_1\boldsymbol{M}_2$, where $\boldsymbol{M}_1$ and $\boldsymbol{M}_2$ are the corresponding staggered magnetizations.

propagation vector, it is then straightforward to obtain an average polarization contribution given by

$$\boldsymbol{P} \propto M_x M_y (\boldsymbol{z} \times \boldsymbol{Q}).\qquad(4.2)$$

This equation shows that a polarization perpendicular to the propagation
vector and to z-axis may occur at least if $\boldsymbol{Q}$ is parallel to the z-axis. Therefore, this result confirms the discussion of the preceding section, showing that
no polarization occurs in the case of an helical magnetic configuration. As
already indicated previously, this class of magnetically induced polarization
seems adequate to explain the existence of multiferroicity in the $TbMnO_3$
compound where magnetic moments localized at Mn^{+3} ions develop a cycloidal spiral lying in the xy-plane with the wave vector along the x-axis as
shown in Figure 4.4.

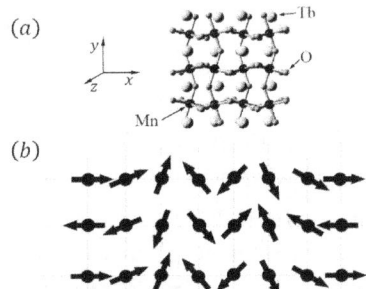

Figure 4.4 (*a*) Crystallographic structure of the $TbMnO_3$ compound. (*b*)
Corresponding magnetic structure in the ferroelectric-antiferromagnetic
phase proposed in Ref. [160]. The cycloidal distribution of the magnetic
moments localized at the Mn ions with the wave vector along the z-axis
is illustrated. The polarization is induced along the x-axis. Adapted from
Ref. [161].

It is interesting to note that there are other combinations of magnetization
and polarization that satisfy the required space and time reversal symmetries
that could, in principle, be considered adequate to describe the magneto-
electric coupling. For instance, terms of the form, $[\boldsymbol{M} \cdot \epsilon(\boldsymbol{r})]^2 [\boldsymbol{P} \cdot b(\boldsymbol{r})]$ or
$[\boldsymbol{M} \cdot \epsilon(\boldsymbol{r})]^2 (\nabla \cdot \boldsymbol{P})$, satisfy this requirement if the function $b(\boldsymbol{r})$ is an odd
function in $\boldsymbol{r}$. Here, the space function $\epsilon(\boldsymbol{r})$ has no special requirements.[8]
Most of these possibilities still remain unexplored from an experimental view
point.

---

[8] This function is a general function of position that should not be confused with strain.

4.2 Electro- and magnetostructural multiferroics

As already indicated previously, the existence of electrostructural and magnetostructural multiferroic materials that are simultaneously ferroelectric and ferroelastic or ferromagnetic and ferroelastic, respectively, is known from quite long time ago. However, the interest in the specific multiferroic features of this class of materials is in fact much more recent and limited to a large extent to magnetic shape memory materials that belong to the class of magnetostructural multiferroics. This section will be devoted essentially to discuss this class of materials but keeping in mind that analogous models can be easily adapted to describe electrostructural materials.

Ni_2MnGa is probably the prototypical example of a magnetostructural multiferroic. On cooling, this Heusler material becomes ferromagnetic at a Curie temperature $T_c \simeq 370$ K and, on further cooling, undergoes a ferroelastic (martensitic) transition at $\sim$200 K. Below this temperature, this material simultaneously shows ferromagnetic and ferroelastic behaviour. This is the multiferroic phase in which magnetization can be induced by stress and large deformations can be induced by application of moderate magnetic fields. This multiferroic phase is characterized by a complex microstructure constituted of twin-related ferroelastic domains with magnetic 180° domains inside twins. Therefore, 90° magnetic domains form between consecutive ferroelastic domains. This complex microstructure is responsible for both elastic and magnetic properties of the multiferroic phase to a large extent. It is interesting that similar microstructures occur in electrostructural multiferroics such as $BaTiO_3$. In this material, ferroelectric domains form inside twin-related ferroelastic domains in its polar tetragonal phase. In Figure 4.5, microstructures reported in multiferroic Ni_2MnGa and $BaTiO_3$ phases are shown. In these materials, often the anisotropy associated with the low-temperature phase is very strong and in the presence of an applied magnetic or electric field, these multiferroic microstructures can be detwinned by twin boundary motion instead of magnetic or polar moment rotation. This allows the possibility of magnetically or electrically inducing large strains or to change the magnetization or polarization by an applied magnetic or electric field. Figure 4.6 shows the giant strain induced by a relatively low magnetic field below 1 T in an off-stoichiometric $Ni_{48.8}Mn_{29.7}Ga_{21.5}$ compound in its multiferroic phase.

Magnetostructural or electrostructural multiferroic domain microstructures can be modelled by generalizing the model presented in Section 3.5, for ferroelastic materials including magnetic or polar degrees of freedom together with the adequate magneto- or electrostructural coupling between

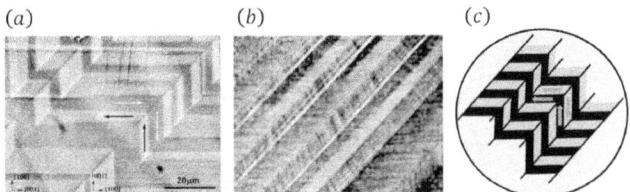

Figure 4.5 (*a*) Domain structure of Ni_2MnGa (from Ref. [162]) and (*b*) $BaTiO_3$ (from Ref. [163]). In both cases twin-related ferroelastic domains are formed with 180° magnetic or polar domains, respectively. In (*c*) the observed domain structure is represented schematically. Ferroelastic twin bands are separated by continuous lines and inside each one polar or magnetic up and down domains are shown with the corresponding magnetization or polarization pointing along the short *c*-lattice parameter of the ferroelastic structure, which determines the magnetic or polar easy-axis (indicated with arrows). Note that 90° magnetic or polar domains form between consecutive ferroelastic domains.

strain and magnetization or polarization. For this purpose, it is particularly interesting to consider the case corresponding to a square-to-rectangle transition. In order to be specific, let us suppose the case of a magnetostructural multiferroic. Then, the free energy functional of the model will have structural, magnetic and magnetostructural contributions

$$F = F_s + F_m + F_{ms} = \int (f_s + f_m + f_{ms})d^2r, \qquad (4.3)$$

where the structural free energy density must be given by Eq. 3.75. According to micromagnetic theory, the pure magnetic part must be of the form

$$F_m = D \int m_x^2(\boldsymbol{r})m_y^2(\boldsymbol{r})d^2r + J \int \int |\nabla \boldsymbol{m}(\boldsymbol{r})|^2 d^2r$$
$$- \mu_0 M_s \int \left(\frac{1}{2}\boldsymbol{H}_d + \boldsymbol{H}_{ext}\right) \cdot \boldsymbol{m}d^2r, \qquad (4.4)$$

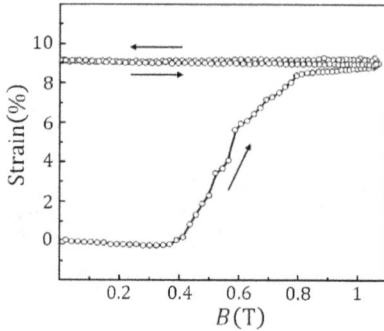

Figure 4.6 Strain associated with the detwinning process induced by an applied magnetic field in $Ni_{48.8}Mn_{29.7}Ga_{21.5}$ Heusler compound at 300 K in its orthorhombic magnetostructural multiferroic phase. The magnetic field was applied along the [100] direction of the high-temperature cubic phase and the strain was measured in a perpendicular direction. Initially the studied sample was in a twinned state. Application of the magnetic field causes detwinning and thus deformation. In the detwinned configuration the sample remains in the same state when the field was removed. Adapted from Ref. [164].

where $m = M/M_s$ is the three-component unit magnetization vector[9] and M_s the magnetization saturation. Here H_d and H_{ext} are the demagnetizing and externally applied magnetic fields. In the preceding equation, the first term corresponds to the anisotropy energy of the square lattice. The parameter D will be assumed positive, which supposes that this term is minimized for the magnetization along the x or y direction.[10] The second term is the exchange energy which accounts for any spatial variation of the magnetization orientation. Therefore, the coefficient J is an exchange stiffness constant. The third term includes the magnetostatic energy associated with the demagnetizing field and the Zeeman energy accounting for the coupling of the magnetization with an externally applied field.

The lowest-order magnetostructural coupling allowed by symmetry is as follows:

$$F_{m-s} = \kappa_1 \int \left[m_x^2(\boldsymbol{r}) + m_y^2(\boldsymbol{r}) \right] e_1(\boldsymbol{r}) \, d^2r + \kappa_2 \int \left[m_x^2(\boldsymbol{r}) - m_y^2(\boldsymbol{r}) \right] e_2(\boldsymbol{r}) \, d^2r$$

$$+ \kappa_3 \int m_x(\boldsymbol{r}) m_y(\boldsymbol{r}) e_3(\boldsymbol{r}) \, d^2r, \tag{4.5}$$

---

[9] Even if the model is a $2d$ model, it is important to assume that the magnetization is a $3d$ vector with two components in-plane in order that a realistic magnetization dynamics can be considered that accounts for the precession of the magnetization vector about an applied magnetic field.

[10] For $D < 0$, this term is minimized when the magnetization is out-of-plane.

where κ_i $(i = 1, 2, 3)$ are magnetostrictive coefficients. If lattice integrity is assumed, it is known that the three strains, e_1, e_2, and e_3 are not independent, which can be taken into account by minimizing the magnetostructural free energy under the Saint-Vénant elastic compatibility condition. Proceeding as discussed in the preceding chapter, in the present case, it is obtained that the ferroelastic non-order parameter free energy terms can be written in Fourier space as,

$$F_{nonOP} = \frac{1}{4\pi^2} \int \hat{U}_2(\boldsymbol{k}) e_2(\boldsymbol{k}) e_2(-\boldsymbol{k}) d^2k$$
$$+ \frac{\kappa_1}{4\pi^2} \int \hat{U}_1(\boldsymbol{k}) e_2(\boldsymbol{k}) \int e^{i\boldsymbol{k}\cdot\boldsymbol{r}} [m_x^2(\boldsymbol{r}) + m_y^2(\boldsymbol{r})] d^2r d^2k$$
$$+ \frac{\kappa_3}{8\sqrt{2}\pi^2} \int \hat{U}_3(\boldsymbol{k}) e_2(\boldsymbol{k}) \int e^{i\boldsymbol{k}\cdot\boldsymbol{r}} [m_x^2(\boldsymbol{r}) m_y^2(\boldsymbol{r})] d^2r d^2k, \quad (4.6)$$

where

$$\hat{U}_1(\boldsymbol{k}) = \frac{(k_x^2 - k_y^2)(k_x^2 + k_y^2)}{k^4 + 8Rk_x^2k_y^2}, \quad (4.7)$$

$$\hat{U}_2(\boldsymbol{k}) = \frac{A_1}{2} \frac{(k_x^2 - k_y^2)^2}{k^4 + \frac{8}{R}k_x^2k_y^2}, \quad (4.8)$$

$$\hat{U}_3(\boldsymbol{k}) = \frac{(k_x^2 - k_y^2)k_xk_y}{\frac{1}{8R}k^4 + k_x^2k_y^2}. \quad (4.9)$$

In the preceding expressions $R = A_1/A_3$ is the ratio of the bulk and the shear moduli. Notice that due to the magnetostructural interaction, in addition to the term $\hat{U}_2$ that controls long-range effects in the pure ferroelastic case discussed in the preceding chapter, now the kernel includes the terms $\hat{U}_1$ and $\hat{U}_3$ that appear due to the magnetostructural coupling. Nevertheless, also in this case the long-range kernel is minimized for $k_x = \pm k_y$ which explains the directionality of long-range correlations along the diagonals of the square lattice. When κ_1 is large, it is expected that the free energy can be minimized for $m_x = \pm 1$ for $e_2 < 0$ and $m_y = \pm 1$ for $e_2 > 0$. In both cases, the minimum energy is found for $e_1 = e_3 = 0$. This situation corresponds in fact to the domain configuration illustrated in Figure 4.5. It is interesting to note that the long-range term essentially determines the magnetocrystalline anisotropy in the ferroelastic rectangular phase.

The model can be solved numerically, which requires discretizing the system into a square mesh with periodic boundary conditions. Starting from a random disordered configuration, the system is then allowed to evolve until it reaches a stabilized configuration. Following the micromagnetic theory, magnetization dynamics must be governed by Gilbert's dynamics (Eq. 3.57) with $\boldsymbol{H}_{eff} = \frac{1}{\mu_0}\frac{\partial F}{\partial \boldsymbol{m}}$, while strain must evolve by means of a pure relaxational dynamics. At temperatures within the multiferroic phase, an example of the obtained domain configuration is shown in Figure 4.7. It compares very well with real microstructures as shown in Figure 4.5.

Figure 4.7 Magnetic and structural domain configuration obtained in the multiferroic phase from numerical simulations of a model for magnetostructural system undergoing a square-to-rectangle transition in a ferromagnetic phase. Twin-related structural domains with 180° magnetic domains inside each twin band are shown. 90° magnetic domains form between neighbouring twins. Adapted from Ref. [165]

Thanks to the possibility of inducing large deformations by application of a magnetic field, some Heusler alloys display *magnetic* shape memory effect in addition to the conventional shape memory properties. This magnetic property is similar to the conventional shape memory effect described in Chapter 3 but in that case the large deformation can also be induced by means of a magnetic field in addition to stress. As in the conventional case, this deformation can be recovered by heating above the ferroelastic transition temperature. In addition to the prototypical Ni_2MnGa, there are other ferromagnetic compounds, including some different Heusler alloys, $Fe_{70}Pd_{30}$ [166] and even some magnetic oxides such as the $La_{2-x}Sr_xCuO_4$ compound [167] that shows a similar behaviour. It is interesting to note that this magnetic shape memory effect can be reproduced by the model previously discussed and the obtained results are illustrated in Figure 4.8.

There is another family of Heusler alloys that also shows magnetic shape memory properties. These compounds are off-stoichiometric and are denoted often as metamagnetic shape memory alloys since the ferroelastic transition is accompanied by a magnetic transition that can be induced not only by stress but also by magnetic field. In most of these materials, the transition occurs from a high-temperature ferromagnetic phase to a ferroelastic, nonmagnetic low-temperature phase, which is followed by a further transition to a ferromagnetic phase that occurs at a lower temperature. Prototypical

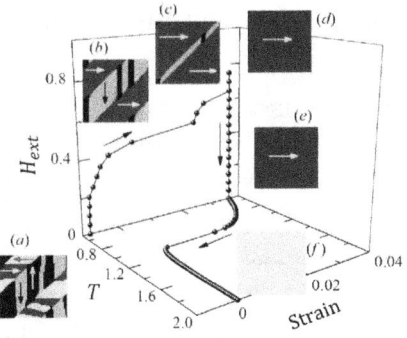

Figure 4.8 Magnetic shape memory effect reproduced from the model of a square-to-rectangle transition in a ferromagnetic system. The figure shows the evolution of the strain throughout the whole process, consisting of loading [(a)-(d)], unloading [(d)-(e)] and heating [(e)-(f)] paths. Snapshots of the magnetostructural microstructure obtained from numerical simulations are shown at representative points of the curves. The two twin-related ferroelastic domains and the austenite parent phase are indicated with different tonalities. Arrows point along the direction of the magnetization. Magnetic field and temperature are given in reduced units. For the model considered the maximum magnetic-field induced deformation is about 4%. Adapted from Ref. [165]

examples are $Ni_2Mn_{1+x}Z_{1-x}$ Heusler compounds with excess of Mn with respect to the 2-1-1 stoichiometry, and with Z being elements such as In, Sn or Sb. The drastic decrease in magnetization at the transition has been associated with the excess of Mn atoms that allows the existence of nearest neighbour Mn-Mn pairs with a tendency to couple antiferromagnetically and to the fact that this effect strengthens in the ferroelastic phase due to the change of lattice parameters [168].

Few materials are known that undergo a direct transition from a high-temperature paraelastic and paramagnetic phase to a ferroelastic and ferromagnetic phase. Among them is Ni-Mn-Ga within a range of compositions with deficit of Mn and electron concentration higher than the concentration corresponding to the 2-1-1 stoichiometry. In all these metamagnetic compounds, the magnetostructural coupling is controlled by the biquadratic coupling term e^2M^2. This coupling is of fourth-order and thus weaker than the third order term discussed previously. Therefore, in these materials, the response in terms of strain or magnetization to the corresponding non-conjugated fields is weak. Nevertheless, large changes of the magnetostructural transition temperature induced by both magnetic field and stress are possible, which provide this class of materials with important technological properties.

4.3 Morphotropic phase boundaries

Originally the term morphotropic phase transition was introduced to refer to a structural transition in a certain compound induced by a change in its composition [169]. Therefore, a morphotropic phase boundary is the boundary between the corresponding two phases in a composition-temperature diagram. At present, the term is more often used to refer to the phase transition induced either by mechanical or, more commonly, chemical pressure or composition between two crystallographically different ferroelectric or ferromagnetic phases. Therefore, in these materials, chemical and mechanical pressure are essential parameters that control phase stability. A well-known example is the lead zirconate-lead titanate solid solution constituted of given fractions of rhombohedral, $PbZrO_3$ (PZO), and tetragonal, $PbTiO_3$ (PTO) ferroelectric compounds. The phase diagram of this system is shown in Figure 4.9. At high temperature, the system shows a paraelectric cubic phase that transforms to a rhombohedral-ferroelectric phase for low PTO content and to a tetragonal-ferroelectric phase for high PTO content. In these materials, the crystal structure imposes a large polar anisotropy and hence the polarization points along $\langle 111 \rangle$ directions in the rhombohedral phase and along $\langle 001 \rangle$ directions in the tetragonal phase.

At a given composition, a triple point exists at which the three phases can coexist. The morphotropic boundary line is a quite-vertical line starting from this point. The interest in mixtures with these properties is due to the fact that across the morphotropic boundary, the crystal structure changes abruptly, which leads to a significant enhancement of the electromechanical response of the solid solution, which provides these mixtures with a potentially interesting behaviour for their use as sensors and actuators. Another interesting solid solution showing similar properties is the $Ba(Ti_{0.8} Zr_{0.2})O_3$-$(Ba_{0.7}Ca_{0.3})TiO_3$ mixture, which has, from the point of view of applications, the advantage of being a lead-free mixture still showing very large piezoelectric response across the morphotropic line [170].

In principle, morphotropic boundary lines are also expected to occur in ferromagnetic solid solutions. Nevertheless, the conditions that allow their existence are more difficult to be satisfied since the relationship between crystallographic structure and the direction of the magnetization vector are often less well defined. Anyhow, similar phase diagrams with morphotropic boundaries have also been reported in a number of ferromagnetic solid solutions. This is, for instance, the case of the solid solution $TbCo_2$-$DyCo_2$, where the rich-$TbCo_2$ rhombohedral solution is separated from the rich-$DyCo_2$

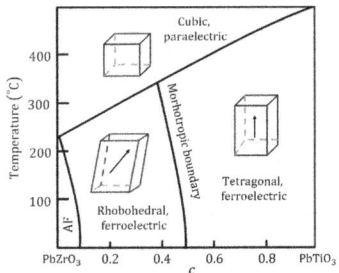

Figure 4.9 Phase diagram of the lead zirconate-lead titanate solid solution. Here c is the molar fraction of $PbTiO_3$ in the solution. Cubic ($Pm3m$), rhombohedral ($R3m$) and tetragonal ($P4mm$) regions are indicated. Arrows indicate the direction of the polarization in the ferroelectric phases. In the AF region, the material shows an antiferroelectric phase ($R3c$).

tetragonal solution by a morphotropic phase boundary where a significant enhancement of magnetostrictive properties has been reported [171].

4.3.1 Modeling morphotropic phase boundaries

Models have been proposed to address systems showing a morphotropic phase boundary. Here, we discuss a Landau model that includes the minimum ingredients that permit to reproduce the phase diagram of the lead zirconate-lead titanate solid solution. The model represents a generalization of the triple-well Landau model proposed for canonical ferroelectrics, such as $BaTiO_3$ or $PbTiO_3$ oxides. Here, however, we need to take into account the expected change of the crystallographic structure with the molar fraction of each component in the solid solution. To this end, ferroelectric phases showing high polarization anisotropy are assumed with an easy polarization direction imposed by the crystal symmetry and the free energy Landau coefficients are taken as composition dependent. The free energy density function f is thus assumed to be of the form [172]

$$f(c, \tau, \boldsymbol{n}, P) = \frac{1}{2} A_2(c)[\tau - \tau_c(c)]P^2 + \frac{1}{4} A_4(c, \tau, \boldsymbol{n})P^4 + \frac{1}{6} A_6(c, \tau, \boldsymbol{n})P^6,$$

(4.10)

where $\tau = T/T_c$ is a reduced temperature relative to the Curie temperature T_c, which in turn is assumed to depend on compostion and the polarization is given by $\boldsymbol{P} = P\boldsymbol{n}$, where $\boldsymbol{n}$ is a unit vector that determines polar anisotropy imposed by crystal symmetry that is expected to result from the coupling between polarization and strain. Therefore, in the strong anisotropy limit imposed by the crystallographic structure, $\boldsymbol{n} = \{0, 0, 1\}$ for a tetragonal phase and $n = \frac{1}{\sqrt{3}}\{1,1,1\}$ for a rhombohedral phase. Considering symmetry-allowed terms up to sixth order, the coefficient A_4 must be of the general form given by the sum of one isotropic and one anisotropic term

$$A_4 = a_4(c,\tau) - a_4'(c,\tau)(n_1^4 + n_2^4 + n_3^4), \tag{4.11}$$

while the sixth-order term must be given by the sum of one isotropic and two anisotropic terms as follows:

$$A_6 = a_6(c,\tau) - a_6'(c,\tau)(n_1^6 + n_2^6 + n_3^6) - a_6''(c,\tau)n_1^2 n_2^2 n_3^2. \tag{4.12}$$

In the preceding expressions, n_i ($i = 1, 2, 3$) are the components of the vector n. In what follows, second- and fourth-order coefficients will be assumed to depend linearly on composition, while the sixth-order coefficient A_6 will be supposed to be a positive constant, independent of composition, temperature, and anisotropy. This is a reasonable approximation since, close to the Curie temperature, higher-order anisotropy terms are expected to be small. Within these assumptions, it is then straightforward to find that for $a_4'(c,\tau) < 0$, the free energy f is minimized either by the vector $n = \frac{1}{\sqrt{3}}\{1,1,1\}$ and, thus, the low-temperature stable phase is rhombohedral, or for $a_4'(c,\tau) > 0$ it is minimized by the vector $n = \{0,0,1\}$ and the low-temperature stable phase is the tetragonal one. The change of sign of the coefficient $a_4'(c,\tau)$ occurs along a line

$$a_4'(c,\tau) = 0, \tag{4.13}$$

in the τ-c diagram, which determines the condition that defines the morphotropic boundary line. Since, in principle, the coefficient a_4' depends on both composition and temperature, the morphotropic line should not be necessarily vertical in this diagram. It is important to note that at the morphotropic line the free energies of the rhombohedral and tetragonal phases become equal, which requires that the crystallographic anisotropy and polarization vanish. Consequently, it is expected that the domain wall energy, σ,[11] vanishes as well. Taking into account that domain width is proportional to $\sqrt{\sigma}$ (see Section 2.2.1), it is expected that near the morphotropic line the characteristic domain size is reduced to the nanoscale. In addition, the symmetry of Eq. 4.10 should change from cubic to spherical since it must reflect the degeneracy with respect to the polarization direction, n. In this situation, the polarization of the ferroelectric domains should be decoupled from the crystal lattice and, consequently, domains may have irregular shapes

---

[11] Note that σ is essentially determined by the energy cost of polarization rotation across a domain boundary between adjacent domains.

with curved boundaries, high mobility, and high sensitivity to external stimuli. This is the basic feature that confers a large electromechanical response across the boundary.

The transition from cubic-to-rhombohedral and cubic-to-tetragonal will be continuous for $A_4 > 0$ and first order for $A_4 < 0$. Therefore, tricritical points could exist given by the condition

$$a_4(c, \tau) - a_4'(c, \tau)(n_1^4 + n_2^4 + n_3^4) = 0. \qquad (4.14)$$

In order to analyse in more detail this condition it is convenient to assume given temperature and composition dependences of the coefficients A_2, τ_c, a_4 and a_4'. For such a purpose, it is reasonable to suppose that $A_2(c)$ and $\tau_c(c)$ are simple linear interpolations between the values associated with the pure compounds in the solid solution corresponding to $c = 0$ and $c = 1$, respectively. On the other hand, the coefficients a_4 and a_4' can be assumed to be of the form[12]

$$a_4(c, \tau) = b_4(c - c_0), \qquad (4.15)$$
$$a_4'(c, \tau) = b_4[\xi(c - c_m) + \zeta(\tau - \tau_m)], \qquad (4.16)$$

where, b_4, c_0, ξ, and ζ are dimensionless constants and c_m and τ_m are the coordinates of the intersection of the morphotropic boundary with the Curie line $\tau = \tau(c)$. Then, from Eqs. 4.14, 4.15, and 4.16 the following equation of the morphotropic boundary is obtained:

$$\tau = \tau_m - \frac{\xi}{\zeta}(c - c_m). \qquad (4.17)$$

Note that this line will be vertical when $\zeta \to 0$, which supposes that the coefficient a_4' is temperature independent.

For $A_4 > 0$ a continuous cubic-to-rhombohedral and cubic-to-tetragonal changes will occur when $A_2 = 0$, which leads to $\tau = \tau_c(c)$. Two tricritical points must be found on this line since the condition $A_4 = 0$ can be satisfied either for $\mathbf{n} = \{1/\sqrt{3}, 1/\sqrt{3}, 1/\sqrt{3}\}$ or $\mathbf{n} = \{0, 0, 1\}$. This leads to the following two conditions, $a_4(c, \tau) - \frac{1}{3}a_4'(c, \tau) = 0$ and $a_4(c, \tau) - a_4'(c, \tau) = 0$, which give tricritical points at $c \leq c_m$ and $c \geq c_m$, respectively. In real systems showing a morphotropic boundary such as the lead zirconate-lead titanate solid solution, there is no clear experimental evidence that confirms the existence of these two tricritical points that might even merge at the triple point [173].

---

[12] The composition and temperature dependence of these coefficients can be approximated by adopting the usual Landau point of view in terms of a Taylor expansion truncated at the lowest order.

The electromechanical response across the morphotropic boundary line can be determined from the calculation of the strain in each phase. If the polarization is assumed homogeneous, the strain tensor components, ε_{ij}, can be obtained as

$$\varepsilon_{ij} = P^2 \sum_{k,l} q_{ijkl} n_k n_l, \tag{4.18}$$

where the indices i, j, k, l can take values 1, 2, and 3 that stand for coordinates x, y, and z. Here q_{ijkl} are the components of the so-called electrostrictive tensor that relates the strain tensor and polarization, which are introduced as phenomenological coefficients. Indeed, this expression is consistent with the assumptions of the preceding Landau model that assumes that polarization is homogeneous in each phase. Once the strain in each phase is known, the piezoelectric coefficients can be obtained from its change under a small applied electric field. For instance, the longitudinal coefficient to an applied electric field in the z-direction, $d_{33}^{[001]}$, will be given by

$$d_{33}^{[001]} = \frac{\partial \varepsilon_{zz}}{\partial E_z}, \tag{4.19}$$

while it will be given by

$$d_{33}^{[111]} = \frac{1}{3} \frac{\partial [\varepsilon_{xx} + \varepsilon_{yy} + \varepsilon_{zz} + 2(\varepsilon_{yz} + \varepsilon_{xz} + \varepsilon_{xy})]}{\partial E}, \tag{4.20}$$

if the field is applied along the [111] direction. Since the strain changes across the morphotropic boundary line, these coefficients are expected to show large values over this line [173].

Rhombohedral and tetragonal phases are ferroelectric as well as ferroelastic and, in general, they are expected to be constituted of polar and ferroelastic domains. In fact, the domain configurations should be similar to those discussed in Section 4.2 with twin-related ferroelastic domains and polar domains inside. Therefore, in this general situation, the calculation of the piezoelectric coefficient should need to take into account such domain configurations and their evolution under an applied electric field.

Exercises

4.1 Consider a material described by the following Landau free energy density function with two order parameters, η and φ, which are biquadratically coupled:

$$f(\eta, \phi, T) = f_0(T) + a_\eta \eta^2 + \frac{1}{2}\eta^4 + a_\phi \phi^2 + \frac{1}{2}\phi^4 + \kappa \eta^2 \phi^2, \quad (4.21)$$

where $a_\eta = T - T_\eta$ and $a_\phi = T - T_\phi$. In addition, T_η and $T_\phi = 2T_\eta$ are the temperatures at which the phases $\eta = 0$ and $\phi = 0$ become, respectively, unstable in the absence of coupling, $\kappa = 0$. If $T_\eta(\kappa)$ is the transition temperature between the phases $(\eta = 0, \phi \neq 0)$ and $(\eta \neq 0, \phi \neq 0)$:

(a) Find an expression for $T_\eta(\kappa)$ as a function of T_η.

(b) Plot the phase diagram of this system in the space T-κ.

(c) Display graphically the expected temperature dependence of the order parameters η and ϕ in the range of coupling $0 < \kappa < 1/2$.

4.2 Consider a ferroelectric thin film described by the following Landau model with a bilinear coupling between the out-of-plane polarization $P_z \equiv P_3$ and the in-plane shear stress, $\sigma_{xy} \equiv \sigma_6$:[13]

$$F(T, P_3, \sigma_6) = F_0(T) + \frac{1}{2}a_0(T - T_0)P_3^2 + \frac{1}{4}a_2 P_3^4 + \frac{1}{6}a_6 P_3^6$$
$$+ \frac{1}{2}S_{66}\sigma_6^2 - b_{36}P_3\sigma_6,$$

where $S_{66} = S_{xyxy}$ is a shear compliance, which is the inverse of the shear elastic constant C_{xyxy}, and b_{36} is the coupling coefficient.

(a) For a film deposited onto a thick enough substrate, the in-plane strain is determined by the misfit strain $\epsilon_6^{(m)}$. In that case, show that the equation of state of a uniform film can be written in the form:

$$E_3 + b_{26}C_{66}\epsilon_6^{(m)} = a_0(T - T_0 - \delta T_0)P_3 + a_4 P_3^3 + a_6 P_3^5,$$

where E_3 is the electric field in the out-of-plane direction and $\delta T_0 = b_{36}C_{66}\varepsilon_0 \mathcal{C}$, with ε_0 being the dielectric constant of vacuum and $\mathcal{C} = 1/\varepsilon_0 a_0$ the Curie–Weiss constant.

(b) Considering now that $\epsilon_6^{(m)} = 0$, up and down domains with polarizations P_+ and P_-, respectively, will form with equal fraction in

[13] This exercise has been adapted from the model introduced in Ref. [174].

the absence of an applied electric field. The applied electric field will unbalance this distribution and, consequently, will induce a stress, $\sigma_6 = -b_{36}C_{66}\langle P_3 \rangle$, where $\langle P_3 \rangle = P_+(1/2 + v) + P_-(1/2 - v) = 2P_s v + (\delta P_+ + \delta P_-)/2$ (v is the deviation of the fraction of positively poled domains from $1/2$). Here, $\delta P_\pm = P_\pm - P_s$, with P_s being the solution of the uniform equation of state with $E_3 = 0$ and $\epsilon_6^{(m)} = 0$. Show that the response of the material to an applied field when $|\delta P_\pm| \ll P_s$, can be obtained from

$$E_3 = \frac{\delta P_\pm}{\chi} + b_{36}^2 C_{66} \left[2v P_s + \frac{1}{2}(\delta P_+ + \delta P_-) \right], \qquad (4.22)$$

where $\chi^{-1} = a_0(T - T_0) + 3a_4 P_s^2 + 5a_6 P_s^5$.

(c) Show that the susceptibility of the polydomain ferroelectric thin film is:

$$\chi_{poly} = \frac{\partial \langle P_3 \rangle}{\partial E_3} = \frac{1}{b_{36}^2 C_{66}}. \qquad (4.23)$$

5

Caloric and multicaloric materials

Any material thermally responds to externally driven changes of its properties. Usually these changes are induced by application or removal of fields thermodynamically conjugated to these properties. Caloric properties arise from the reversible contribution to this thermal response. There are two interesting limiting situations that correspond to either varying the external field isothermally or adiabatically. In the first case, the field induces a change of entropy and, thus, heat is reversibly exchanged with the surroundings. In the second, the system responds with a temperature change. These reversible isothermal change of entropy and adiabatic change of temperature are commonly used to quantify the caloric response of a given material. However, in general, both entropy and temperature will change when the external field is modified. Often these changes are small and strongly influenced by the presence of non-equilibrium dissipative effects. Materials denoted as caloric materials are those that display large caloric effects. This class of materials is nowadays acknowledged to be potentially important for refrigeration, heat pumping, and energy harvesting applications.

In the mid nineteenth century, James Prescott Joule already realized that small changes in temperature may be induced in deformable materials such as natural rubber by application or removal of stress [175]. Later, a similar effect was found to occur induced by a magnetic field in some magnetic materials [176]. In the mid twenties of the last century, Debye [177] and Giauque independently predicted that this effect might permit reaching very low temperatures below 1 K in paramagnetic salts. This magnetocaloric effect was experimentally corroborated shortly after by Giauque and MacDougall [179] and became the basis of the adiabatic demagnetization cooling technology that was exploited in cryogenic applications. In 1930, an electrically analogous, electrocaloric effect was reported in Rochelle salt [180].

For many years, little extra effort was devoted to the development of new caloric materials because it was considered that the caloric response would be, in general, very small, especially at high temperatures. The situation changed in 1976 when the possibility of sustained magnetic cooling about room temperature was demonstrated by making use of the magnetocaloric response of Gd in the vicinity of its Curie temperature [181]. This achievement revived the interest of the magnetic refrigeration technology community, which lead to the discovery of materials such as the intermetallic compound $Gd_5(Si_2Ge_2)$ [182] that displays giant magnetocaloric effects close to room temperature. More recently, the interest has been extended to other caloric effects induced by electric and mechanical fields [183, 184]. At present, materials displaying giant caloric effects are foreseen as efficient and environmentally clean solid-state alternatives to the standard fluid compression technology commonly used in refrigeration applications.

Giant caloric effects occur in the vicinity of phase transitions where thermodynamic properties show a strong temperature dependence. Especially interesting are materials undergoing a first-order transition that can be field induced. From this viewpoint, ferroic materials are especially attractive since a caloric effect is expected to occur associated with the appearance of the ferroic property that, in the absence of applied external fields, may emerge spontaneously upon cooling from high temperature through a phase transition. Depending on specific symmetry-dictated conditions, and possible coupling of the ferroic property to a secondary order parameter, this transition can be either continuous or first order. In the latter case, the transition should involve latent heat that provides the basis for the expected large caloric response. As a consequence, ferroelectrics, ferromagnets, and ferroelastics are expected to be good candidates to display giant electrocaloric, magnetocaloric, and mechanocaloric effects, respectively [185]. A caloric effect associated with toroidization, called toroidocaloric effect, has also been predicted but not yet experimentally realized [186].

In multiferroic materials, two or more phase transitions are expected to occur either at different or the same temperature. In the latter case, the different ferroic orders emerge simultaneously and, therefore, caloric effects associated with each ferroic property are expected to occur in an interdependent manner. Since the strength of the interplay between ferroic properties is strongly enhanced at the lower temperature phase transition, multicaloric effects are expected to occur in its vicinity due to the possibility of cross-response to multiple fields.

In this chapter, we will study basic aspects of the caloric and multicaloric response of materials in the vicinity of phase transitions. This represents the study of thermal response associated with a phase transition, which, for the sake of generality, must take into account the influence of non-equilibrium effects that may affect the reversible response to applied external fields.

5.1 Caloric response

Let us consider a generic multiferroic material characterized by ω ferroic properties $\{X_i\}$. As discussed in Chapter 1, this supposes that the material considered can exchange ω different kinds of work with the surroundings associated with the coupling of its ferroic properties with their corresponding thermodynamically conjugated fields, $\{y_i\}$. Pairs of these thermodynamically conjugated variables can be polarization and electric field, magnetization and magnetic field, or strain and stress. Variables in each pair must have the same tensorial order so that their tensorial product is a scalar with units of energy density. Isothermal, $\boldsymbol{\xi}_T^i$, and adiabatic, $\boldsymbol{\xi}_S^i$, thermal response functions to an applied field $\boldsymbol{y}_i$ can be defined, respectively, as

$$\boldsymbol{\xi}_T^i \equiv \left(\frac{\partial S}{\partial \mathbf{y}_i}\right)_T, \tag{5.1}$$

$$\boldsymbol{\xi}_S^i \equiv \left(\frac{\partial T}{\partial \mathbf{y}_i}\right)_S, \tag{5.2}$$

where S is entropy and T temperature. These response functions have, in principle, the same tensorial order as the fields $\{\mathbf{y}_i\}$. In general, only the response to a given component (or combination of components) of the fields is considered and, thus, scalar functions can be used to characterize the response to such field components. This provides an adequate determination of the caloric response in the case of isotropic materials when the field has a vector-like character (rank-1 tensor). Considering that only one ferroic property, $\boldsymbol{X}$, is relevant and taking into account Maxwell relations introduced in Chapter 1, it is straightforward to show that

$$\xi_T = \left(\frac{\partial X}{\partial T}\right)_y, \tag{5.3}$$

$$\xi_S = \left(\frac{\partial X}{\partial S}\right)_y = -\frac{T}{C_y}\left(\frac{\partial X}{\partial T}\right)_y = -\frac{C_y}{T}\xi_T, \tag{5.4}$$

where now X is the component (or combination of components) of the ferroic property corresponding to those of the field, and C_y is the heat capacity. Indeed, X and y are the magnetization and magnetic field in the case of the magnetocaloric effect, the polarization and electric field in the case of the electrocaloric effect, and strain and stress in the case of the mechanocaloric effect. In this last case, the caloric effect is usually denoted with different names depending on the applied stress mode. When it is induced by a hydrostatic pressure, it is denoted as barocaloric and as elastocaloric effect when it is induced by a uniaxial stress.[1]

It is worth noting that this approach provides a complete characterization of the electro- and magnetocaloric properties of magnetic and polar isotropic materials. In the case of the mechanocaloric effect, even in elastically isotropic materials, a complete characterization requires the determination of caloric properties induced by hydrostatic pressure and shear stress. Usually, instead of pure shear, the elastocaloric response induced by uniaxial stress is studied.

In all these cases, both the isothermal and adiabatic responses are, essentially, controlled by the corresponding response coefficients, $\xi_T(T, y)$. For a finite change of the field, for instance, from 0 to y, the reversible isothermal field-induced change of entropy, $\Delta S(T, y)$, and reversible adiabatic field-induced change of temperature, $\Delta T(S, y)$, are, respectively, given by

$$\Delta S(T, y) = S(T, y) - S(T, 0) = \int_0^y \xi_T(y)dy, \tag{5.5}$$

$$\Delta T(S, y) = T(S, y) - T(S, 0)$$
$$= T(0, y)\left[\exp\left[-\int_0^y \frac{1}{C_y}\xi_T(T, y)dy\right] - 1\right]. \tag{5.6}$$

Figure 5.1 schematically illustrates, in an *S-T* diagram, the change of entropy induced by isothermal application of a field y and the subsequent change of temperature that occurs when the field is adiabatically removed. It is interesting to note that, in general, $\xi_T(y) < 0$, which supposes that the entropy decreases ($\Delta S(T, y) < 0$) and the temperature increases ($\Delta T(S, y) > 0$) when the field is applied. This is the so-called conventional caloric effect, that gives rise, for instance, to the possibility of cooling by adiabatic demagnetization. Nevertheless, it is not forbidden that in certain regions of the thermodynamic variables, $\xi_T(y) > 0$, which leads to the so-called inverse caloric effect. In this case, application of a field gives rise to an isothermal increase of entropy and an adiabatic decrease in temperature.

Often, the temperature change is small and the heat capacity can be assumed independent of the field and, then, from Eq. 5.6, the adiabatic change of temperature is given as $\Delta T(S, y) \simeq -T\Delta S(T, y)/C_y$ to a reasonably good approximation.

---

[1] It is referred to as flexocaloric or twistocaloric effect when it is induced by bending or twisting, respectively. In these two cases, it must be noted that the strain gradient has to be taken into account to quantify the caloric effect.

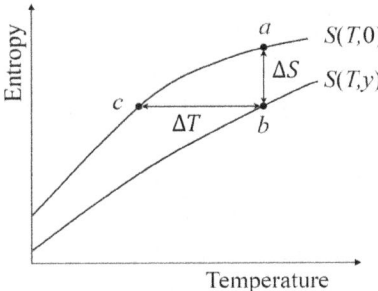

Figure 5.1 Schematic representation in an entropy-temperature diagram (far from the absolute zero temperature) of the changes of entropy and temperature induced, respectively, by isothermal and adiabatic application/removal of a field y. The two curves represent the entropy as a function of temperature at zero and at an applied field y. In the state a the system is at temperature T and zero applied field. When a field y is applied the system reaches the state b with an entropy decrease $-\Delta S$. Then, when the field is adiabatically removed, the state c is reached. In this last process the system cools down to a temperature $T - \Delta T$.

To analyse further the preceding expressions giving ΔS and ΔT, it is useful to express the variable X as a Taylor expansion about $y = 0$. This gives

$$X(T, y) = X_0 + \left(\frac{\partial X}{\partial y}\right)_0 y + \frac{1}{2}\left(\frac{\partial^2 X}{\partial y^2}\right)_0 y^2 + ..., \qquad (5.7)$$

where the first term, X_0 is X in the absence of an applied field at temperature T, and the second, $(\partial X/\partial y)_0 = \chi_0$, is the linear susceptibility. Then, a good approximation to the isothermal field-induced entropy can be given as

$$\Delta S(T, y) = \alpha y + \frac{1}{2}\frac{\partial \chi_0}{\partial T}y^2 + \frac{1}{6}\frac{\partial}{\partial T}\frac{\partial \chi_0}{\partial y}y^3 + ..., \qquad (5.8)$$

where a generalized *thermal expansion* coefficient has been defined as $\alpha \equiv (\partial X_0/\partial T)$. Usually, the coefficient α is small and shows little dependence on temperature and field. In this case, the caloric response, given by the first term of the preceding expansion, is weak. This is not the case in the neighbourhood of phase transitions, either continuous or first order, where the property X is expected to strongly depend on temperature and field.[2] Therefore, the behaviour of the caloric response near continuous and first-order transitions is analysed next in detail.

---

[2] The interesting magnetocaloric properties of paramagnetic salts at low temperature that have been used to produce millikelvin temperatures are not related to a phase transition. They are a consequence of the strong variation in temperature of the linear susceptibility that goes as T^{-1} (Curie law) at low temperature, together with the low value of the heat capacity (that goes to zero at absolute zero as required by the third law of thermodynamics).

Continuous transition

To analyse the behaviour of ΔS and ΔT close to a critical point, we take into account that the property X is the appropriate order parameter for the phase transition. In the critical region, thermodynamic variables show power-law behaviour and, in particular, we know that $X_0 \sim |1 - T/T_c|^\beta$ and $(\partial X/\partial y)_0 \sim |1 - T/T_c|^{-\gamma}$, which leads to $y \sim |1 - T/T_c|^{\beta+\gamma}$. Therefore, in the (X, y, T) space, the thermodynamic behaviour near a critical point can be assumed to be described by the equations

$$X_0 = a \left| 1 - \frac{T}{T_c} \right|^\beta , \tag{5.9}$$

$$y = b \left| 1 - \frac{T}{T_c} \right|^{\beta+\gamma} , \tag{5.10}$$

where a and b are constants. Taking into account that at the critical isotherm $X_0 \sim y^{1/\delta}$, within this framework, it is obtained that the expansion of X can be written in the form [187]

$$X = y^{1/\delta}[1 \pm a|T_c - T|y^{-1/(\beta+\gamma)} + b|T_c - T|^2 y^{-2/(\beta+\gamma)} \pm ...], \tag{5.11}$$

where the signs $\pm$ stand for temperatures $T > T_c$ and $T < T_c$, respectively. Therefore, ΔS is obtained after temperature differentiation and integration over y. Then, in the neighbourhood of the critical point, the following expansion for the isothermal field-induced entropy change is obtained

$$\Delta S = A y^{[\beta(1+\delta)-1]/\beta\delta} \mp B(T_c - T) y^{[\beta(1+\delta)-2]/\beta\delta} + ..., \tag{5.12}$$

where the scaling relation $\gamma = \beta(\delta - 1)$ has been used. Here A, B, ..., are temperature and field-independent coefficients. In the preceding expression, the first term describes the dominant critical behaviour characterized by an exponent $[\beta(1 + \delta) - 1]/\beta\delta$. Note that in mean-field $[\beta(1 + \delta) - 1] = 1$ and $\beta\delta = 3/2$, while the exponent of the second term of the expansion vanishes. Thus, in this approximation, it is expected that $\Delta S \sim y^{2/3}$. Instead, in systems belonging to the Heisenberg universality class, an exponent 0.64 is expected, which is quite similar to the mean-field value and therefore difficult to discriminate from experimental data.

To obtain the critical behaviour of ΔT, the critical behaviour of C_y is first needed. We take into account that the heat capacity satisfies the following thermodynamic expression:

$$\left(\frac{\partial C_y}{\partial y} \right)_T = T \frac{\partial}{\partial y} \left(\frac{\partial S}{\partial T} \right)_y = T \left(\frac{\partial^2 X}{\partial T^2} \right)_y , \tag{5.13}$$

where the Maxwell relation $(\partial S/\partial y)_T = (\partial X/\partial T)_y$ has been used. Then, taking into account Eq. 5.11, it is found that

$$C_y \sim y^{-[2-\beta(1+\delta)]/\beta\delta}, \tag{5.14}$$

if the exponent $[2 - \beta(1 + \delta)]/\beta\delta \geq 0$. Finally, from Eq. 5.6, it is obtained that

$$\Delta T \sim y^{1/\beta\delta}. \tag{5.15}$$

Note that in mean-field, both ΔS and ΔT are expected to display the same critical behaviour. Figure 5.2 (*a*) shows experimental results giving the change of temperature ΔT induced by adiabatic application of a magnetic field in the region of the paramagnetic-ferromagnetic critical point of Gd. In Ref. [189], it has been reported that critical behaviour of Gd is mean-field to a good approximation due to the influence of long-range dipolar effects. Consistently, Figure 5.2 (*b*) corroborates that close to the critical temperature, experimental data nicely satisfy the expected mean-field scaling, $\Delta T \sim h^{2/3}$.

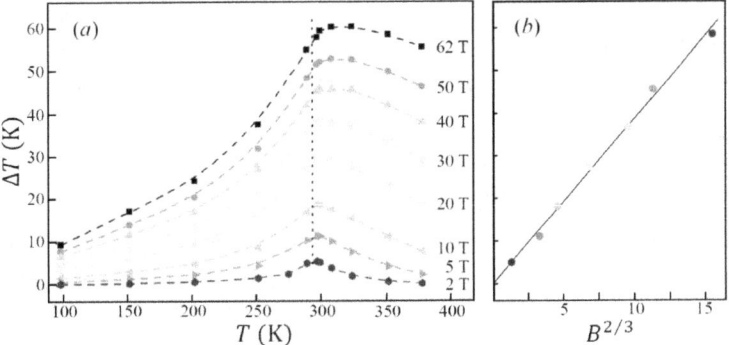

Figure 5.2 (*a*) Temperature change ΔT as a function of the initial temperature T of the material, induced by adiabatic application of pulsed magnetic fields in the region of the paramagnetic-ferromagnetic critical point of Gd crystal. The magnetic induction field, B, is applied in the $\langle 0001 \rangle$ crystallographic direction. The value of the applied field is indicated in each curve. The dashed lines connecting experimental data points are guides to the eye. The vertical dashed line indicates the critical temperature of Gd, $T_c = 294$ K. (*b*) Temperature change for an initial temperature T_c as a function of $B^{2/3}$ that confirms that mean-field scaling is satisfied to a good approximation. Adapted from Ref. [188].

First-order transition

First-order phase transitions are characterized by the discontinuity of extensive properties such as X and the entropy. Therefore, associated with this discontinuity, the transition involves latent heat that, together with the fact that the transition can be field induced, provides the large caloric response to the applied field. For this class of transitions, it is clear from Eq. 5.7 that the relevant term of the expansion is the first term that already shows the discontinuity at the transition temperature, even in the absence of an applied field.

With the aim of studying caloric effects associated with first-order transitions, we assume that the dependence of the property X in the neighbourhood of the transition is of the type:

$$X(T,y) = \tilde{X}_0(T,y) + \Delta X(y)\mathcal{F}\left[\frac{T_t(y) - T}{\delta_T(y)}\right], \tag{5.16}$$

where $\tilde{X}_0$ is a smooth function of T and y, which is supposed to describe the temperature and field dependence of X outside the transition region. Since we are interested in the caloric response associated with the transition, we will, for the time being, not consider this term. Here $\mathcal{F}$ is assumed to be an arbitrary function, not necessarily analytic, that varies from 0 to 1 within a range $\delta_T(y)$. The equation describes an ideal first-order transition taking place in strict equilibrium in the limit $\delta_T(y) \to 0$. In this case, X shows a discontinuity $\Delta X(y)$ at $T_t(y)$, which is the transition temperature. In this case

$$\lim_{\delta_T(y) \to 0} \mathcal{F} = \mathsf{h}[T_t(y) - T], \tag{5.17}$$

where h is the Heaviside step function. Then, taking into account Eq. 5.5, the isothermal entropy change corresponding to a change in the field from 0 to y can be obtained as

$$\Delta S(T,y) = \int_0^y \Delta X(y)\left[\frac{\partial \mathsf{h}[T_t(y) - T]}{\partial T}\right]_y dy, \tag{5.18}$$

where the derivative of the Heaviside function is the Dirac delta function. Then, assuming that ΔX is independent of y, the result of the preceding integral is

$$\Delta S(T,y) = \begin{cases} -\frac{\Delta X}{\vartheta} & \text{for } T \in [T_t(0), T_t(y)], \\ 0 & \text{for } T \notin [T_t(0), T_t(y)], \end{cases} \tag{5.19}$$

where $\vartheta = dT_t(y)/dy$ has also been assumed independent of y. Taking into account the Clausius–Clapeyron equation, $\vartheta = -\Delta S_t/\Delta X$, where ΔS_t is the transition change of entropy. Therefore, as expected, $\Delta S(T, y) = \Delta S_t$. The maximum adiabatic temperature change corresponds to the shift of transition temperature induced by the field, $\Delta T = T_t(y) - T_t(0)$.

It is important to note that when $\Delta X > 0$ ($\vartheta > 0$), the caloric response is conventional, while it is inverse when $\Delta X < 0$ ($\vartheta < 0$). This last situation can occur when the property X in the low-temperature phase is lower than in the high-temperature phase and leads to an inverse caloric effect. This situation may arise due to frustration effects or due to the interplay of property X with secondary degrees of freedom. The behaviour in the conventional and inverse caloric effect cases is illustrated in Figure 5.3.

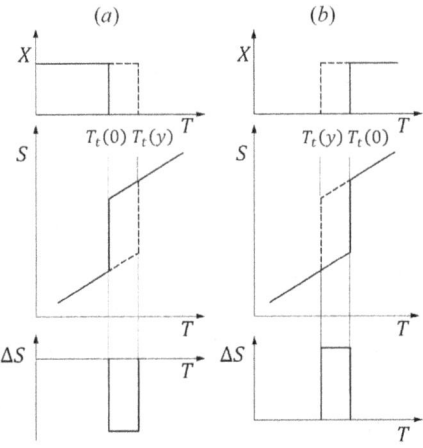

Figure 5.3 Schematic representation of the behaviour of the property X, entropy S and isothermal field-induced entropy change in the region of a first-order ideal transition for systems showing conventional (*a*) and inverse (*b*) caloric effects. In the X vs. T and S vs. T plots, continuous lines indicate the zero-field behaviour, while discontinuous lines correspond to an applied field y. The slope of the curves S vs. T is C/T, where the heat capacity C is assumed to be independent of y. In the conventional case, an applied field shifts the transition to a higher temperature and $\Delta S < 0$. In the inverse case, the field shifts the transition to a lower temperature and $\Delta S > 0$.

Often, in real materials, first-order transitions spread over a certain range of temperatures. There are several factors that can explain this fact including the existence of composition gradients in the studied specimens, defects, etc. In the case of transitions that involve strain, the temperature spread is essentially a consequence of the influence of elasticity that can lead to a thermoelastic equilibrium that renders possible the coexistence of phases in a finite temperature range. In this case, taking into account Eq. 5.16, the general expression giving the isothermal entropy can be expressed as

$$\Delta S(T, y) = \int_0^y \Delta X(y) \left[\frac{\partial \mathcal{F}\{[T_t(y) - T]/\delta_T(y)\}}{\partial T} \right]_y dy + \alpha y, \qquad (5.20)$$

where now a dependence of ΔX on y has been considered and the contribution that describes the temperature and field dependence of X outside the transition region has been included assuming that it is of the form, $\tilde{X}_0(T, y) = X_0(T)$. Then, an average value of the isothermal change over the effective temperature range is obtained as

$$\langle \Delta S(y) \rangle = -\frac{1}{\delta_T(y)} \int_0^y \Delta X(y) dy + \alpha y, \qquad (5.21)$$

with $\delta_T(y) \simeq \delta_T(0) + \Delta T_t(y)$, where $\delta_T(0)$ is the transition range in the absence of applied field, and $\Delta T_t(y)$ represents the transition temperature shift induced by the field.

Experimental determination of caloric effects

In the case of both, first-order and continuous transitions, experimental determination of the isothermal field-induced entropy change can be undertaken based on Eq. 5.5. This is especially adequate in the case of the magnetocaloric effect, which requires measuring magnetization *vs.* magnetic field isotherms at close enough temperature intervals. Then, integral in Eq. 5.5 is computed numerically. When this procedure is used to study the caloric effect near a first-order transition, in addition to the contribution associated with the transition, the obtained $\Delta S(T, y)$ includes the contribution from a possible field and temperature dependence of the magnetization outside the transition. For high-enough fields, this contribution can yield a field-induced entropy change larger than the transition entropy change depending on the sign of α and on whether the caloric effect is conventional or inverse.

The isothermal field-induced entropy change can also be obtained from measurements of the heat capacity at selected values of the magnetic field from which the entropy can be computed as

$$S(T, y) = \int_0^T \frac{C_y(T, y)}{T} dT, \qquad (5.22)$$

from which, in turn the entropy change at temperature T associated with a change of the field from 0 to y can be computed as, $\Delta S(T, y) = S(T, y) - S(T, 0)$.

Heat flux scanning calorimeters specifically designed to work at constant temperature by sweeping the field are especially suitable to provide a direct determination of $\Delta S(T, y)$. These devices are well adapted to study caloric effects associated with first-order transitions. In this case, integration of the calorimetric signal gives the heat exchanged, $q(T, y)$, when the transition is isothermally field induced. Then $q(T, y)/T$ represents a good estimation of $\Delta S(T, y)$ if dissipative effects are sufficiently weak.

The adiabatic field-induced temperature change is usually determined from direct thermometric measurements. Nevertheless, it is often determined from the isothermal entropy change as $\Delta T(y) \simeq T\Delta S(T, y)/C$ which at high enough temperatures provides a sufficiently good estimation.

Metamagnetic materials and the Deformable Ising model

Large caloric effects are expected to occur in materials undergoing a first-order transition with an associated high latent heat, which can be field induced. Therefore, this class of materials is desirable for applications that take advantage of a large caloric response. Ferroelastic and ferroelectric transitions are often first order basically due to symmetry reasons in the former case and due to the interplay of polar and lattice degrees of freedom in the latter. Instead, in most ferromagnetic materials, the magnetization decreases continuously and vanishes at the Curie point. Nevertheless, first-order magnetic transitions occur in a number of materials where the magnetization shows a strong coupling to the lattice. This is the class of magnetic materials that are expected to display a large magnetocaloric effect. An important example is the $La(Fe_xSi_{1-x})_{13}$ compound, which shows a cubic $Fm\bar{3}c$ space group crystallographic structure ($NaZn_{13}$-like structure). For high Fe content, the ground state is antiferromagnetic, but becomes ferromagnetic for $x > 0.89$. In a relatively narrow range of composition, close to the boundary that separates ferromagnetic and antiferromagnetic ground states, the transition from the paramagnetic phase to ferromagnetic one is first-order and for $x = 0.86$ this line ends at a tricritical point. Instead, for lower Fe contents, the transition is continuous. This behaviour stems from the fact that the increase of the Fe content induces a decrease of the lattice parameter with no change of symmetry, which leads to a reduction of Fe-Fe distances and is accompanied by a corresponding increase of the coordination number [190]. This effect gives rise to a magnetostructural coupling responsible for the change of order of the transition.[3] The first-order transition can be induced

---

[3] At a microscopic level, the metamagnetic behaviour of this compound has its origin in the magnetic field-dependent change of the band structure of the 3d electrons that show a very marked peak of the electronic density of states just below the Fermi level.

by the application of moderate magnetic fields, which have the effect of considerably increasing ($\sim 4\mathrm{K/T}$) the transition temperature. Therefore, in this composition regime, $\mathrm{La(Fe}_x\mathrm{Si}_{1-x})_{13}$ compounds show very interesting magnetocaloric effects. The phase diagram of this compound and magnetocaloric properties are illustrated in Figure 5.4.

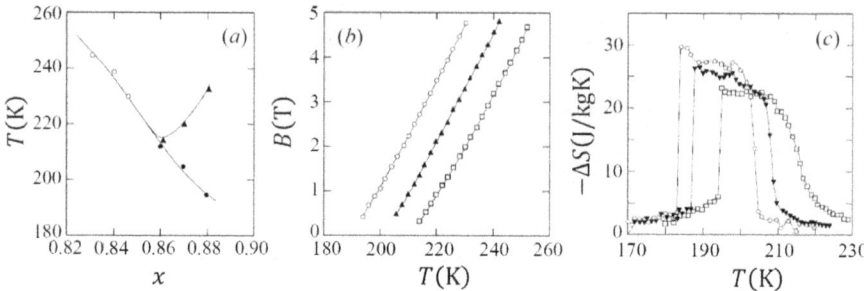

Figure 5.4 (*a*) Phase diagram of the $\mathrm{La(Fe}_x\mathrm{Si}_{1-x})_{13}$ compound. Continuous para-ferromagnetic transition occurs for low x (empty symbols). The line ends at a tricritical point at $x \simeq 0.85$. For larger values of x, the transition is first order. The line of solid circles corresponds to transition temperatures at zero field while the line of solid triangles corresponds to the transition under a finite applied magnetic field. (*b*) Variation of the transition temperature as a function of the applied field (magnetic induction, B) for three different values of x corresponding to the first-order transition region. From left to right, $x = 0.88$, 0.89 and 0.90. (*c*) Isothermal entropy changes corresponding to a magnetic field change of 2 T for the same compositions as in (*b*). Adapted from Refs. [190] and [191].

To address metamagnetic materials that can show first-order and continuous transitions, Bean and Rodbell proposed a generalization of the Weiss model of ferromagnetism that includes the dependence of the exchange coupling on lattice spacing [192]. The model is essentially equivalent to the *Deformable Ising model* introduced in Chapter 1. Therefore, this model is suitable to provide a simple description of caloric effects near first-order and continuous transitions. The region of interest is the one with $J > 0$ in the vicinity of the tricritical point that occurs for a value of the coupling constant $\lambda = 1/3$ (see Figure 1.2). The Gibbs free energy density function is of the general form

$$G(m, \delta, T, h) = f_m(m, \delta, T) + f_v(m, \delta, \{\omega\}, T) - hm, \qquad (5.23)$$

where f_m and f_v are the free energy densities given by Eqs. 1.94 and 1.98, respectively, with $m = m_\alpha + m_\beta$, and $\delta = m_\alpha - m_\beta$ being the ferromagnetic and antiferromagnetic order parameters, and $\{\omega\}$ is the set of normal frequency vibrations. The Zeeman term $-hm$ has been included to account for the coupling of m with an applied external field, h. In the region of interest, minimization of G with respect to m and δ gives $\delta = 0$ [193]. Therefore, the entropy is given by

$$S(m, T, h) = - \left(\frac{\partial G}{\partial T} \right)_h, \quad (5.24)$$

where m is a solution of the equation,

$$m = \tanh \left(\frac{zJm + h}{k_B T} + \frac{3\lambda m}{1 + \lambda m^2} \right). \quad (5.25)$$

Results obtained from this equation are depicted in Figure 5.5. Panel (*a*) shows h *vs.* T coexistence lines for selected values of the parameter λ. These lines end at the corresponding critical point. Magnetization and entropy as a function of temperature for $\lambda = 0.65$ are shown in panels (*b*) and (*c*), respectively, for selected values of the applied field. The isothermal field-induced entropy change is determined as

$$\Delta S(T, h) = S(T, h) - S(T, 0). \quad (5.26)$$

Results for $\lambda = 0.65$ are shown in Figure 5.5(*d*).

Comparison of Figures 5.4 and 5.5 clearly shows that the Deformable Ising model provides a good description of the magnetocaloric properties of metamagnetic materials. In the model, the parameter λ controls the strength of the magnetostructural coupling and when it is used to describe the properties of a metamagnetic material such as the $La(Fe_x Si_{1-x})_{13}$ compound, this parameter must be related to the Fe content, x, that controls the strength of the magnetoelastic interplay in the experiments.

5.2 Multicaloric effects

Multicaloric effects occur when more than one external fields are applied (or removed) either simultaneously or sequentially. For the sake of clarity and without loss of generality, we will assume that the system considered responds to only two fields that we will denote $\boldsymbol{y}_1$ and $\boldsymbol{y}_2$, while the corresponding conjugated displacements (ferroic properties) are $\boldsymbol{X}_1$ and $\boldsymbol{X}_2$, respectively. Assuming isotropy, the change of entropy induced by the fields $\boldsymbol{y}_1$ and $\boldsymbol{y}_2$ can be expressed as [194]

$$\begin{aligned} \Delta S\, [T, (0,0) &\rightarrow (y_1, y_2)] \\ &= \Delta\, S[T, (0,0) \rightarrow (y_1, 0)] + \Delta S[T, (y_1, 0) \rightarrow (y_1, y_2)], \quad (5.27) \end{aligned}$$

where the first term on the right-hand side is the isothermal entropy change induced by the application of the field y_1 and, thus, quantifies the single

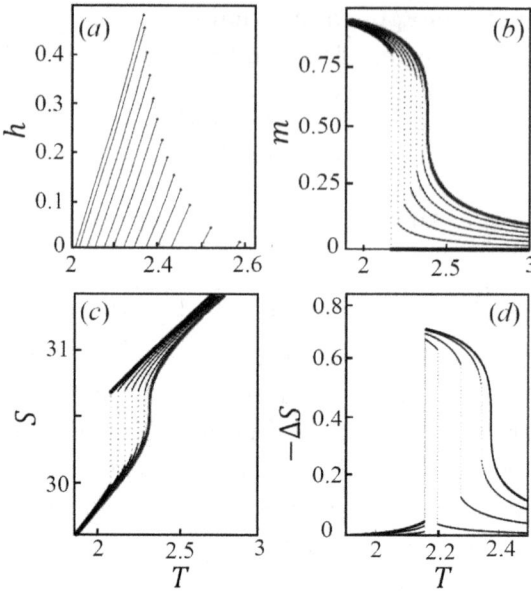

Figure 5.5 (*a*) Paramagnetic to ferromagnetic h-T transition lines for values of the coupling parameter λ ranging from 0.35 (left) to 0.738 (right). Dots indicate the critical end points of the transition lines. (*b*) Magnetization as a function of temperature in the region of the first-order transition for $\lambda = 0.65$ and values of the magnetic field ranging from 0 (left) to the critical field $h_c = 0.05$ (right curve). (*c*) Corresponding entropy vs. temperature curves. (*d*) Isothermal field-induced entropy change as a function of temperature corresponding to field changes $0 \to h$, with (from left to right) $h = 0.05$, 0.15, 0.25 and 0.295. Magnetic field and temperature are given in reduced units.

caloric effect associated with the property X_1, $\Delta S[T, (0,0) \to (y_1,0)] = \Delta S(T, y_1) = \int_0^{y_1} (\partial X_1/\partial T)_{y_1} dy_1$. The second term is the isothermal entropy change induced by the field y_2 keeping the field y_1 constant. This second term can be expressed as

$$\Delta S \left[T, (y_1, 0) \to (y_1, y_2) \right]$$
$$= \int_0^{y_2} \left(\frac{\partial X_2}{\partial T} \right)_{y_2, y_1=0} dy_2 + \int_0^{y_1} \frac{\partial}{\partial y_1'} \left[\left(\frac{\partial X_2}{\partial T} \right)_{y_1', y_2} dy_2 \right] dy_1'.$$

Finally, the multicaloric entropy change is given as

$$\Delta S \quad [T, (0,0) \to (y_1, y_2)]$$
$$= \Delta S \left(T, y_1 \right) + \Delta S(T, y_2) + \int_0^{y_2} \int_0^{y_1} \frac{\partial \chi_{12}}{\partial T} dy_1 dy_2, \qquad (5.28)$$

which states that it is given as the sum of the single caloric entropy changes plus a term associated with the cross-response to both fields. This last term is related to the temperature derivative of the cross-coefficient χ_{12}, which can also be expressed as $\partial^2 S / \partial y_1 \partial y_2$. Of course, in the absence of interplay between the ferroic properties X_1 and X_2, the multicaloric effect is simply the sum of the single caloric effects associated with each ferroic property. Schematically, this is illustrated in Figure 5.6. In the figure, the effect of the interplay is taken into account through the curvature of the entropy surface. It is worth noting that the entropy curve will always show a curvature in the neighbourhood of a phase transition.

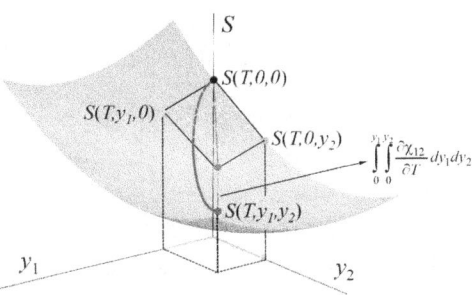

Figure 5.6 Schematic representation of the entropy curve as a function of fields y_1 and y_2 at a given temperature T. The three contributions to the isothermal multicaloric entropy change are indicated.

It is also possible to obtain an expression similar to Eq. 5.28 for the entropy change that gives the multicaloric adiabatic temperature change induced by the two fields. It is obtained that this temperature change is the sum of the contributions associated with each single caloric effect plus an extra term, ΔT_{inter}, which results from the interdependence between the properties X_1 and X_2. This term has the form

$$\Delta T_{inter} = \int_0^{y_2} \int_0^{y_1} \frac{\partial}{\partial y_1} \left[\frac{T}{C} \left(\frac{\partial X_2}{\partial T} \right)_{y_1, y_2} \right] dy_1 dy_2, \qquad (5.29)$$

where C is the heat capacity.

Multicaloric effects in a Landau model with biquadratic coupling

It is useful to analyse the preceding general treatment of multicaloric effects within the framework of a Landau model for a system with two order parameters with biquadratic coupling. As an example, we will consider a magnetoelectric multiferroic material for which the relevant order parameters are magnetization, M, and polarization, P. Compared to the previously discussed Deformable Ising model which includes the effect of magnetoelastic coupling, the present model is more complete in the sense that it takes into account the existence of phase transitions associated with both order parameters.

As has been discussed in Chapter 4, a biquadratic coupling between polarization and magnetization necessarily exists in any magnetoelectric multiferroic. In any case, the strength of this interplay is usually weak and has a small effect on the multicaloric interplay. Nevertheless, it is important to analyse this case to illustrate the consequences of the coupling on the multicaloric properties. Therefore, we consider a Gibbs free energy, $\mathcal{G}$, of the general form

$$\mathcal{G} = G_p(T, P) + G_m(T, M) + G_{p-m}(P, M). \tag{5.30}$$

For symmetry reasons, the expansion of the pure polar, G_p, and magnetic, G_m, contributions may include only even terms in the corresponding order parameters. It will be assumed that, on cooling, the polar transition occurs first and is continuous and that the magnetic transition takes place at a lower temperature and, depending on the strength of the magnetoelectric interplay, might be first order or continuous. Therefore, the simpler choice of the free energy contributions is

$$G_p(T, P) = \frac{1}{2} a_p(T) P^2 + \frac{1}{4} b_p P^4 - EP, \tag{5.31}$$

$$G_m(T, M) = \frac{1}{2} a_m(T) M^2 + \frac{1}{4} b_m M^4 + \frac{1}{6} c_m M^6 - HM, \tag{5.32}$$

$$G_{p-m}(P, M) = \frac{1}{2} \kappa P^2 M^2. \tag{5.33}$$

where $a_p(T) = a_p^0(T - T_c^p)$ and $a_m(T) = a_m^0(T - T_c^m)$ are the inverse of the pure paraelectric and paramagnetic susceptibilities, which are assumed to display Curie–Weiss linear temperature dependence. The limits of stability of the pure phases satisfy that $T_c^p > T_c^m$. All parameters are assumed to be positive. The energies that couple the order parameters with external electric, E, and magnetic, B, fields have also been included.

Minimization of $\mathcal{G}$ with respect to P leads to the following saddle path relationship between P and M:

$$b_p P^3 + a_p(T) P + \kappa P M^2 = E. \tag{5.34}$$

In the absence of applied electric and magnetic fields, in the ferroelectric phase, $P^2 = -(a_p + \kappa M^2)/b_p$. In this case, substitution in $\mathcal{G}$ leads to the following effective free energy:

$$\mathcal{G}_{eff} = G_0(T) + \frac{1}{2}A(T,\kappa)M^2 + \frac{1}{4}B(\kappa)M^4 + \frac{1}{6}c_m M^6, \qquad (5.35)$$

where $A(T,\kappa) = a_m(T) - \kappa a_p(T)/b_p$ and $B(\kappa) = b_m - \kappa^2/b_p$. Therefore, the magnetoelectric transition predicted by the model will be continuous for weak enough coupling such that $b_m > \kappa^2/b_p$ and first order for $b_m < \kappa^2/b_p$. In the former case, the transition will occur at a critical temperature, T_c^{mp}, at which $A(T_c^{mp}, \kappa) = 0$, and in the latter, at $T_0^{mp} = T_c^{m-p}(\kappa) + 3B(\kappa)^2/16$. The obtained phase diagram is shown in Figure 5.7.

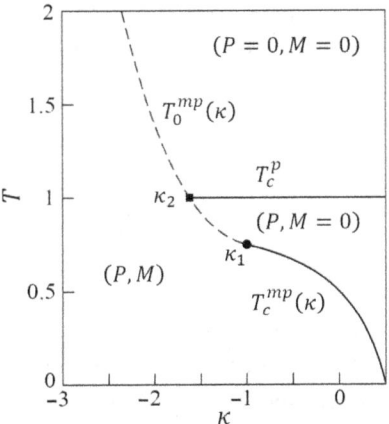

Figure 5.7 Phase diagram of the model giving the transition lines as a function of the coupling parameter in the region $\kappa < 0.5$. Continuous lines indicate continuous transitions and dashed lines correspond to first-order transitions. It is represented in reduced units $b_m = b_0 = c_m = 1$ and $a_p^0 = a_m^0 = 1$ as well as $T_c^p = 1$. Note that $\kappa_1 = -1$ and $\kappa_2 = -1.623$ locate tricritical points. For $\kappa_1 < \kappa < 0.5$, the model predicts two consecutive continuous phase transitions from paraelectric to ferroelectric at T_c^p and from ferroelectric to magnetoelectric multiferroic at $T_c^{mp}(\kappa)$ during cooling from high temperature. For $\kappa < -1$, this transition is first order and the two transitions occur simultaneously for $\kappa < -1.623$. This is the strong coupling region. Adapted from Ref. [194].

The transition boundaries between the different phases shown in Figure 5.7 are affected by applied electric and magnetic fields. The study of the effect of these fields is especially interesting in the strong coupling region corresponding to $\kappa < \kappa_2$ where the model shows a single transition from a paraelectric-paramagnetic phase to a multiferroic magnetoelectric phase. In this region, the first-order transition can be induced by both electric and magnetic fields. Therefore, this is the region where the study of the multicaloric effect associated with the simultaneous application of electric and magnetic fields is more interesting. The corresponding isothermal entropy change that quantifies the multicaloric response can be decomposed as the sum, $\Delta S(T, 0 \rightarrow E, 0 \rightarrow B) = \Delta S(T, 0 \rightarrow B, E = 0) + \Delta S(T, B, 0 \rightarrow E)$, where the first term on the right hand side can be expressed as

$$\Delta S(T, E = 0, 0 \rightarrow B) = \Delta S_p[T, E = 0, P(0) \rightarrow P(B)].$$
$$+ \Delta S_m[T, E = 0, M(0) \rightarrow M(B)]. \quad (5.36)$$

This expression takes into account that due to the magnetoelectric interplay, application of a magnetic field not only induces a change of magnetization but also a change of polarization. In the preceding equation, the polarization will be a solution of Eq. 5.34 with $E = 0$ and M a solution of $\partial G_{eff}/\partial M = 0$, where $G_{eff} = F_{eff} - BM$. Therefore, $\Delta S(T, E = 0, 0 \rightarrow B)$ can be expressed as

$$\Delta S(T, E = 0, 0 \rightarrow B) = -\frac{1}{2} \left(a_m^0 - \frac{\kappa}{b_m} a_p^0 \right) [M^2(T, B) - M^2(T, 0)]. \quad (5.37)$$

This is illustrated in Figure 5.8 that shows the polar and magnetic contributions to the isothermal entropy change induced by application of a magnetic field in the absence of an applied electric field.

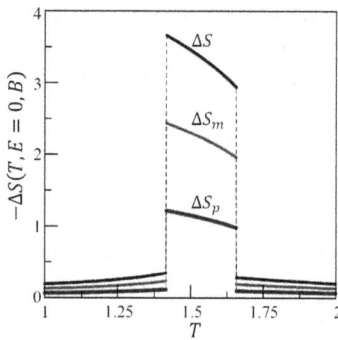

Figure 5.8 Entropy change induced by isothermal application of a magnetic field $B = 0.6$ in the absence of an applied electric field as a function of temperature for $\kappa = -2$. Total, magnetic and polar contributions are indicated. Adapted from Ref. [194].

The calculation of the second contribution of the multicaloric entropy change, $\Delta S(T, B, 0 \rightarrow E)$, requires considering the effective free energy G_{eff} obtained in the presence of an electric field. Therefore, in this case, the relationship between P and M must be obtained from Eq. 5.34 with $E \neq 0$. Then,

$$\Delta S(T, B, 0 \to E) = -\frac{\partial}{\partial T} \{\mathcal{G}_{eff}[T, M(B, E)] - \mathcal{G}_{eff}[T, M(B, E = 0)]\},$$
(5.38)

where M is a solution of $\partial \mathcal{G}_{eff}/\partial M = 0$. In Figure 5.9, the multicaloric entropy change corresponding to application of both magnetic and electric fields in the region of the first-order transition from paraelectric-paramagnetic phase to the multiferroic magnetoelectric phase is shown. The obtained result indicates that while the gain of applying two fields is weak in relation to the strength of the caloric response, the most noteworthy effect is the widening of the temperature range where the material displays a large caloric response.

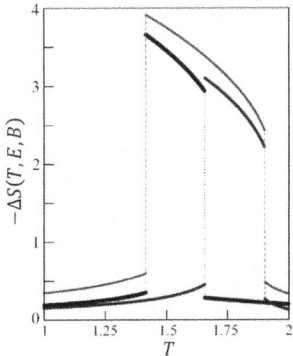

Figure 5.9 Multicaloric entropy change (upper curve) induced by isothermal application of a magnetic field $B = 0.6$ and an electric field $E = 0.4$ in the case of $\kappa = -2$. The contributions $\Delta S(T, E = 0, 0 \to B)$ (lower-left curve) and $\Delta S(T, B, 0 \to E)$ (lower-right curve) are shown. Adapted from Ref. [194].

So far, multicaloric effects induced by electric and magnetic fields in magnetoelectric mutiferroics have not been reported, probably due to the fact that this class of materials is scarce. One of the few materials where multicaloric effects have been studied in detail is the metamagnetic Fe-Rh compound. This material shows a first-order transition from a low-temperature antiferromagnetic phase to a high-temperature ferromagnetic phase, which is accompanied by a significant volume change. Thus, the transition can be induced both by application of a magnetic field and hydrostatic pressure.

As shown in Figure 5.10, pressure favours the stability of the low-temperature phase thus increasing the transition temperature, while the magnetic field has an opposite effect. Therefore, this material shows conventional barocaloric effect and inverse magnetocaloric effect. Figure 5.10 also shows the entropy change induced by application of pressure at selected values of the applied field and the entropy change induced by magnetic field at selected values of the hydrostatic pressure. It confirms that application of pressure shifts the magnetocaloric response to higher temperatures, while application of magnetic field shifts the barocaloric response to lower temperatures. In contrast, the maximum values are affected very little by the application of the secondary field. This behaviour is in agreement with the results of the model discussed above in the sense that in spite of the magnetostrictive interplay, a secondary field does not affect the multicaloric strength but it allows a fine-tuning of the operation range where the material displays a large response (see Ref. [195]).

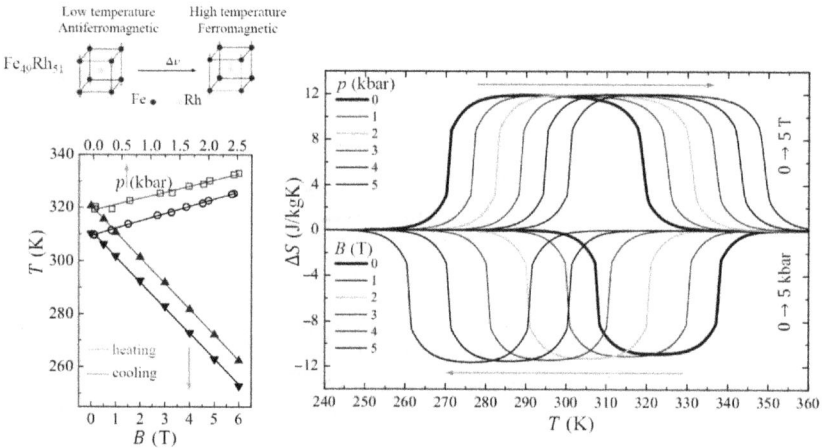

Figure 5.10 Multicaloric effect in $Fe_{49}Rh_{51}$ induced by a magnetic field and hydrostatic pressure. The left panel shows the transition temperature of the first-order ferro-antiferromagnetic transition as a function of magnetic field and pressure. Data correspond to application (up triangles) and removal (down triangles) of the field. The right panel shows, as a function of temperature, the entropy change induced by application of a magnetic field of 5 T at the indicated values of pressure (top curves) and the entropy change induced by pressure at the indicated applied magnetic fields (bottom curves). The magnetic field-induced entropy change is positive, which confirms its inverse character while the pressure induced entropy change is negative as expected for a system showing conventional barocaloric effect.

5.3 Influence of hysteresis

Caloric response is the reversible contribution of the thermal response of a given material to an externally applied field. So far, in the present chapter, we have studied the caloric response by strictly assuming equilibrium conditions. In real materials, however, this response is affected by non-equilibrium dissipative effects which manifest themselves by the presence of hysteresis effects, which is particularly important in the vicinity of first-order transitions where a large caloric response is expected. In practice, this is an important question since the existence of these effects causes a reduction in the efficiency of devices utilizing caloric materials. This can be illustrated by considering a Carnot cycle that works in the region of a first-order transition. This is schematically illustrated in Figure 5.11. Thus, we consider a generic material that shows conventional caloric effect associated with a first-order magnetic phase transition, which can be induced with a field y. We can assume that initially the material is at temperature T_2 under an applied field, y. The field is then adiabatically removed and the material cools down to T_1. Then, isothermally, it undergoes a phase transition at constant field to the high-temperature phase. In this process, the material absorbs from the cold reservoir at T_1 the latent heat $q_1 = T_1 \Delta S$. In this transition, the property X, thermodynamically conjugated to y, undergoes a change ΔX. Then the field is adiabatically applied and the material heats up to T_2. Finally, the cycle is closed when isothermally at constant field, the material transforms back to the low-temperature phase. For a field y, the refrigerant capacity of this cycle is given as[4] $|\Delta S \Delta T| \simeq |y \Delta X|$, where $\Delta T = T_2 - T_1$ is the shift of the transition temperature induced by the applied field y and ΔX is the change of magnetization of the material at the transition. If the transition occurs with hysteresis, for the same maximum value of the field, y, the forward transition will occur at a temperature $T_2' < T_2$ and the reverse transition at a temperature $T_1' > T_1$. Therefore, even if the entropy change is approximately the same as in the equilibrium case, for a given y, $|\Delta S \Delta T| > |\Delta S' \Delta T'|$, which assumes a loss of the refrigerant capacity. Note that if the shift of transition temperature was lower than the transition hysteresis, the refrigerant capacity would vanish.

In principle, any refrigeration device can work reversibly if the corresponding cycle is driven quasistatically. In this case, the heat removed per unit time from the cold reservoir will vanish, which would make the device

---

[4] This equation simply expresses that the area of the Carnot cycle in the S-T space (which represents the total heat exchanged with reservoirs in the cycle) is equal to the area of the cycle represented in the y-X space (which gives the total work exchanged during the cycle) and assumes that ΔS and ΔX are independent of temperature and field, respectively.

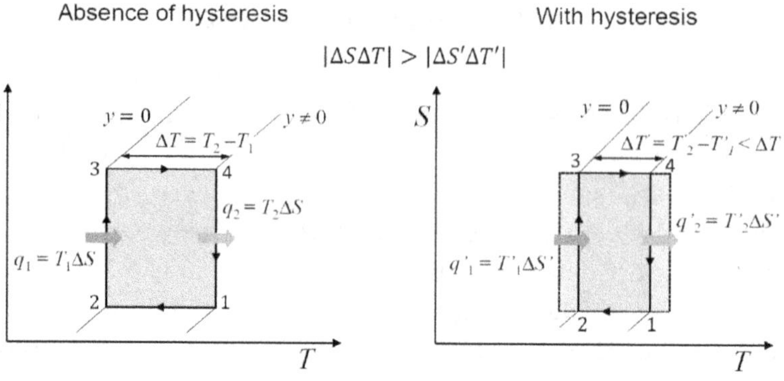

Figure 5.11 Schematic representation in an S-T space of a refrigeration Carnot cycle working in the region of a first-order transition. 1 → 2: adiabatic removal of the applied field. 2 → 3: reverse transition from the low to the high-temperature phase and the caloric material absorbs the latent heat from the low-temperature reservoir. 3 → 4: adiabatic application of the field. 4 → 1: forward transition from the high to the low-temperature phase and the caloric material delivers the latent heat to the high-temperature reservoir. In the presence of hysteresis, the forward transition occurs at a lower temperature, while the reverse transition occurs at a higher temperature than in equilibrium. Therefore, for a given field y, $|\Delta S \Delta T| < |\Delta S' \Delta T'|$ and the refrigeration efficiency is reduced due to hysteresis.

unuseful for practical applications. In general, this is so for any device used as a refrigeration, heat pump, or harvesting device. Instead, for an arbitrary cycle that runs in finite time, Clausius inequality must be satisfied, that is,

$$\oint \frac{\delta q}{T_{res}} \leq 0, \tag{5.39}$$

where T_{res} are the temperatures of the external reservoirs and the equality only applies when the cycle is reversibly performed. Therefore, this means that $\delta q / T_{res} = dS - \delta S_i$, where dS is the reversible contribution of the differential entropy change and $\delta S_i \geq 0$ is the entropy production due to dissipative effects. When an external field is adiabatically applied or removed, $\delta q = 0$ and the induced change of temperature is approximately given by

$$\Delta T \simeq \frac{T}{C}(-\Delta S + S_i) = \Delta T_{rev} + \frac{T S_i}{C}, \tag{5.40}$$

where C is the heat capacity and $T S_i$ is the dissipated energy due to non-equilibrium effects. While the sign of ΔT_{rev} depends on whether the field is applied or removed, $T S_i$ is always positive. Therefore, this term quantifies the irreversible contribution to the thermal response induced by an external field.

It is important to indicate that the multicaloric effect associated with the response to multiple fields can be utilized to eliminate hysteresis in one field by transferring it to another field. This is an important issue that should permit controlling hysteresis by judicious manipulation of more than one field. In practice, this strategy may allow reaching an apparent perfect reproducibility under the application and removal of one field.

Exercises

5.1 The following model is proposed to account for the behaviour of the magnetization in the region of a metamagnetic transition. The model includes both temperature and magnetic field dependence of the magnetization together with the possibility that the transition extends in a given temperature range.[5] The magnetization in the low-temperature phase is assumed to decrease as:

$$M(T) = \Delta M_0 (1 - bT),$$

where ΔM_0 and b are constants, and $M = 0$ in the high-temperature region. The transition is assumed to occur at a temperature $T_t(H)$ within a range ΔT_t which is assumed independent of the applied magnetic field. For each magnetization curve, T_t is defined at the centre of the region ΔT_t and is assumed to increase linearly with the applied magnetic field in such a way that $dT_t/dH = \alpha$, with α being a positive constant.

(a) Find the isothermal field-induced entropy change, $\Delta S(T, 0 \to H)$ as a function of T and H.

(b) Compare the contribution to $\Delta S(T, 0 \to H)$ associated with the transition, obtained after subtracting the contribution outside the transition region, with the entropy change estimated using the Clausius–Clapeyron equation.

5.2 Consider a ferromagnetic material with localized moments, which arise from a spin S described by the equation of state (Eq. 3.33)

$$\sigma = \mathcal{B}_S(x),$$

where $\mathcal{B}_S$ is the Brillouin function, $x = g\mu_B S(B + \lambda M)/k_B T$ and $\sigma = M/ng\mu_B S$. Show that, close to the Curie point, the field induced change of temperature caused by adiabatic application of a field B is of the form:

$$\Delta T \simeq aB^{2/3} - bB^{4/3} + ...,$$

and find expressions for the coefficients a and b.

---

[5] The fact that a transition occurs in a finite temperature range may be caused by the existence of internal elastic stresses that locally modify equilibrium conditions.

5.3 Consider a magnetic material described by the following free energy function:

$$F = \frac{1}{2}A(T)m^2 + \frac{1}{4}B(\eta)m^4 + \frac{1}{6}Cm^6,$$

where m is the magnetization, $A(T) = a[T - T_c(\eta)]$ and $B(\eta) > 0$ for $\eta > \eta_t$, vanishes at $\eta = \eta_t$, and becomes negative for $\eta < \eta_t$. Show that:

(a) For $\eta > \eta_t$, close to the critical point $T_c(\eta)$, the entropy change induced by the application of a magnetic field h behaves as:

$$\Delta S_c(0 \to h) \sim h^{2/3}.$$

(b) In the vicinity of the tricritical point $T_t = T_c(\eta_t)$, the entropy change should depend on the magnetic field as

$$\Delta S_t(0 \to h) \sim h^{2/5}.$$

Note that both critical and tricritical behaviour of the magnetocaloric effect are consistent with the general result given by Eq. 5.12, with mean-field critical exponents, $\beta = 1/2$ and $\delta = 3$ and tricritical exponents $\beta = 1/4$ and $\delta = 5$, respectively.

5.4 Consider the model for a paraelectric-antiferroelectric transition in Exercise 3.8 and discuss under what circumstances it can describe an inverse electrocaloric effect.

6

Liquid crystals and topological defects

Liquid crystals are complex materials that possess some of the mechanical properties that characterize conventional liquid and, at the same time, show physical properties similar to those of crystalline solids [196]. For instance, these materials do not support shears and can flow like a liquid but can also show anisotropic optical, electrical or magnetic properties like crystalline solids. This peculiar behaviour is often considered to define a state of matter, the liquid crystalline state [197], whose properties are between those of isotropic liquids and crystalline solids. This behaviour is mainly a consequence of the fact that liquid crystalline compounds consist of highly anisotropic molecules that can show different combinations of the ordering of their mass centres and their orientation. Therefore, liquid crystalline phases are determined by given combinations of some long-range positional and orientational orders. Depending on specific characteristics of these two classes of orders, many different liquid crystalline phases have been identified. In fact, in addition to the melting from the crystalline solid phase to the liquid crystal state and from this state to the isotropic liquid state, other intermediate phase transitions between different liquid crystalline structures are expected to occur. In the present chapter, we will mainly focus on the transition from the nematic phase with only orientational order to the isotropic liquid phase. This transition is especially relevant since it illustrates most of the peculiarities of phase transitions in liquid crystals that arise from the intrinsic anisotropy of the molecules, which is reflected in the resulting anisotropy of the physical properties of this class of materials. In this respect, especially interesting are elastic properties, sometimes denoted as orientational elasticity that allow liquid crystals to transmit torques. The Fréedericksz transition, which occurs under an applied electric or magnetic field in a confined liquid crystal, is a consequence of this peculiar behaviour. Another interesting aspect related to this behaviour is the possibility of existence in liquid crystals of a great variety of topological defects. In reality, liquid crystals offer a unique opportunity to study the interplay among topology, elasticity and anisotropy, an issue which is also very relevant in relation to other classes of materials such as magnetic materials as well as the recently emerging class of topological materials. In this book, this interplay will be discussed in the present chapter.

It must be mentioned that in addition to the interest concerning basic aspects of condensed matter and phase transitions, liquid crystals are also relevant in relation to their technological applications. From this point of view, main applications such as displays that are based on the fact that liquid crystals are optically active materials, which is, indeed, a consequence of the strong anisotropy of the molecules and of the existence of orientational long-rage order [198]. This property is used, for instance, to tune the polarization of light passing through them.

6.1 General features

In general, liquid crystals are organic compounds constituted of anisotropic molecules of many chemical types such as acids, azo- or azoxy-compounds, and cholesteric ester [199]. Liquid crystals are classified into two broad classes denoted as thermotropic and lyotropic liquid crystals [200]. In thermotropic liquid crystals, the liquid crystalline state appears when the solid melts and, as temperature increases different liquid crystalline phases usually occur in a sequence of phase transitions. At high enough temperature, they behave as isotropic liquids. This is the class of liquid crystalline materials that will be considered in the present chapter. Lyotropic liquid crystals are found in a number of colloidal solutions and certain polymers. In this class of materials, the liquid crystalline state occurs by controlling the concentration of the solution, which is the main parameter while temperature and pressure act as secondary control parameters.

Examples of thermotropic liquid crystals are para-azoxyanisole ($C_{14}H_{14}N_2O_3$, abbreviated as PAA) and p-methoxybenzylidene p-n-butylaniline ($C_{18}H_{21}NO$, abbreviated as MBBA), while a colloidal aqueous solution of tobacco mosaic virus is a well-known example of lyotropic liquid crystal. The molecular structure of the first two molecules is given in Figure 6.1.

All classes of liquid crystals may exhibit a quite large diversity of phases, which form by combining some long-range order of their mass centres and of their orientations. Indeed, the actual macroscopic symmetry of each of these phases will strongly depend on the microscopic shape of the molecules, which

para-azoxyanisole (PAA)

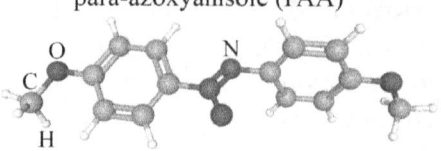

p-methoxybenzylidene *p*-*n*-butylaniline (MBBA)

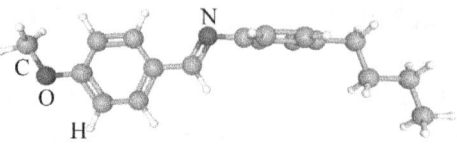

Figure 6.1 Molecular structures of PAA and MBBA liquid crystals.

can exhibit a great variety. The simplest liquid crystals are those consisting of rod-like and disk-like molecules but liquids crystals with more complex molecules, including bend-shaped molecules, are also known [201].

As an example we can consider the case of a thermotropic liquid crystal comprising rod-like molecules. At low temperature, these compounds display a solid crystalline phase. As temperature increases, at melting, it may happen that positional ordering is not completely lost as occurs in many crystalline solids, but rather some partial long-range positional ordering remains together with the orientational ordering of the molecules. This positional ordering occurs often in 1*d* as happens, for instance, in smectic phases in which the mass centres of the molecules are, on average, in layers that can slide freely over one another. A great variety of smectic phases have been identified depending on the orientation of the molecules in each layer. In the smectic A phase, the molecules are aligned with the long axis perpendicular to the layers. The smectic B phase is similar but the organization of the mass centres of the molecules is not random but rather shows some structure, which is usually hexagonal. In contrast with these two smectic phases in the smectic C phase, the preferred axis is not perpendicular to the layers so that this phase is characterized by exhibiting biaxial symmetry. Usually, the macroscopically preferred axis of the molecules is characterized by a vector field $n(r)$ that defines the local orientation of the molecules. This vector is called the *director*, and since its magnitude is not important, it is defined to be unity.

At a higher temperature, in this class of compounds, long-range positional ordering completely disappears but still the orientational ordering of the molecules survives. The obtained phases are denoted as nematic phases in which the molecules are aligned with their long axis parallel to each other. In nematic phases an important feature is the fact that n and $-n$ orientations are indistinguishable, even in the case that the molecules are polar. This issue is essential in order to understand most of the properties of nematic liquid crystals. Another interesting feature is the fact that the absence of long-range positional order explains the high fluidity of these phases, which still show highly anisotropic optical properties associated with the orientational ordering. As temperature is further increased, at the nematic–isotropic transition these systems lose the orientational order and the conventional liquid phase is recovered. The sequence of phases described above is schematically represented in Figure 6.2. It must be noted that similar sequences might occur in the case of disk-like molecules. For instance, in nematic discotic liquid crystals molecules tend to align along a common director axis perpendicular to the plane of the molecules. This phase is schematically represented in Figure 6.3. On the other hand, in the so called columnar phases, in addition to the orientational order, the molecules form stacks that provide some $1d$ positional long-range order of the mass centres of the molecules. These phases can be understood as the equivalent of smectic phases in discotic liquid crystals. Examples of typical discotic molecules are derivatives of triphenylene, phthalocyanine or hexabenzocoronene [202].

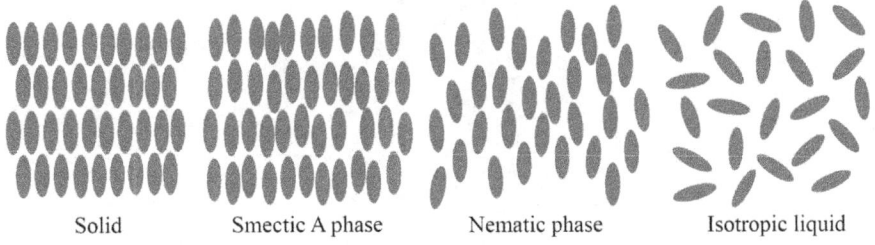

| Solid | Smectic A phase | Nematic phase | Isotropic liquid |

Figure 6.2 Schematic representation of the arrangement of molecules in an increasing temperature sequence from the low temperature solid to the high temperature isotropic liquid. The intermediate phases are the smectic A and the nematic liquid crystal phases.

The so-called cholesteric phases are similar to nematic phases in the sense that they display orientational but not positional long-range order but differ in that the director varies in space in a regular way. The resulting configuration may be understood as the one induced by twisting about one axis perpendicular to the director of an initially aligned nematic material. This leads to the formation of a structure which can be visualized as a stack of very thin $2d$ nematic-like layers with the director in each layer twisted with

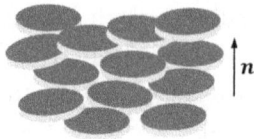

Figure 6.3 Schematic representation of the arrangement of disk-like molecules in a nematic phase, also denoted as discotic nematic. In this phase the disks tend to be oriented along the axis perpendicular to the plane of the molecules, which defines the director.

respect to those above and below. This structure is illustrated in Figure 6.4. Therefore, this phase has a helical axis perpendicular to the director. In this sense, it is similar to helical magnetic phases with the difference that in liquid crystals n and $-n$ orientations along the succession of staking layers resulting from the rotation of the director are indistinguishable. These phases are also denoted as chiral nematic phases.

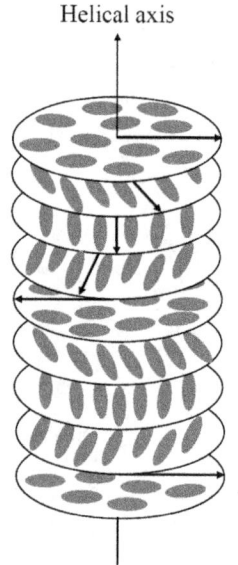

Figure 6.4 Schematic representation of a cholesteric liquid crystal consisting of nematic layers with the corresponding director rotating along the helical axis, perpendicular to the layers. In each layer the director is indicated.

Note that in some compounds when increasing the temperature from the solid crystalline state, at melting, only some positional ordering survives. In these cases, a plastic crystal phase instead of a liquid crystal is said to form. The name plastic crystal refers to the mechanical softness of such phases, which can be easily deformed. As the liquid crystal state, the plastic crystal state can also be considered as an intermediate state between the solid and the isotropic liquid state. The difference between the two classes of phases originates from the corresponding shape of the molecules that, to be distinguished, provides the strong anisotropy to liquid crystals but not to plastic crystals that are often comprising almost spherical molecules, which, in any case, still possess orientational degrees of freedom. In this respect, liquid crystals and plastic crystals can be seen as opposites. This latter class of compounds will not be discussed in the present chapter.

6.2 The orientational order parameter

For the sake of simplicity, we will start by assuming a nematic liquid crystal consisting of rigid molecules of rod-like shape as illustrated in Figure 6.5. Molecules possess a centre of symmetry and thus, as already discussed, opposite directions of the director n cannot be distinguished, which supposes that its average value must vanish. Consequently, for liquid crystals, it is not possible to define an order parameter with vectorial character as done in ferromagnets. Therefore, it is necessary to consider an order parameter of tensorial nature. It can be introduced as follows [196]. For each molecule i, a unit vector $\boldsymbol{\nu}^{(i)}$ is considered along its long axis, which describes its orientation. Then a natural order parameter adequate to describe orientational ordering in nematic and cholesteric liquid crystals is the following second-rank tensor

$$S_{\alpha\beta}(\boldsymbol{r}) = \frac{1}{N} \sum_{i=1}^{N} \left(\nu_\alpha^{(i)} \nu_\beta^{(i)} - \frac{1}{3}\delta_{\alpha\beta} \right), \qquad (6.1)$$

where $\nu_{\alpha(\beta)}$ is the $\alpha(\beta)$ component of $\boldsymbol{\nu}^{(i)}$ referred to as a set of laboratory fixed axes. The sum is over the N molecules in a small but still macroscopic volume located at the point $\boldsymbol{r}$, and $\delta_{\alpha\beta}$ is the Krönecker delta. This order parameter is a symmetric traceless tensor of rank 2, which has, in general, five independent components. As expected, at very high temperature, when the orientations of the molecules are randomly distributed, in the isotropic liquid state, this tensor is expected to vanish. Note that often some short-range order will still occur in the isotropic liquid phase and then this parameter will not strictly vanish.

Maier and Saupe [203] proposed an alternative order parameter for uniaxial liquid crystals with a single preferred direction along the director consisting of molecules with a well-defined long axis. It has the form

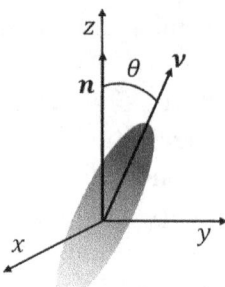

Figure 6.5 Example of a molecule with a well-defined long axis. The vector $\boldsymbol{\nu}$ determines its orientation, which may be different than the direction defined by the director $\boldsymbol{n}$. In the figure θ is the angle between the two vectors. Uniaxial symmetry has been assumed.

$$S_{\alpha\beta} = \left\langle \cos\phi_\alpha \cos\phi_\beta - \frac{1}{3}\delta_{\alpha\beta} \right\rangle, \tag{6.2}$$

where cartesian coordinates that coincide with the molecular fixed axes are considered and $\phi_{\alpha(\beta)}$ stands for the angle between the $\alpha(\beta)$ axis and the preferred direction. In the preceding expression, the angular brackets indicate an average over the molecules in a small but macroscopic volume as in Eq. 6.1. It is worth noting that for molecules with a well-defined long axis about which they rapidly rotate, both definitions given by Eqs. 6.1 and 6.2 must be equivalent.

In real liquid crystals, molecules are not rigid. Flexibility can be taken into account by describing different parts of molecules with different $S_{\alpha\beta}$ tensors. This situation indeed renders much more complex treatment and suggests that the order parameter should be chosen following a different strategy. With this idea in mind, de Gennes [204] proposed to define the amount of orientational order in terms of a macroscopic property, which is independent of any assumption about the rigidity of the molecules. This point of view is adequate for a thermodynamic study of the orientational order in liquid crystals. He suggested to define the order parameter in terms of the anisotropic part of the magnetic susceptibility tensor as

$$Q_{\alpha\beta} = \chi_{\alpha\beta} - \frac{1}{3}\sum_\gamma \delta_{\alpha\beta}\chi_{\gamma\gamma}, \tag{6.3}$$

where $\chi_{\alpha\beta}$ is the magnetic susceptibility tensor per unit volume.[1] Note that just like $S_{\alpha\beta}$, here $Q_{\alpha\beta}$ is also a traceless symmetric rank-2 tensor.

---

[1] Often, liquid crystals are diamagnetic and, thus, $\chi_{\alpha\beta}$ is the diamagnetic susceptibility.

In liquid crystals, the magnetic exchange between molecules is usually very weak and, thus, the susceptibility can be expressed as a sum of the susceptibilities of individual molecules. In this case, $Q_{\alpha\beta}$ should be simply related to the order parameter $S_{\alpha\beta}$ introduced for rigid rod-like molecules. Thus, suppose that the susceptibilities along the principal axes of a molecule are χ_1, χ_2, and χ_3 respectively. If fixed coordinates, x', y', z', are chosen that coincide with the principal axes, it can be easily shown that

$$
\begin{aligned}
Q_{xx} &= Q_{yy} \\
&= 2N[(S_{y'y'} + S_{z'z'})\chi_1 + (S_{z'z'} + S_{x'x'})\chi_2 + (S_{x'x'} + S_{y'y'})\chi_3], &(6.4) \\
Q_{zz} &= N[S_{x'x'}\chi_1 + S_{y'y'}\chi_2 + S_{z'z'}\chi_3], &(6.5)
\end{aligned}
$$

where N is the number of molecules per unit volume. The traceless character of $S_{\alpha\beta}$ supposes that $S_{x'x'} + S_{y'y'} + S_{z'z'} = 0$, which implies that there are only two independent parameters on the right-hand sides of the preceding equations. Therefore, for molecules with an axial symmetry, the tensors $Q_{\alpha\beta}$ and $S_{\alpha\beta}$ must be proportional. Therefore,

$$
Q_{\alpha\beta} = N\chi_a S_{\alpha\beta}, \tag{6.6}
$$

where $\chi_a = \chi_{\ell\ell} - \chi_\perp$, with $\chi_{\ell\ell}$ and $\chi_\perp$ being the susceptibilities along and perpendicular to the main axis. Therefore, χ_a represents a measure of the amount of anisotropy in the susceptibility.

It is worth noting that any other second-rank tensor property could be chosen instead of the magnetic susceptibility to define the order parameter. For instance, $Q_{\alpha\beta}$ could be defined in terms of the electric susceptibility. However, in that case, the simple relationships 6.4 and 6.5 cannot be maintained since the electric susceptibility cannot be expressed as the sum of susceptibilities of individual molecules. Anyhow, relation 6.6 can still be used considering that χ_a is an average susceptibility.

Uniaxial symmetry

We consider now the cases of uniaxial and biaxial symmetry in more detail. In the first case, the preferred direction of the molecules is along the director and the order parameter can be written as $S_{\alpha\beta} = \mathcal{S}(n_\alpha n_\beta - \frac{1}{3}\delta_{\alpha\beta})$, where $n_{\alpha(\beta)}$ are the components of the director $\boldsymbol{n}$. Taking the director along the z-axis then, the three non-zero components of $S_{\alpha\beta}$ are, $S_{xx} = S_{yy} = -\frac{1}{3}\mathcal{S}$ and $S_{zz} = \frac{2}{3}\mathcal{S}$, where the scalar $\mathcal{S}$ represents a measure of the degree of alignment of the molecules along the director. Supposing that $\varphi(\theta)\sin\theta d\theta$ is the fraction of molecules with long axes making angles between θ and $\theta + d\theta$ with the director, then $\mathcal{S}$ can be expressed as

$$\mathcal{S} = \int_0^\pi \left(1 - \frac{3}{2}\sin^2\theta\right)\varphi(\theta, T)\sin\theta d\theta, \tag{6.7}$$

where $\varphi(\theta, T)$ is an angular distribution function, independent of the azimuthal angle but that should depend on temperature. In the isotropic phase, $\mathcal{S} = 0$ while $\mathcal{S} = 1$ when all molecules are aligned with the director. Note that $\mathcal{S}$ vanishes as well in a cholesteric phase. This order parameter is usually expressed in the form

$$\mathcal{S} = \left\langle \frac{1}{2}(3\cos^2\theta - 1) \right\rangle = \langle P_2(\cos\theta) \rangle, \tag{6.8}$$

where $P_2(\cos\theta)$ is the second Legendre polynomial.

In this uniaxial case, if the macroscopic order parameter given by Eq. 6.3 is considered, it must be written as, $Q_{\alpha\beta} = Q(n_\alpha n_\beta - \frac{1}{3}\delta_{\alpha\beta})$, with $Q = \chi_a$. Therefore, in uniaxial liquid crystals, the order parameter can be specified by the magnitude of Q and the director $\boldsymbol{n}$.

Biaxial symmetry

The smectic C phase occurs in liquid crystals with biaxial symmetry.[2] In this case,

$$Q_{xx} = \chi_{xx} - \frac{1}{3}\sum_\gamma \chi_{\gamma\gamma} = -\frac{1}{2}(P - R), \tag{6.9}$$

$$Q_{yy} = \chi_{yy} - \frac{1}{3}\sum_\gamma \chi_{\gamma\gamma} = -\frac{1}{2}(P + R), \tag{6.10}$$

$$Q_{zz} = \chi_{zz} - \frac{1}{3}\sum_\gamma \chi_{\gamma\gamma} = P, \tag{6.11}$$

which depends on the two scalar quantities P and R. Actually, this is the most general form that can take a symmetric, rank-2 traceless tensor in a diagonalized form.

The Maier-Saupe mean-field statistical model

Maier and Saupe [203] proposed a simple mean-field model for the isotropic–nematic transition in a uniaxial liquid crystal based on the order parameter

---

[2] Compared with smectic A phases, where the director is perpendicular to the layers, in the smectic C phase the director is tilted with respect to the perpendicular direction.

$\mathcal{S}$ given by Eq. 6.8. In this model, a nematic potential $V(\theta)$ is assumed to be

$$V(\mathcal{S}, \theta) = -v\mathcal{S}P_2(\cos\theta), \qquad (6.12)$$

where v is a positive parameter that is taken independent of temperature. This potential depends on the actual orientation of a given molecule, which is assumed to interact with the average orientation of the remaining molecules and, therefore, it is assumed proportional to the degree of order of the system, $\mathcal{S}$, as expected in a mean-field approximation. This is a purely geometric argument that ignores, for instance, any particular dipolar interactions or the specific shape of the molecules.

Within a variational approach, a free energy functional per molecule, $f = u - Ts$, can be obtained [206] taking into account that the internal energy u can be expressed as

$$u = \frac{1}{2}\int_0^\pi V(\mathcal{S}, \theta)\varphi(\theta)\sin\theta d\theta, \qquad (6.13)$$

and the entropy s as

$$s = -k_B\int_0^\pi \ln\varphi(\theta)\varphi(\theta)\sin\theta d\theta. \qquad (6.14)$$

Note that in the expression of the internal energy, the factor $1/2$ is introduced in order to avoid considering twice the interaction of pairs of molecules. Therefore, the free energy function per molecule can be written as

$$f = -\frac{1}{2}v\mathcal{S}^2 + k_BT\langle\ln\varphi(\theta)\rangle. \qquad (6.15)$$

Minimization under the normalization condition, $\int_0^\pi \varphi(\theta)\sin(\theta)d\theta = 1$, leads to, $\delta[f - \lambda(1 - \int_0^\pi \varphi(\theta)\sin(\theta)d\theta)] = 0$, where λ is a Lagrange multiplier and δ indicates functional derivative with respect to φ. A straightforward calculation gives the following orientational distribution function

$$\varphi(\theta) = \frac{\exp[-V(\mathcal{S}, \theta)/k_BT]}{\int_0^\pi \exp[-V(\mathcal{S}, \theta)/k_BT]\sin\theta d\theta}. \qquad (6.16)$$

Therefore, the equilibrium order parameter can be obtained by solving the transcendental equation

$$\mathcal{S} = \frac{\int_0^\pi \exp[-V(\mathcal{S}, \theta)/k_BT]P_2(\cos\theta)\sin\theta d\theta}{\int_0^\pi \exp[-V(\mathcal{S}, \theta)/k_BT]\sin\theta d\theta}. \qquad (6.17)$$

It is clear that $S = 0$ is a solution of this equation, which corresponds to the isotropic phase. To find non-trivial solutions, it is convenient to integrate Eq. 6.17 by parts. Then, it is obtained that

$$S = \frac{3}{4}\left[\frac{\exp(x^2)}{xD(x)} - \frac{1}{x^2}\right] - \frac{1}{2}, \tag{6.18}$$

where $x = (\frac{3}{2}vS/k_BT)^{1/2}$ and $D(x) = \int_0^x \exp(y^2)dy$. A numerical solution of the preceding equation permits to obtain S as a function of T. The model predicts that the isotropic–nematic transition is first order and occurs at $T_0 \simeq 0.220v/k_B$. At this temperature, the order parameter of the nematic phase is $S_0 \simeq 0.43$, independent of v.

The Maier-Saupe theory considers only an effective attractive long-range intermolecular potential and ignores the effect of short-range forces.[3] This is indeed a crude approximation but, in spite of that, the theory is very successful in accounting for the first-order character of the nematic transition and for the orientational properties of real nematics [207]. In fact, the agreement with experiments can be improved even further by including terms in the nematic potential that takes into account higher-rank interactions and deviations from strict uniaxial symmetry.

6.3 The Landau-de Gennes model

The Landau approach to the isotropic–nematic transition is useful because it takes explicitly into account the relevance of symmetries in the transition. Therefore, in order to propose a Landau model for an isotropic–nematic transition, the starting point is the fact that the free energy function must be invariant under rotations, which supposes that all terms in the expansion must be scalar functions of the order parameter $Q_{\alpha\beta}$ (or $S_{\alpha\beta}$) [205]. In order to proceed in a systematic manner, it must be reminded that the general result that shows that all the rotational invariants of a given tensor $\mathbf{Q}$, with components $S_{\alpha\beta}$, can be expressed as, $\text{Tr}(\mathbf{Q})^n$ with $n = 1, 2, 3, ...$[4] Since in our case $\mathbf{Q}$ can be represented as a 3×3 matrix, it follows that all $\text{Tr}(\mathbf{Q})^n$ can be expressed as products of $\text{Tr}(\mathbf{Q})$, $\text{Tr}(\mathbf{Q}^2)$ and $\text{Tr}(\mathbf{Q}^3)$, only. The traceless character of $\mathbf{Q}$ leads to a general Landau expansion in terms of

---

[3] The theory also assumes that the configuration of the centres of mass of the molecules is not affected by the orientational order.

[4] The demonstration of this general result can be found in Ref. [205].

the two non-vanishing terms $\mathrm{Tr}(\mathbf{Q}^2)$ and $\mathrm{Tr}(\mathbf{Q}^3)$ only. Therefore, the singular part of free energy density function must have the following general form:

$$
\begin{aligned}
f(Q_{\alpha\beta}, T) = \frac{1}{2}A \left[\mathrm{Tr}(\mathbf{Q}^2)\right] &+ \frac{1}{3}B \left[\mathrm{Tr}(\mathbf{Q}^3)\right] + \frac{1}{4}C \left[\mathrm{Tr}(\mathbf{Q}^2)\right]^2 \\
&+ \frac{1}{5}D \left[\mathrm{Tr}(\mathbf{Q}^2)\mathrm{Tr}(\mathbf{Q}^3)\right] + \frac{1}{6}E \left[\mathrm{Tr}(\mathbf{Q}^2)\right]^3 \\
&+ \frac{1}{6}E' \left[\mathrm{Tr}(\mathbf{Q}^3)\right]^2 + \dots \, .
\end{aligned}
\tag{6.19}
$$

This expansion is usually denoted as the Landau-de Gennes model. It is insightful to note that odd terms of order 3, 5, ..., are allowed due to the symmetry associated with the non-polar character of the molecules and is responsible for the fact that the isotropic–nematic transition shows first-order character. It is interesting to note as well that there are two independent 6-order terms in the expansion. As will be seen later, the second one, with coefficient E', is important as it makes possible the existence of a biaxial nematic phase.

In a general case, we have already seen that the diagonalized order parameter is of the form

$$
\mathbf{Q} = \begin{pmatrix} -\frac{1}{2}(P - R) & 0 & 0 \\ 0 & -\frac{1}{2}(P + R) & 0 \\ 0 & 0 & P \end{pmatrix}.
\tag{6.20}
$$

Therefore, the invariants of $\mathbf{Q}$ are, $\mathrm{Tr}(\mathbf{Q}^2) = \frac{1}{2}(3P^2 + R^2)$ and $\mathrm{Tr}(\mathbf{Q}^3) = \frac{3}{4}P(P^2 - R^2)$. Note that in the uniaxial case with the main axis along the z-axis, $R = 0$, while when it is along the x- or y-axis, $R = -3P$ or $R = 3P$, respectively. In the first case, P must be $\frac{2}{3}\mathcal{S}$, where $\mathcal{S}$ is the order parameter given by Eq. 6.8. Therefore, in this case, the resulting Landau expansion reads

$$
\begin{aligned}
f_u(\mathcal{S}, T) = \frac{1}{3}A\mathcal{S}^2 &+ \frac{2}{27}B\mathcal{S}^3 + \frac{1}{9}C\mathcal{S}^4 \\
&+ \frac{2}{45}D\mathcal{S}^5 + \frac{4}{81}E\mathcal{S}^6,
\end{aligned}
\tag{6.21}
$$

where the term with coefficient E' has been omitted, which is adequate as the considered liquid crystal shows uniaxial symmetry. As it is usual within the framework of the Landau theory, the coefficient A is assumed to linearly depend on temperature as $A = a(T - T_c)$, and the remaining parameters are supposed to be temperature independent. In the preceding expression, a is a positive constant and, in general, T_c represents the low temperature stability limit of the isotropic phase. In the simplest model dealing with the isotropic–nematic transition $D = E = 0$ and $C > 0$ can be assumed. Then for a conventional nematic of rod-like molecules, the order parameter $\mathcal{S} > 0$ in the nematic phase and, thus, B must be negative. Instead, in the case of a discotic nematic phase, since $\mathcal{S} < 0$, B must be positive.

The equilibrium temperature of the first-order transition is $T_0 = T_c + \frac{1}{27} \frac{B^2}{aC}$ and the value of the order parameter of the nematic phase at the transition temperature is given by $\mathcal{S}_0 = \frac{|B|}{3C}$. The transition latent heat can be obtained as $\ell = T_0 \Delta s$, where

$$\Delta s = -\left(\frac{\partial F}{\partial T}\right)_{\mathcal{S}=\mathcal{S}_0} = -\frac{2}{3} a \mathcal{S}_0^2 = -\frac{2B^2}{81C^2} \qquad (6.22)$$

is the entropy discontinuity at the transition or entropy change.

The preceding simple model for the isotropic–nematic transition of uniaxial liquid crystals gives results that compare well with those obtained with the Maier-Saupe theory. The phenomenological parameters of the free energy expansion can be estimated from experimental data, which gives rise to a semi-quantitative description of the behaviour of thermodynamic quantities in the vicinity of the phase transition. For most liquid crystals, such as MBBA ($C_{18}H_{21}NO$) and BMAB ($C_{17}H_{20}N_2O_2$) [210], it is found that the parameters B and C are very small. This suggests that the transition might occur close to a tricritical point determined by $A = C = 0$, which are coefficients of terms with the same symmetry in the Landau expansion. These results suggest an alternative model that assumes that $C = 0$. In this case, a positive stabilizing term E must be considered in the free energy expansion, and one can yet suppose that $D = 0$ since, in the presence of a cubic term, this term leads only to minor corrections. In this case, the transition temperature is given as $T_0 = T_c + \frac{11}{54a} \sqrt[3]{\frac{3B^4}{8E}}$, and the value of the order parameter at the transition temperature is $\mathcal{S}_0 = \sqrt[3]{-\frac{3B}{8E}}$. It is not clear which of the two models describes better uniaxial liquid crystals. On the one hand, the significant discontinuity of the order parameter at the isotropic-nematic transition seems to be better accounted for by the first (critical) model. Nevertheless, the increase of fluctuations as the transition is approached, that is observed in the behaviour of response functions and light scattering, seems to be in favour of the tricritical model.

In the biaxial case, the general form of the tensor $\mathbf{Q}$ given by Eq. 6.20 must be considered. Therefore, the two parameters, P and R are needed to describe the orientational order. The formation of a biaxial phase may be not only a consequence of a non-uniaxial molecular symmetry, but can occur as well due to the existence of strong correlations that may give rise to aggregates of molecules that do not have uniaxial symmetry. In fact, it can also be a consequence of an external applied field.

The simplest Landau expansion for biaxial liquid crystals neglects the terms with coefficients D and E. However, the term E' is essential to make possible a biaxial phase. Stability then requires that $C > 0$ and $E' > 0$. From the expressions of $\text{Tr}(\mathbf{Q}^2)$ and $\text{Tr}(\mathbf{Q}^3)$ given as functions of P and R, a free energy function of the form,

$$f_b(P, R, T) = a(P) + b(P)\,R^2 + c(P)\,R^4, \qquad (6.23)$$

is obtained. In the preceding expression, $a(P) = \frac{3}{4}AP^2 + \frac{1}{4}BP^3 + \frac{9}{16}CP^4 + \frac{3}{32}E'P^6$, $b(P) = \frac{1}{4}A - \frac{1}{4}BP + \frac{3}{8}CP^2 - \frac{3}{16}E'P^4$ and $c(P) = \frac{1}{8}C + \frac{3}{16}E'P^2$. Minimization with respect to R gives the two solutions, $R = 0$ and $R = -b(P)/c(p)$. The first case corresponds to the uniaxial solution and then $f_b = a(P) = f_u(P)$ given from Eq. 6.21 after taking into account that $P = \frac{2}{3}\mathcal{S}$. In the second case, $f_b = a(P) - b^2(P)/2c(P)$. Taking into account that $c(P)$ must always be positive this biaxial solution is only allowed when $b(P) < 0$. Note also that in this case, $f_b(P) < f_u(P)$. Then, the general problem reduces to finding the minima of the function,

$$f(P) = \begin{cases} f_u(P) = a(P) & \text{for } b(P), \geq 0 \text{ (uniaxial case)} \\ f_b(P) = a(P) - b^2(P)/2c(P) & \text{for } b(P), < 0 \text{ (biaxial case).} \end{cases}$$
$$(6.24)$$

It is interesting to indicate that a biaxial nematic phase has not been reported in thermotropic liquid crystals. Nevertheless, biaxial nematic phases have been observed in lyotropic systems constituted of amphiphilic molecules in aqueous solution[5] associated with a combined effect related to the strong correlation between the molecules and the shape of the resulting aggregates. Actually, in these solutions, the molecules tend to cluster into aggregates so that hydrophilic groups locate close to the surface in order to optimize the interaction with water, while the lipophilic molecules tails occupy the inner part of the aggregates. These aggregates or micelles, which are usually of spherical shape, grow when the concentration increases and become either rod-like or disk-like. Therefore, these micelles behave as large-scale molecules that have a tendency to display orientational order [208]. Depending on the actual shape of the micelles, uniaxial and biaxial nematic phases can form [209].

6.3.1 Fluctuation effects, distortion and effect of external fields

The Landau-deGennes model given by Eq. 6.19 is only adequate to deal with homogeneous liquid crystals. While this is the stable configuration, there are

---

[5] These solutions are similar to soap-like solutions.

non-uniform configurations that cannot relax. This may lead, for instance, to the existence of domains separated by domain walls where the local preferred axis or director changes. In general, spatial inhomogeneities originate from thermal fluctuations. These effects are worth considering when, for instance, pretransitional effects become very relevant and a proper study of response functions close to the phase transition must be done in terms of correlation functions and thus fluctuations cannot be neglected. This is especially important to address the influence of orientational order on light scattering. On the other hand, inhomogeneities also occur associated with the distortion of liquid crystals. Therefore, the study of this diversity of problems requires a generalization of the Landau-de Gennes model given by Eq. 6.19 that properly takes into account space inhomogeneities of molecule orientations associated with distortions. To that purpose, within the spirit of the Ginzburg-Landau theory, we need to include in the model an explicit dependence on spatial derivatives of $\mathbf{Q}$. Therefore, the general Landau-de Gennes free energy functional must be of the form

$$\mathcal{F}(\mathbf{Q}, \nabla\mathbf{Q}, T) = \int \{f[\mathbf{Q}(r, T] + f_d[\nabla\mathbf{Q}(r)]\} \, dr, \qquad (6.25)$$

where the integral is performed over the volume of the system and $f(\mathbf{Q}, T)$ has the form of free energy density given by Eq. 6.19 and $f_d(\nabla\mathbf{Q})$ is the density term that brings the dependence on the space derivatives of $\mathbf{Q}$. This term, that represents the distortional or elastic free energy, must be invariant under translations and rotations. Then, only the combinations of $Q_{\alpha\beta}$ and its derivatives that satisfy these symmetries are allowed in the resulting expansion. Taking only the lowest-order terms, the elastic energy density has the form [213]

$$f_d = \frac{1}{2} \sum_{\alpha,\beta,\gamma} \left[L_1 \left(\frac{\partial Q_{\alpha\beta}}{\partial x_\gamma} \right)^2 + L_2 \frac{\partial Q_{\alpha\beta}}{\partial x_\beta} \frac{\partial Q_{\alpha\gamma}}{\partial x_\gamma} + L_3 \frac{\partial Q_{\alpha\gamma}}{\partial x_\beta} \frac{\partial Q_{\alpha\beta}}{\partial x_\gamma} \right]$$
$$+ \frac{1}{2} \sum_{\alpha,\beta,\gamma,\delta} L_4 \epsilon_{\delta\alpha\gamma} Q_{\delta\beta} \frac{\partial Q_{\alpha\beta}}{\partial x_\gamma}, \qquad (6.26)$$

where x_i with $i = \alpha, \beta, \gamma, \delta$ that takes values 1, 2, and 3, are cartesian coordinates, $Q_{\alpha\beta}$ are the components of the tensor $\mathbf{Q}$ and $\epsilon_{\delta\alpha\gamma}$ is the Levi-Civita symbol. Note that the first term is the usual gradient square term of the most common Ginzburg-Landau theory.

The preceding distortional free energy is directly connected with the Oseen-Frank free energy [211, 212], which is a vector representation formulated in terms of the director and its spatial derivatives, based on a generalization of Hooke's law which assumes that curvature stresses are proportional to curvature strains. It has the following form:

$$f_{OF} = \frac{1}{2}K_{11}(\nabla \cdot \boldsymbol{n})^2 + \frac{1}{2}K_{22}(\boldsymbol{n} \cdot \nabla \times \boldsymbol{n})^2 + \frac{1}{2}K_{33}(\boldsymbol{n} \times \nabla \times \boldsymbol{n})^2$$
$$- \frac{1}{2}(K_{22} + K_{24})\nabla \cdot [\boldsymbol{n}\nabla \cdot \boldsymbol{n} + \boldsymbol{n} \times \nabla \times \boldsymbol{n}]$$
$$- q_0 K_{22}(\boldsymbol{n} \cdot \nabla \times \boldsymbol{n}), \tag{6.27}$$

where the coefficients K_{ij} are the Frank elastic constants that measure the response in deformation to given forces. For instance, K_{11}, K_{22} and K_{33} measure the response to splay, twist and bending forces, respectively. The corresponding defomation modes are illustrated in Figure 6.6. Note that q_0 is denoted as the chirality index and it is a coefficient that measures the degree of chirality[6] of the liquid crystal.

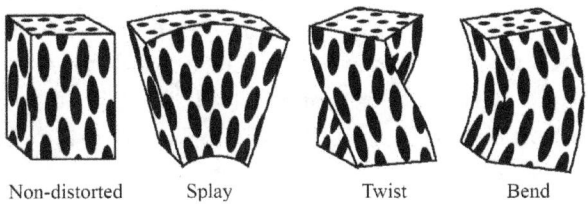

Non-distorted   Splay   Twist   Bend

Figure 6.6 Schematic representation of splay, twist and bend distortions of a nematic liquid crystal. Adapted from Ref. [214].

Considering a nematic liquid crystal with uniaxial orientational order characterized by an order parameter $\mathcal{S}$, the relationship between these elastic constants and the coefficients L_i of the tensor representation Eq. 6.26 are [213]

$$L_1 = \frac{1}{6\mathcal{S}^2}(K_{33} - K_{11} + 2K_{22}), \tag{6.28}$$

$$L_2 = \frac{1}{\mathcal{S}^2}(K_{11} - K_{22} + 2K_{24}), \tag{6.29}$$

$$L_3 = \frac{1}{\mathcal{S}^2}K_{24}, \tag{6.30}$$

$$L_4 = \frac{2}{\mathcal{S}^2}q_0 K_{22}. \tag{6.31}$$

The Oseen-Frank representation of the distortion energy is given in terms of the five independent parameters, K_{ij}, that represent elastic curvature moduli and the chirality index q_0. However, the tensorial expansion of f_d has

---

[6] The concept of chiral symmetry will be discussed Section 6.4.

only four independent L_i coefficients since it has been limited to symmetry allowed second-order terms. Therefore, in this case, the mapping of both models is not bijective, that is not one-to-one related. The degeneracy can be removed by adding the third-order term $L_5 \sum_{\alpha\beta\gamma\delta} Q_{\delta\gamma} \left(\frac{\partial Q_{\alpha\beta}}{\partial x_\delta} \right) \left(\frac{\partial Q_{\alpha\beta}}{\partial x_\gamma} \right)$ in the expansion given by Eq. 6.26, which leads to the relation $L_5 = (K_{33} - K_{11})/2\mathcal{S}^3$. Then, both vectorial and tensorial representations become equivalent. Nevertheless, it must be noted that the tensor formulation has the advantage of removing the problem of the apparent difference between $\boldsymbol{n}$ and $-\boldsymbol{n}$ directors in the vectorial formulation [213].

Light scattering

Light scattering anomalies precede the occurrence of the isotropic–nematic transition that, in spite of being first order, usually takes place on cooling quite close to the limit of stability, T_c, of the isotropic phase where the intensity of light scattering should diverge. This anomalous behaviour originates from correlations between fluctuations that extend over larger and larger regions as the temperature T_c is approached. In reality, this behaviour is related to the critical opalescence phenomenon that, is expected to occur at critical points.

The physical quantity relevant to address light scattering is the optical dielectric tensor $\boldsymbol{\varepsilon}$. This tensor can be split into isotropic and anisotropic contributions. The components of the isotropic contribution are, $\varepsilon_{\alpha\beta}^{iso} = \frac{1}{3}\delta_{\alpha\beta}\mathrm{Tr}(\boldsymbol{\varepsilon})$, and the anisotropic part, $\Delta\boldsymbol{\varepsilon} = \boldsymbol{\varepsilon} - \boldsymbol{\varepsilon}^{iso}$, which must be proportional to the orientational order parameter $\boldsymbol{Q}$. In a typical elastic scattering experiment, if $\boldsymbol{k}$ and $\boldsymbol{k}'$ are the wave vectors of the incident and scattered beams, energy conservation supposes that $|\boldsymbol{k}| \cong |\boldsymbol{k}'|$ and the transferred momentum $\boldsymbol{q} = \boldsymbol{k}' - \boldsymbol{k}$ satisfies, $|\boldsymbol{q}| = 2|\boldsymbol{k}|\sin(\theta/2)$, where θ is the angle formed by the incident and scattered beams. It is then found that the scattering intensity $I(\boldsymbol{q})$ is given as

$$I(\boldsymbol{q}) \propto \langle |e'_\alpha \varepsilon_{\alpha\beta}(\boldsymbol{q}) e_\beta|^2 \rangle = e'_\alpha e'_\gamma e_\beta e_\delta \langle Q^*_{\alpha\beta}(\boldsymbol{q}) Q_{\gamma\delta}(\boldsymbol{q}) \rangle, \qquad (6.32)$$

where $\boldsymbol{e}$ and $\boldsymbol{e}'$ are the unit vectors with components e_α and e'_β giving the polarization direction of the incident and scattered beams. In the preceding expression, it is assumed that the scattering intensity is only associated with orientational fluctuations and density fluctuations are not considered. Two interesting cases are, $\boldsymbol{e}\|\boldsymbol{e}'$, and $\boldsymbol{e}\|z, \boldsymbol{e}'\|y$, which lead, respectively, to

$$I_\| \propto \langle |Q_{zz}(\boldsymbol{q})|^2 \rangle, \qquad (6.33)$$

$$I_\perp \propto \langle |Q_{yz}(\boldsymbol{q})|^2 \rangle. \qquad (6.34)$$

Proceeding as in Section 1.5.1, correlation functions can be obtained within the gaussian approximation after linearizing the local term, f, in the free energy functional given by Eq. 6.25. Keeping only the gradient terms with coefficients L_1 and L_2, in Fourier space, the free energy functional takes the form

$$\mathcal{F} = \int_{q<q_D} \frac{d\boldsymbol{q}}{2\pi^3} \left[\frac{1}{2}(A + L_1 q^2)|Q_{\alpha\beta}(\boldsymbol{q})|^2 + \frac{1}{2}L_2 q_\alpha q_\beta Q_{\alpha\gamma}^*(\boldsymbol{q})Q_{\beta\gamma}(\boldsymbol{q}) \right], \quad (6.35)$$

where $q = |\boldsymbol{q}|$ and $q_D = (6\pi\rho)^{1/3}$ is the radius of the Debye sphere which, in this case, plays a role analogous to the Brillouin zone in a crystalline solid. Without loss of generality, we can take $\boldsymbol{q} = (0, 0, q)$ and it is obtained that

$$I_\| \propto \frac{k_B T}{[A + (L_1 + \frac{2}{3}L_2)q^2]}, \quad (6.36)$$

$$I_\perp \propto \frac{k_B T}{[A + L_1 q^2]}. \quad (6.37)$$

Therefore, order parameter correlations and thus light scattering follow the Ornstein-Zernike law, which means that close to a critical point where $T \to T_c$ and $A \to 0$, light scattering is expected to diverge in the optical region, $q \approx 0$. In the case of the isotropic–nematic transition in a liquid crystal, the intensity is large but still remains finite at the transition, which is first-order. Experimental results confirm that the inverse of the intensity decreases linearly with temperature down to the transition temperature where it is still finite as predicted by the preceding model [215].

Effect of an external field

A complete formulation of the Landau-de Gennes model requires to take into account the coupling of external fields with the orientational order parameter. The Landau-de Gennes model is especially adequate for this aim since the order parameter is taken proportional to the anisotropic part of the magnetic susceptibility. Liquid crystals are diamagnetic and thus in the presence of a static magnetic field $\boldsymbol{H}$ the extra energy density term that must be considered in the functional $\mathcal{F}$ given by Eq. 6.25 is

$$f_f = -\frac{1}{2}\sum_{\alpha\beta} H_\alpha H_\beta \chi_{\alpha\beta}. \quad (6.38)$$

Then, expressing the susceptibility as $\chi_{\alpha\beta} = \Delta\chi Q_{\alpha\beta} + \delta_{\alpha\beta}\bar{\chi}$, where $\bar{\chi} = \frac{1}{3}\sum_\gamma \chi_{\gamma\gamma}$, leads to

$$f_f = -\frac{1}{2}\bar{\chi}H^2 - \frac{1}{2}\Delta\chi \sum_{\alpha\beta} H_\alpha H_\beta Q_{\alpha\beta}. \quad (6.39)$$

The first term in the preceding equation is, for a given field, a constant independent of the molecular orientational order and can thus be omitted. Note that for uniaxial molecules, the relevant term giving the coupling with the field can be written in terms of the director as $-\frac{1}{2}\chi_a(\boldsymbol{n}\cdot\boldsymbol{H})^2$ where similarly to $\Delta\chi$, $\chi_a = \chi_\| - \chi_\perp$ measures the degree of anisotropy of the molecules. In fact, depending on the shape of the molecules, $\Delta\chi$ and χ_a can be positive or negative. For uniaxial elongated molecules, they are positive, which supposes that molecules tend to align the long axis parallel to the applied field. Instead, when $\Delta\chi$ or χ_a are negative, molecules will tend to align perpendicular to the field. It is worth noting that the situation is analogous in the case of an applied electric field but then it must be reminded that the electrical susceptibilities are related to average values over the whole system.

In the uniaxial case corresponding to an homogeneous system, in the presence of a magnetic field, a term $-h\mathcal{S}$ should be added to Eq. 6.21, with $h = \frac{1}{2}\Delta\chi H^2$. Then, the free energy density function has the form

$$f_u(\mathcal{S}, T, H) = \frac{1}{3}A\mathcal{S}^2 + \frac{2}{27}B\mathcal{S}^3 + \frac{1}{9}C\mathcal{S}^4 - h\mathcal{S}. \qquad (6.40)$$

Note that the simplest model with $D = E = 0$ and $C > 0$ has been considered. Minimization with respect to the order parameter $\mathcal{S}$ yields the following equation of state

$$h = \frac{2}{3}a(T - T_c)\mathcal{S} + \frac{2}{9}B\mathcal{S}^2 + \frac{4}{9}C\mathcal{S}^3. \qquad (6.41)$$

From this equation it is obtained that, in the presence of an applied magnetic field, the system will still display a first-order transition at a temperature that increases with h or H^2. For low fields, the transition temperature varies linearly with the field as $T_t(h) = T_0 + 3h/a\mathcal{S}_0$, where $T_0 = T_c + B^2/27aC$ is the transition temperature at zero field and $\mathcal{S}_0 = -B/3C$ the corresponding order parameter of the nematic phase. The coexistence line ends at a field-induced critical point, (T_{cp}, h_{cp}), which is given by the conditions

$$\frac{\partial f_u}{\partial \mathcal{S}} = \frac{\partial^2 f_u}{\partial \mathcal{S}^2} = \frac{\partial^3 f_u}{\partial \mathcal{S}^3} = 0, \qquad (6.42)$$

and which, as seen in Chapter 1 (Eq. 1.27), indicates that the critical point is an inflection point of the critical isotherm. These conditions lead to

$$h_{cp} = -\frac{B^3}{486C^2}, \qquad (6.43)$$

$$T_{cp} - T_0 = \frac{B^2}{18aC}. \qquad (6.44)$$

Note that $(T_{cp} - T_c)/(T_0 - T_c) = 3/2$. Isofield curves in the order parameter-temperature diagram showing the critical point are depicted in Figure 6.7.

At temperatures above the critical point, in the paranematic phase,[7] the field induces a weak orientational order that can be obtained disregarding non-linear terms in the order parameter in the equation of state. That is, at $T > T_{cp}$, $\mathcal{S} \cong h/a(T - T_c)$.

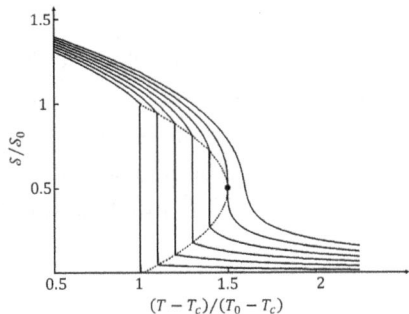

Figure 6.7 Isofield curves corresponding to the equation of state Eq. 6.41 in a reduced order parameter, $\mathcal{S}/\mathcal{S}_0$, versus reduced temperature space. Here $\mathcal{S}_0$ is the order parameter of the nematic phase at the transition in the absence of an applied field. From left to right, the field h increases from 0 to 1.2. The dot locates the critical point.

The dependence of $T_t - T_0$ on the square of the applied field has been confirmed in the electric [216] and magnetic [217] cases. In the electric case, the existence of a critical point has been observed in MBBA liquid crystals [218]. Nevertheless, in thermotropic liquid crystals, a magnetic-induced critical point is not reachable since critical magnetic fields have been estimated to be of about 10^3 T, which is an enormous field, not available experimentally.

The Fréedericksz transition

An interesting phenomenon related to the effect of a magnetic field on a liquid crystal is the Fréedericksz transition. To describe this transition, consider a nematic liquid crystal consisting of rod-like molecules confined between two glass slides. The interaction between the liquid crystal and the glass is assumed to be such that the director is constrained to lie perpendicular to the glass at the boundaries. When a magnetic or electric field, applied perpendicular to the director, exceeds a certain threshold value, H_{tr}, the optical properties of the system are observed to change abruptly

---

[7] This phase is not strictly isotropic due to the small orientational order and is often denoted as the paranematic phase.

due to a distortion effect experienced by the molecules. This distortion is a consequence of the fact that both the magnetic field and the boundaries exert torques on the molecules that compete, and when the field exceeds the threshold value, it becomes energetically favourable for the molecules in the bulk of the sample to turn in the direction of the field as illustrated in Figure 6.8 (a). This effect, which was first observed by Fréedericksz and Zolina [219], is, strictly speaking, not a phase transition since the crossover occurs with no singularity. Notwithstanding, it is interesting to analyse this phenomenon since it nicely illustrates the relevance of the distortion energy. It is worth noting that there are indeed other geometries in which the competition between torques produced by a magnetic or an electric field and the boundaries can be studied. Among them, the case of molecules that have the long axis parallel to the boundaries and the field is perpendicular to it is also worth mentioning. This case is also illustrated in Figure 6.8 (b).

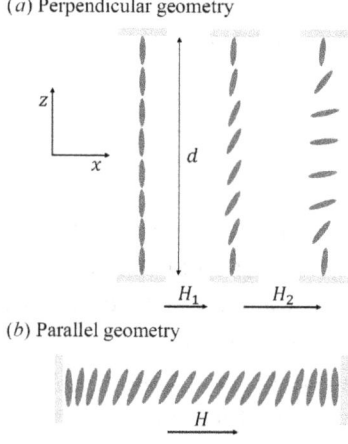

(a) Perpendicular geometry

(b) Parallel geometry

Figure 6.8 Schematic illustration of the Fréedericksz transition induced by a magnetic field in the perpendicular, (a), and parallel, (b), geometry cases. In the former case the figure shows the distortion with a field H_1 which is close to the threshold value and already induces a distortion. Distortion increases by application of a larger field $H_2 > H_1$.

As an example, let us analyse the case of uniaxial molecules with the long axis perpendicular to the glass slides. Consider that the z-axis is taken perpendicular to the slides and that the applied field lies along the x-direction as indicated in Figure 6.8 (a). In general, the director will be of the form $n = (n_x, n_z) = (\cos\theta, \sin\theta)$, where θ is the angle between the director and the z-axis, which will depend on z only. In the nematic phase only the first three terms in Eq. 6.27 are relevant[8] and thus the energy accounting for the coupling with the field in terms of the director leads to the following free energy functional

$$F = \frac{1}{2} \int_{-d/2}^{d/2} \left[(K_{11} \sin^2 \theta + K_{33} \cos^2 \theta) \left(\frac{\partial \theta}{\partial z} \right)^2 - \chi_a H^2 \sin^2 \theta \right] dz, \quad (6.45)$$

where d is the distance between the glass slides. Now, for the sake of simplicity, we will assume that $K_{11} = K_{33} = K$ which allows writing the preceding functional in the form

$$F = \frac{1}{2} \frac{K}{\xi^2} \int_{-d/2}^{d/2} \left[\xi^2 \left(\frac{\partial \theta}{\partial z} \right)^2 - \sin^2 \theta \right] dz, \quad (6.46)$$

where $\xi = \sqrt{(K/\chi_a H^2)}$. Minimization of this functional with respect to $\theta(z)$ leads to a differential equation that must be solved with the boundary condition $\theta(d/2) = \theta(-d/2) = 0$. It is given by

$$\xi^2 \frac{\partial^2 \theta}{\partial z^2} + \sin \theta \cos \theta = 0. \quad (6.47)$$

Indeed $\theta = 0$ is a trivial solution of this equation that satisfies boundary conditions. In any case, a non-trivial general solution can be found by taking into account that the preceding differential equation has a first integral of the form [9]

$$\xi^2 \left(\frac{d\theta}{dz} \right)^2 + \sin^2 \theta = \sin^2 \theta_m, \quad (6.48)$$

where the constant of integration $\sin^2 \theta_m$ can be identified taking into account that $d\theta/dz = 0$ where the distortion θ takes the maximum value that presumably occurs at $z = 0$, which is halfway between the two boundaries. Then integration leads to

$$z = \frac{1}{2}d - \xi \int_0^\theta \frac{d\theta}{\sqrt{\sin^2 \theta_m - \sin^2 \theta}} = \xi \csc \theta_m E(\csc \theta_m, \theta), \quad (6.49)$$

where it has been imposed that $\theta = 0$ at $z = d/2$. Here, $E(\csc \theta_m, \theta)$ is the incomplete elliptic integral of the second kind. If the maximum distortion θ_m is small, integration leads to the solution $\theta = \theta_m \sin[(\pm d - 2z)/2\xi]$, where the positive sign holds for $z > 0$ and the negative sign for $z < 0$. Therefore,

---

[8] These three terms are adequate to describe distortions in non-chiral liquid crystals. For stability reasons, the elastic moduli K_{11}, K_{22} and K_{33} may be positive.

[9] As in the Example 2.1.1, this first integral can be obtained by taking into account that the differential equation has the form of a one-dimensional mechanical equation of motion of a particle after identifying θ with the position x, z with time and ξ^2 with the mass. Then, $-\sin \theta \cos \theta$ plays the role of a force that only depends on position.

this requires that at $z = 0$, $d = \pi\xi$, which supposes that a reasonably good estimation of the threshold field is given as

$$H_{tr} = \frac{\pi}{d}\sqrt{\frac{K_{33}}{\chi_a}}. \tag{6.50}$$

This field is given in terms of K_{33} since the distortion is largely of bend type, with very little splay. Moreover, this value of the threshold field holds even in the case $K_{11} \neq K_{33}$. In fact, this indicates that the experiment provides a useful method to measure the elastic modulus K_{33}.

Note that the analysis of the parallel case with $K_{11} = K_{33}$ differs only from the perpendicular case by a rotation of axes by $\pi/2$. The corresponding distortion is a splay to a large extent and in this case $H_{tr} = \frac{\pi}{d}\sqrt{K_{11}/\chi_a}$ and thus can be used to measure the elastic modulus K_{11}.

6.3.2 The nematic to smectic transition

It has been already indicated that on cooling from high temperature, liquid crystals are expected to exhibit a sequence of temperature-induced phase transitions, first from an isotropic liquid to a nematic state breaking rotational symmetry and then from the nematic to a smectic state breaking, partially, translational symmetry, usually along one axis. Note that at a lower temperature, the remaining translational symmetry could be broken through a phase transition towards a crystalline state. In this section, we briefly consider the transition from the nematic to a smectic transition.

A suitable order parameter for the nematic to smectic transition can be introduced from the distribution of the mass centres of the molecules. Therefore, an order parameter can be defined in terms of the local density that is expected to show a modulation along a given direction. Then, a convenient choice consists of taking a complex scalar quantity such as $\psi(\boldsymbol{r}) = \psi_0 \exp(-i\phi_0)$, where the modulus ψ_0 is defined as the amplitude of a $1d$ density wave characterized by the phase ϕ [220]. In the case of a smectic A phase, the wave vector of the order parameter, $\nabla\phi$, should be parallel to the director $\boldsymbol{n}$. Then, the layer spacing in the smectic A phase should be given as $d = 2\pi/k_0$, where $k_0 = |\nabla\phi|$.

An important point is the fact that the partial breaking of translational symmetry at the smectic transition might affect the orientational order and, consequently, any realistic model for the nematic–smectic transition should take into account the coupling of both orientational and positional order parameters. The general form of the free energy density function to address the isotropic–nematic and nematic–smectic phases that include the lowest-order symmetry allowed positional and coupling terms, should be of the form:

$$f_{n-s} = f + f_d + \frac{1}{2}A_s|\psi|^2 + \frac{1}{4}C_s|\psi|^4 + \frac{1}{2}\Lambda_1(\nabla\psi)^2 + \frac{1}{2}\Lambda_2(\nabla^2\psi)^2$$
$$+ \frac{1}{2}K_1|\psi|^2[\text{Tr}(\mathbf{Q}^2)] + \frac{1}{2}K_2\sum_{\alpha,\beta}Q_{\alpha\beta}\frac{\partial\psi}{\partial x_\alpha}\frac{\partial\psi}{\partial x_\beta}, \qquad (6.51)$$

where $f_n(Q_{\alpha\beta}) = f + f_d$ and f is given by Eq. 6.19 and f_d by Eq. 6.26. The last two terms in the preceding expression are the terms coupling orientational and positional order parameters.

In the case of a uniaxial liquid crystal with spatial uniform order, the free energy density can be written in the simpler form

$$f_{n-s} = \frac{1}{3}A\mathcal{S}^2 + \frac{2}{27}B\mathcal{S}^3 + \frac{1}{9}C\mathcal{S}^4$$
$$+ \frac{1}{2}A_s\psi_0^2 + \frac{1}{4}C_s\psi_0^4 + \frac{1}{2}\Lambda_1k_0^2\psi_0^2 + \frac{1}{2}\Lambda_2k_0^4\psi_0^2$$
$$+ \frac{1}{3}K_1\psi_0^2\mathcal{S}^2 + \frac{3}{8}K_2k_0^2\mathcal{S}\psi_0^2, \qquad (6.52)$$

where, as in Eq. 6.19, the coefficients D and E have been taken to be zero. Note that the main term coupling orientational and positional orders is biquadratic in the corresponding two order parameters. The coupling term linear in $\mathcal{S}$ depends also on k_0^2 and, thus, vanishes in non-smectic phases. As usual, only the coefficients $A = a(T - T_c)$ and $A_s = a_s(T - T_c^s)$ are assumed to be temperature dependent.

Minimization of the preceding free energy density function with respect to $\mathcal{S}$, ψ_0 and k_0 yield the following phases:

Isotropic phase: $\mathcal{S} = 0$, $\psi_0 = 0$ and $k_0 = 0$.

Nematic phase: $\mathcal{S} = \mathcal{S}_n = \frac{1}{4C}[B + \sqrt{B^2 - 24AC}] > 0$, $\psi_0 = 0$ and $k_0 = 0$.

Smectic A phase: $\mathcal{S} = \mathcal{S}_s > 0$, $\psi_0 = -\frac{1}{\bar{C}_s}(\bar{A}_s - \bar{K}_2\mathcal{S}_f + \frac{3}{2}\bar{K}_1\mathcal{S}_s^2)$, $k_0 = -\frac{1}{2\Lambda_2}(\Lambda_1 + K_2\mathcal{S}_s)$.

In the preceding expressions, $\mathcal{S}_s$ is the solution of, $2\bar{A}_s\bar{K}_2/3C_s + 2\bar{A}\mathcal{S}_s + 3\bar{C}\mathcal{S}_s^2 = 0$, and $\bar{A}_s = A_s - \Lambda_1^2/4\Lambda_2$, $\bar{K}_1 = K_1 - K_2^2/6\Lambda_2$, $\bar{K}_2 = \Lambda_1K_2/2\Lambda_2$, $\bar{C} = C - \bar{K}_1/C_s$, $\bar{B} = B - 3\bar{K}_2\bar{K}_1/C_s$, $\bar{A} = A - K_1A_s/C_s - \bar{K}_2^2/3C_s$.

It is important to note that both isotropic to nematic and nematic to smectic A transitions are first-order transitions which means that a temperature range can exist where the three phases might overlap, in stable or metastable states. In addition, it is interesting to point out that for given values of model parameters, a direct transition from the isotropic to the smectic phase can occur.

6.4 Topological defects

When orientational order in liquid crystals is broken, defects topologically related to the actual symmetry of the system may appear [221]. These defects are qualified as topological because they represent properties that remain unchanged when the space is smoothly deformed. Therefore, when a topological property changes, it occurs by integer steps and not gradually. In liquid crystals, singular and non-singular topological defects can occur. Singular defects are related to a discontinuity in the orientational order and the main examples are disclinations.[10] These line defects occur typically in nematic liquid crystals and create an internal elastic field in the liquid crystal. They are not limited to liquid crystals and can be observed also in other media with non-solid-crystalline symmetries. On the other hand, non-singular defects are especially interesting in the case of chiral liquid crystals. These defects are related to objects such as skyrmions and merons that have attracted a lot of attention in the field of magnetism but can also form in liquid crystals. In this context they are important in order to understand the existence of the so-called blue phases.

Disclinations

Disclinations are defect lines where the preferred direction or director changes discontinuously. Therefore, they are topological defects in orientational order that were introduced in analogy with dislocations in crystalline solids, which are defects in positional order [212]. Beginning with a uniform configuration, a disclination L can be understood using the Volterra construction process that yields the same main conclusions as the topological stability theory [222]. The following operations are needed: (*i*) A cut along a surface A limited by a line is made in the liquid crystal.[11] (*ii*) The upper and lower surfaces of A, denoted as A_1 and A_2, respectively, are then rotated relative to each other about an axis a by a certain finite angle, Ω. Indeed, the symmetry of the liquid crystal requires that the rotation must be a multiple of π. That is, $\Omega = m\pi$ where m is the Frank index of the disclination. (*iii*) Empty spaces resulting from the preceding operation are filled with undistorted liquid crystal or conversely, any extra material is removed. The whole

---

[10] This word indicates discontinuity in the inclination of the molecules.

[11] It is assumed that the molecules are not perturbed by this cut and that on the upper and lower sides of A, A_1 and A_2, the molecules remain in place.

structure is then allowed to elastically relax to a stationary state. The index m is chosen as positive if the material is removed and negative if it is added.

The change of the director can be detected by considering a closed path through the liquid crystal that encircles the disclination by means of a unit vector that changes continuously along the path. Then it may be found that after the complete path, this vector has returned to its original orientation or, rather, has the reverse orientation. This second possibility cannot be ruled out since the sign of the director has no physical meaning. In this case, a branch plane, that contains the disclination line, should be introduced in the medium at which the sign of the vector is reversed. This should correspond to disclinations of odd-index that, in principle, could be detected in this way. Note that there is no test to detect even-index disclinations.

Since disclinations are structures that involve changes of the director over short distances, they appear as energetically very unfavourable and it would seem advantageous to eliminate the singularity and spread out the disclination over a smooth strain pattern consistent with boundary conditions. However, for topological reasons associated with the required branch plane, this is not possible in the case of odd-index disclinations. In contrast, even-index disclinations can always relax into a non-singular strained configuration.

The rotation axis in the Volterra construction can be taken perpendicular to the director but not parallel since, in that case, the system could elastically relax to the unperturbed state. While the rotation can have any arbitrary orientation relative to the disclination line, perpendicular and parallel cases corresponding to twist and wedge disclinations are especially interesting in uniaxial nematics. They are illustrated in Figure 6.9.

Twist disclination   Wedge disclination

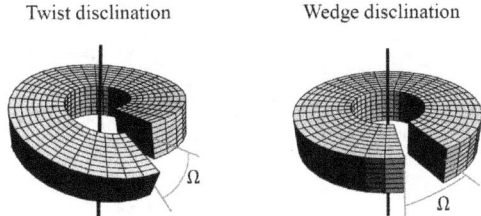

Figure 6.9 Illustration of the Volterra process to form twist and wedge disclinations. In both cases, the vertical line represents the disclination line. Rotation Ω is performed about an axis perpendicular to the disclination line in a twist disclination, while it is parallel to the line in the case of a wedge disclination.

In twist or perpendicular disclinations, the rotation axis is perpendicular to the disclination line. Considering that the z-axis is the rotation axis and the disclination line is taken along the y-axis, then the director should be parallel to the x-y plane. This kind of disclination involves twist distortion and the director can be assumed of the form $\boldsymbol{n} = [\cos\phi(x,z), \sin\phi(x,z)]$. The elastic energy must be of the form

$$F_e = \frac{1}{2}\int \left[(K_{11}\sin^2\phi + K_{33}\cos^2\phi)\left(\frac{\partial\phi}{\partial x}\right)^2 + K_{22}\left(\frac{\partial\phi}{\partial z}\right)^2 \right] d^3r, \quad (6.53)$$

which has been obtained assuming that only the first three terms in Eq. 6.27 are relevant. Assuming that $K_{11} = K_{22} = K_{33} = K$, minimization of the preceding functional leads to the two-dimentional Laplace equation $(\partial^2/\partial x^2 + \partial^2/\partial z^2)\phi = 0$. Then, it is obtained that the solution representing the disclination is of the form, $\phi = \frac{1}{2}m\psi + \phi_0$, where m is the Frank index that can be a positive or negative integer and ψ the angle between the x-axis and a line joining the origin and the point (x,z). In the plane $z = 0$, assuming that $\phi_0 = 0$, the configuration is the following. For $x > 0$, $\phi = 0$, while for $x < 0$, $\phi = \pi/2(\text{mod }\pi)$ if m is odd and $\phi = 0(\text{mod }\pi)$ if m is even. Note that disclinations with indices $\pm m$ are obtained from each other by reflection over the x-y plane.

In the case of wedge or axial disclinations, the rotation axis is parallel to the disclination line. As in the previous case, taking the z-axis as the rotation axis, the disclination line must also be oriented along the z-axis and the director $\boldsymbol{n}$ must lie parallel to the x-y plane. In this case, the disclination only involves splay and bend distortions and the director can be taken of the form $\boldsymbol{n} = [\cos\phi(x,y), \sin\phi(x,y)]$. The corresponding elastic energy is of the form

$$F_e = \frac{1}{2}K\int \left[\left(\frac{\partial\phi}{\partial x}\right)^2 + \left(\frac{\partial\phi}{\partial y}\right)^2 \right] d^3r, \quad (6.54)$$

where $K_{11} = K_{33} = K$ has been assumed again. Then, minimization leads to $(\partial^2/\partial x^2 + \partial^2/\partial y^2)\phi = 0$, and the solution representing disclination lines is of the form $\phi = \frac{1}{2}m\psi + \phi_0$, where the Frank index m may be positive or negative. Then, as ψ varies from 0 to 2π, the angle ϕ between the director and the x-axis must change from ϕ_0 to $\phi_0 + 2\pi$.

It is interesting to note that this type of wedge disclinations also exist in cholesteric liquid crystals. In that case, the obtained solution must be superposed to the cholesteric twist [223].

Skyrmions and merons

Interesting non-collinear topological defects occur in chiral liquid crystals. In general, an object is said to show chiral[12] symmetry if it is not distinguishable from its mirror image. This means that the mirror image cannot superpose onto itself. From this point of view, chirality should be considered as a property of asymmetry.

When chiral liquid crystals such as cholesteric liquid crystals are confined in an anisotropic environment, they experience geometric frustration originating from the fact that while chirality favours a twist of the director field, anisotropy favours a director orientation which is not compatible with the twist. As a consequence of this competition chiral liquid crystals form complex patterns of topological defects, with regions of twist separated by regions of the favoured orientation [224].

In this section, we focus on a class of non-singular defect structures such as skyrmions which are characterized by the fact that the orientation of the molecules varies in a topological configuration that cannot be annealed away but for which the magnitude of the order parameter remains constant in magnitude and its orientation varies smoothly. Examples of skyrmions are represented in Figure 6.10. The configuration of these defects is different from topological vortices in the sense that there is no singularity where the magnitude of the order parameter goes to zero or changes from the bulk value. These non-singular defects are also observed in chiral magnets. In the last part of this section, we will discuss similarities and differences of these classes of defects in both kinds of chiral materials. A further interesting aspect is the similarity of skyrmions and blue phases. These phases, that form in chiral liquid crystals, consist of a periodic array of double-twist tubes separated by disclination lines. In each tube, the director follows a $2d$ variation going outward from its centre thus giving rise to a structure similar to a half-skyrmion, which is called a meron.[13] Therefore, it is interesting to understand the crossover between skyrmions and blue phases, which require the occurrence of singularities associated with the disclination lines.

A simple model suitable to take into account these kinds of defects starts with the general free energy functional given by Eq. 6.25. The local term is limited to fourth order with $A = a(T - T_c)$ and only the terms with coefficients L_1 and L_4 of the elastic contribution are considered. Following Ref. [225], it can be assumed that $L_1 = L$ and $\frac{1}{4}L_4 = q_0 L$ and then this elastic free energy density is given as

---

[12] The words chiral and chirality are derived from the Greek $\chi\varepsilon\iota\rho$ (kheir) that means hand, which is a chiral object.

[13] The word meron is also derived from the Greek $\mu\varepsilon\rho o\varsigma$ (meros) that means part or portion.

Hedgehog skyrmion   Spiral skyrmion

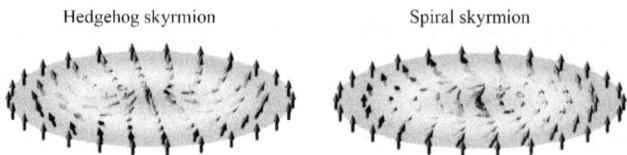

Figure 6.10 Vector field variation of $2d$ examples of skyrmions. The direction of this vector represents the director in liquid crystals and the magnetic moment in magnetic materials.

$$f_d = \frac{1}{2}L \sum_{\alpha,\beta,\gamma} \left(\frac{\partial Q_{\alpha\beta}}{\partial x_\gamma}\right)^2 - 2Lq_0 \sum_{\alpha,\beta,\gamma,\delta} \epsilon_{\alpha\beta\gamma} Q_{\alpha\delta} \frac{\partial Q_{\beta\delta}}{\partial x_\gamma}. \tag{6.55}$$

The first term is the only term that accounts for the excess of energy associated with splay, twist, or bend deformations. The second term is the essential term that favours a chiral twist. Here q_0 is the index of chirality that is associated with an inverse length characterizing molecular chirality.

To proceed further with this model, it is useful to show that after rescaling the free energy density depends only on two parameters. Reminding that the equilibrium order parameter $\mathcal{S}_0$ of the nematic phase at the first-order isotropic–nematic transition is proportional to $|B|/C$, it turns out that the free energy density of the nematic phase relative to the isotropic phase goes as B^4/C^3. On the other hand, the core radius of disclinations in the nematic phase goes as $\xi \sim \sqrt{(LC/B^2)}$. Therefore, the Q-tensor, the free energy density and all lengths can be rescaled by these characteristic values. Rescaling then shows that the model is controlled by two dimensionless parameters, t and κ. The first, given as

$$t = \frac{27AC}{B^2} \tag{6.56}$$

is a dimensionless temperature, and the second

$$\kappa = \sqrt{\frac{108CLq_0^2}{B^2}} \tag{6.57}$$

is a dimensionless chirality coefficient that represents the natural twist q_0 relative to the disclination core radius, ξ. In the nematic phase, in terms of these two dimensionless parameters, the free energy density associated with the chiral twist is $L\mathcal{S}^2 q_0^2$, while the free energy density corresponding to a disclination core is $L\mathcal{S}^2/\xi^2$. Therefore, κ^2 can also be interpreted as the energy scale that favours the chiral twist relative to the energy scale associated with the change of the director orientation inside a disclination core. Therefore, a liquid crystal is denoted as low- or high-chirality liquid crystal depending on whether κ is low or high, respectively.

For a complete analysis of the problem an extra dimensionless parameter representing the degree of anisotropy is still needed. In experiments, a liquid crystal is usually placed in an environment that favours some actual alignment of the nematic director. In analogy with magnetism, the axis along which the alignment is favoured is denoted as the *easy axis*. Instead, when perpendicular alignment is favoured it is called the *easy plane*. The Fréedericksz transition discussed in Section 6.3.1 provides a nice example of an environment that induces anisotropy. In reality, this experiment indicates that there are two mechanisms to create anisotropy. The first is associated with an external applied electric or magnetic field. The second occurs when a liquid crystal is anchored between the two surfaces of a narrow cell. In the first case, if the field is applied along the z-axis, we have seen that according to Eq. 6.39, the free energy density has an additional term of the form $\sim -\Delta\chi H^2 Q_{zz}$, which induces an easy axis for $\Delta\chi > 0$, and an easy plane for $\Delta\chi < 0$. Then, a dimensionless anisotropy α can be introduced as

$$\alpha = \frac{\Delta\chi H^2 C^2}{|B|^3}. \tag{6.58}$$

In the case of a surface-induced anisotropy, if d is the thickness of the cell LS/d^2 will play the role of $\Delta\chi H^2$ in the field-induced anisotropy case and, thus, the dimensionless anisotropy will have the form $\alpha = LC/d^2 B^2$.

In order to determine the structures, modulated or not, that minimize the free energy functional defining the model, it is convenient, for the sake of simplicity, to address the question considering a 3d nematic order tensor that depends only on two spatial coordinates. Therefore, we consider that $\mathbf{Q} = \mathbf{Q}(x, y)$. This model has been studied in detail in Ref. [225]. The following different phases can be accounted for by the model.

Isotropic phase. This is the disordered phase with $\mathbf{Q} = 0$ and with no anisotropy contribution.

Vertical nematic phase. This is an homogeneous nematic phase with the director parallel to the z-axis. The free energy density is the one given by Eq. 6.21, with $D = E = 0$. Note that in the presence of anisotropy, an additional term $-\alpha Q_{zz} = -\alpha S$ should be added.

Planar nematic phase. In this phase, the director is aligned perpendicular to the z-axis, for instance, along the y-axis. The situation is almost identical to the preceding vertical case, but the anisotropy contribution should be of the form, $-\alpha Q_{zz} = \frac{1}{2}\alpha\mathcal{S}$.

Cholesteric phase or lattice of walls. This phase is shown in Figure 6.11 and can be described as a periodic lattice of twisted walls that separate regions in which the director field has an orientation imposed by the anisotropy. It can be considered that the director is of the form, $\boldsymbol{n} = -\hat{\boldsymbol{y}}\sin(\pi x/d) + \hat{\boldsymbol{z}}\cos(\pi x/d)$, with d being the periodicity of the modulation along the z-axis, and $\hat{\boldsymbol{y}}$ and $\hat{\boldsymbol{z}}$ are unit vectors along y- and z-axes. Then, the homogeneous part of the free energy is of the same form as in the preceding two cases but now an extra term associated with the modulation, $\pi^2 L\mathcal{S}^2/d^2 - 2\pi^2 Lq_0\mathcal{S}^2/d$, must be included. The contribution originating from the anisotropy is of the form $-\alpha\langle Q_{zz}\rangle = -\frac{1}{4}\alpha$.

Blue phase or meron lattice. This phase has a structure consisting of a hexagonal lattice of double-twist tubes that can be viewed as half skyrmions or merons. It is shown in Figure 6.11. In each meron, the director twists through an angle of $\pi/2$ from a vertical orientation at the centre of the tube to a horizontal orientation at the edge. In cylindrical coordinates, the director can be assumed to be of the form $\boldsymbol{n}(r) = -\hat{\boldsymbol{\phi}}\sin(\pi r/d) + \hat{\boldsymbol{z}}\cos(\pi r/d)$, for $0 \leq r \leq d/2$, where d is the diameter of the tube. The tubes are separated by disclinations at which the director changes orientation while lying within the x-y plane. The average free energy density of this phase, $\langle f \rangle$, can be estimated assuming that the unit cell, comprising one meron and two disclinations, has an area $A = \sqrt{3}d^2/2$ and that the area of one meron is $A_m = \pi d^2/2$. Then, $\langle f \rangle = f_m A_m + 2f_{dis}A_{dis}/A$, where f_m and f_{dis} are, respectively, the free energy densities of merons and disclinations and the area of a disclination is $A_{dis} = A - A_m$. Then, the resulting average free energy density is $\langle f \rangle = \frac{\sqrt{3}\pi A\mathcal{S}^2}{18} + \frac{\pi B\mathcal{S}^3}{27\sqrt{3}} + \frac{\sqrt{3}\pi C\mathcal{S}^4}{54} + \frac{15L\mathcal{S}^2}{d^2} - \frac{\sqrt{3}(4+\pi^2)Lq_0\mathcal{S}^2}{3d}$.

Skyrmion lattice. This structure is depicted in Figure 6.11. In the skyrmions, the director twists by an angle π, from vertical in its centre, to horitzontal and back to vertical at the edge. Similarly to the case of merons, now a simple assumption consists of taking the director of the form, $\boldsymbol{n}(r) = -\hat{\boldsymbol{\phi}}\sin(2\pi r/d) + \hat{\boldsymbol{z}}\cos(2\pi r/d)$, for $0 \leq r \leq d/2$. In this case, however, in the regions between the tubes the director field is vertical with no singularity associated with disclinations. As in the preceding case, an average free energy density can be estimated as $\langle f \rangle = \frac{\sqrt{3}\pi A\mathcal{S}^2}{18} + \frac{\pi B\mathcal{S}^3}{27\sqrt{3}} + \frac{\sqrt{3}\pi C\mathcal{S}^4}{54} + \frac{15L\mathcal{S}^2}{d^2} - \frac{\sqrt{3}(4+\pi^2)Lq_0\mathcal{S}^2}{3d}$.

The phase diagram can be determined by comparing rescaled thermodynamic free energy densities of the different phases. Note that the

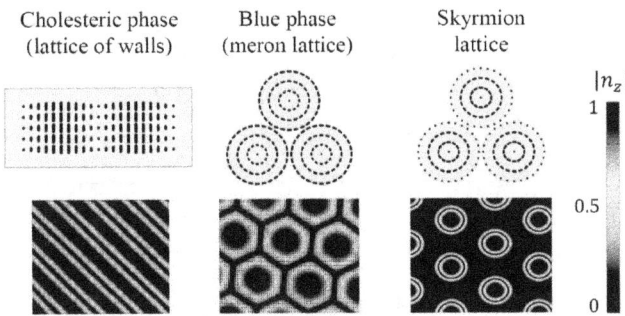

Figure 6.11 Two-dimensional illustration of the cholesteric phase (or lattice of walls), of the blue phase or meron lattice, and of the skyrmion lattice. The top row has schematic representations of the director modulation of these structures. The bottom row shows Monte Carlo numerical simulations of the model described in this section. The vertical bar on the right-hand side indicates the value of the z-component of the director. Adapted from Ref. [225].

thermodynamic free energy density is the free energy corresponding to the values of S and d that minimize the free energy functions. In the case with no anisotropy, the obtained expressions after rescaling depend only on the parameters t and κ. They are listed in Table 6.1, which also includes the corresponding anisotropy contributions.

The phase diagram for $\alpha = 0$ is shown in Figure 6.12 in the t-κ space. At high temperature, the isotropic phase is, as expected, the stable phase. At low temperature and high chirality, the system forms the blue phase in spite of the unfavourable contribution from the disclinations forming between merons. It is interesting to note that the energy cost of these disclinations can still be assumed finite but small, thanks to the fact that nematic order is soft due to the high chirality. Instead, at low temperature and low chirality, the stable phase is the cholesteric phase. In that case, the energy cost of disclinations is high since the nematic order is stiff due to the low chirality.

The anisotropy indeed affects the phase diagram. For high $|\alpha|$, the isotropic phase transforms first on cooling to the vertical nematic phase if $\alpha > 0$ or to the planar nematic phase if $\alpha < 0$. At lower temperature, these phases transform to modulated cholesteric or blue phases depending on the balance between α and κ.

An interesting aspect is the fact that the skyrmion lattice is not the stable phase in any region of the phase diagram. In any case, this phase cannot be strictly excluded since skyrmion defects might form as metastable defects.

Table 6.1 *Thermodynamic free energy densities and anisotropy contributions of the phases that can form in a chiral liquid crystal*

Phase	Thermodynamic free energy density	Anisotropy contribution
Isotropic[a]	0	0
Vertical nematic	$-\dfrac{(3+\sqrt{9-8t})^2(3+\sqrt{9-8t}-4t)}{93312}$	$-\dfrac{\alpha(3+\sqrt{9-8t})}{18}$
Planar nematic	$-\dfrac{(3+\sqrt{9-8t})^2(3+\sqrt{9-8t}-4t)}{93312}$	$\dfrac{\alpha(3+\sqrt{9-8t})}{36}$
Cholesteric	$-\dfrac{(3+\sqrt{9-8t+6\kappa^2})^2(3-4t+3\kappa^2+\sqrt{9-8t+6\kappa^2})}{93312}$	$-\dfrac{\alpha(3+\sqrt{9-8t+6\kappa^2})}{72}$
Blue (meron lattice)[b]	$-\dfrac{9700(3+\sqrt{9-8t+6\kappa^2})^2(3-4t+4.1\kappa^2+\sqrt{9-8t+6\kappa^2})}{1000}$	$\dfrac{27\alpha(\sqrt{9-8t+6\kappa^2})}{100}$
Skyrmion lattice	$-\dfrac{(3+\sqrt{9-8t+6\kappa^2})^2(3-4t+1.37\kappa^2+\sqrt{9-8t+6\kappa^2})}{93312}$	$-\dfrac{\alpha(8-\pi\sqrt{3})(3+\sqrt{9-8t+6\kappa^2})}{144}$

[a] Any effect of the anisotropy is neglected in the isotropic state.
[b] In the blue phase, it is assumed that the order parameter S has the same value as the order parameter in the cholesteric phase.

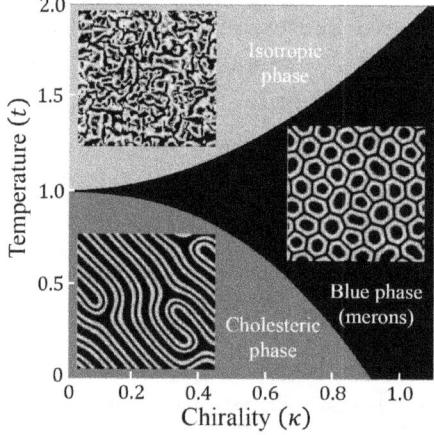

Figure 6.12 Phase diagram for chiral liquid crystals with no anisotropy in a reduced temperature (t) reduced chirality (κ) space. The snapshots of the different phases were obtained from Monte Carlo simulations of the model described in the text. Adapted from Ref. [225].

This is likely to occur for low chirality near the transition from the vertical nematic phase to the cholesteric phase for $\alpha > 0$.

Chiral magnets

Topological defects such as skyrmions and merons also form in chiral magnets. These defects have raised a lot of interest due to the fact that they represent highly stable discrete magnetic states that might be used as bits to store information. Therefore, this confers materials with this class of defects with a lot potential for applications related to magnetic memories, information technology and spintronics. Skyrmions have been reported to form both in bulk magnetic materials such as in MnSi [226] or in magnetic thin films [227]. Nevertheless, it is important to take into account that compared to liquid crystals, there is an important difference which arises from the symmetry of the orientational order parameter, which has a tensorial nature in liquid crystals while it is a vector in magnets.

The simplest free energy density adequate for chiral magnets, analogous to the model discussed for chiral liquid crystals, that takes into account the proper symmetries of magnets has the following form:

$$
f_{cm} = \frac{1}{2} A |\boldsymbol{M}|^2 + \frac{1}{4} C |\boldsymbol{M}|^4 + \frac{1}{2} K \sum_{\alpha\beta} \left(\frac{\partial M_\alpha}{\partial x_\beta} \right)^2
$$

$$
+ K q_0 \sum_{\alpha\beta\gamma} \epsilon_{\alpha\beta\gamma} M_\alpha \left(\frac{\partial M_\beta}{\partial x_\gamma} \right) - H M_z - \mathcal{A} M_z^2, \qquad (6.59)
$$

where $\boldsymbol{M} = (M_x, M_y, M_z)$ is the magnetization vector, $\boldsymbol{H}$ is a magnetic field applied along the z-axis and, as in the liquid crystal case, q_0 is an index of chirality. Note that the homogeneous contribution to the free energy density includes only even terms in the magnetization, which is consistent with the polar character of the order parameter. As usual, the quadratic coefficient is assumed to depend linearly on temperature as, $A = a(T - T_c)$, and the rest of the parameters are taken constant with respect to temperature. The third term accounts for the energy cost associated with any variation of the magnetization as a function of position and the fourth term is the Dzyaloshinskii-Moriya interaction that favours certain twist modulations of the magnetization originate from spin–orbit coupling. The last two terms are, respectively, the Zeeman term and the magnetocrystalline anisotropy term, which is assumed uniaxial. It may be easy-axis for $\mathcal{A} > 0$ or easy-plane for $\mathcal{A} < 0$.

Following the analogy with the chiral liquid crystal case, this free energy can also be rescaled. In this case, the characteristic values of the order parameter, free energy density of the ferromagnetic phase relative to the disordered phase and the core radius of a vortex in magnetic order are, $M \sim \sqrt{|A|/C}$, $f_{cm} \sim A^2/C$ and $\xi \sim \sqrt{K/|A|}$, respectively. Hence, magnetization, free energy density and all lengths can be rescaled by these characteristic values and then the model depends only on the three dimensionless parameters,

$$\kappa = q_0 \sqrt{\frac{K}{|A|}}, \qquad h = H \sqrt{\frac{C}{A^3}}, \qquad \alpha = \frac{A}{|A|}, \tag{6.60}$$

which play a similar role as the dimensionless parameters defined for liquid crystals in Eq. 6.56, 6.57 and 6.58. Actually, κ is a dimensionless chirality given as the ratio between the disclination core radius ξ and the natural twist length q_0^{-1}. Here α is a dimensionless anisotropy analogue to the dimensionless anisotropy in the liquid crystal case. In contrast, the parameter h does not exist in the liquid crystal, while a reduced temperature is not necessary in the magnetic case since the system is supposed to be in the low temperature phase, below T_c. The following magnetic structures can then be accounted for by the model.

Vertical ferromagnetic phase. In this phase, the magnetization is uniform along the z-axis, $\boldsymbol{M} = \hat{\boldsymbol{z}} M$.

Tilted ferromagnetic phase. The magnetization is of the form, $\boldsymbol{M} = -M$ $(\hat{\boldsymbol{x}} \sin \theta + \hat{\boldsymbol{z}} \cos \theta)$, where θ is the tilt angle determined by the competition between h and α.

Spiral phase. This phase is analogous to the cholesteric phase in liquid crystals. For $h = 0$ and $\alpha = 0$, $\boldsymbol{M} = -M[\hat{\boldsymbol{x}} \sin(\pi x/d) - \hat{\boldsymbol{z}} \cos(\pi x/d)]$. For small h and α, this structure is only slightly distorted so that the expression of $\boldsymbol{M}$ can still be used.

Skyrmion lattice. This structure is constituted of disks with a skyrmion structure that forms a hexagonal lattice. Therefore, in each disk, the magnetization that points downwards at its centre, twists through an angle of π so that it points upwards at the edge of the disk. It can be supposed that this variation is described by the expression, $\boldsymbol{M} = -M[\hat{\boldsymbol{\phi}} \sin(2\pi r/d) - \hat{\boldsymbol{z}} \cos(2\pi r/d)]$, for $0 \leq r \leq r/2$. Within the regions that separate disks, the magnetization is uniform and points upwards.

Meron lattice. This structure can also be modelled as a lattice of disks with a meron structure. Therefore, in this case, the magnetization only twists through an angle of $\pi/2$ from the centre of the disk. However, these disks cannot be arranged in a hexagonal lattice since the magnetization would be incompatible at the point where two disks with this structure meet. This situation, that in liquid crystals forces merons to be separated by disclinations in the blue phase, is forbidden in magnetism due to the vector character of

the order parameter. The alternative is to arrange disks with meron structure in a square lattice. In this structure, there is a regular alternation of
merons with the magnetization at their centre pointing downwards or upwards. In this structure, the magnetization in each meron can be assumed
to vary as, $\boldsymbol{M} = -M[\hat{\boldsymbol{\phi}}\sin(\pi r/d) - \hat{\boldsymbol{z}}\cos(\pi r/d)]$, for $0 \leq r \leq r/2$. Then,
within the regions between disks, the magnetic order must have a vortex,
which can be modelled as a disordered isotropic region.

The scaled thermodynamic free energies, obtained after minimization of the
corresponding free energy densities with respect to M and d, are listed in
Table 6.2. The phase diagram is then determined by comparing these thermodynamic free energy densities for different values of the dimensionless
parameters. An example of a cross-section of this phase diagram for a given
value, $\kappa = 0.5$, of chirality is depicted in Figure 6.13.

Table 6.2 *Thermodynamic free energy densities and anisotropy*
contributions of the phases of a chiral magnet

Phase	Thermodynamic free energy density
Vertical ferromagnetic	$-\frac{1}{4} - h - \alpha$
Tilted ferromagnetic[a]	$-\frac{1}{4} - h\cos\theta - \alpha\cos^2\theta$
Spiral	$-\frac{1}{4}(1+\kappa^2)^2 - \frac{1}{2}\alpha(1+\kappa^2)$
Skyrmion lattice	$-\frac{1}{4} - 0.36\kappa^2 - 0.15\kappa^4 - 0.093h - 0.37h\sqrt{(1+0.8\kappa^2)}$ $-0.55\alpha - 0.36\alpha\kappa^2$
Meron lattice	$-\frac{1}{4} - 0.36\kappa^2 - 0.11\kappa^4 - 0.23\alpha - 0.28\alpha\kappa^2$

[a] θ is the tilt angle determined by the competition between h and α.

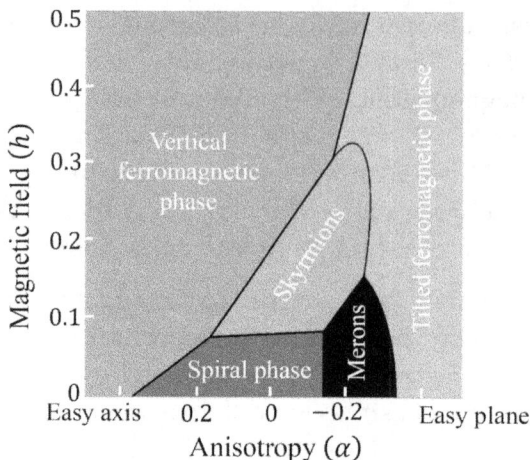

Figure 6.13 Cross-section of the phase diagram of a chiral magnet in a reduced magnetic field (h), reduced anisotropy (α) space corresponding to a value $\kappa = 0.5$ of the reduced chirality. For high positive α, the anisotropy is easy-axis, while for high negative α, it is easy-plane. Adapted from Ref. [225].

Exercises

6.1 Consider the Maier-Saupe mean-field model for an isotropic–nematic transition in a uniaxial liquid crystal. Suppose now that the liquid crystal is constituted of polar molecules and assume that the nematic potential is now of the form:

$$V(\mathcal{S}, \eta, \theta) = -v_{\mathcal{S}}\mathcal{S}P_2(\cos\theta) - v_\eta\eta P_1(\cos\theta),$$

where $\mathcal{S} = \langle P_2(\cos\theta)\rangle = \langle \frac{1}{2}(3\cos^2\theta - 1)\rangle$ and $\eta = \langle P_1(\cos\theta)\rangle = \langle\cos\theta\rangle$. The parameters $v_{\mathcal{S}}$ and v_η measure the effective orientational and polar interactions between molecules, respectively.

(a) Show that the mean-field free energy density, $f = u - Ts = \langle V(\mathcal{S}, \eta, \theta)\rangle - k_BT\langle\ln\varphi(\theta)\rangle$, where $\varphi(\theta)$ is the angular distribution function, can be written in the form

$$f = \frac{1}{2}v_{\mathcal{S}}\mathcal{S}^2 + \frac{1}{2}v_\eta\eta^2 - k_BT\ln Z(\mathcal{S}, \eta),$$

where $Z(\mathcal{S}, \eta)$ is given by

$$Z = \int_0^\pi \exp\left[\frac{v_{\mathcal{S}}\mathcal{S}P_2(\cos\theta) + v_\eta\eta\cos\theta}{k_BT}\right]\sin\theta d\theta.$$

(b) Show that the equilibrium order parameters can be obtained by minimizing f with respect to $\mathcal{S}$ and η or by solving the self-consistent equations:

$$\mathcal{S} = \int_0^\pi \frac{1}{Z}\exp\left[\frac{v_{\mathcal{S}}\mathcal{S}P_2(\cos\theta) + v_\eta\eta\cos\theta}{k_BT}\right]P_2(\cos\theta)\sin\theta d\theta,$$

$$\eta = \int_0^\pi \frac{1}{Z}\exp\left[\frac{v_{\mathcal{S}}\mathcal{S}P_2(\cos\theta) + v_\eta\eta\cos\theta}{k_BT}\right]\cos\theta\sin\theta d\theta. \quad (6.61)$$

6.2 Consider a uniaxial liquid crystal subjected to an applied magnetic field, H, described by a free energy density (Eq. 6.40):

$$f_u(\mathcal{S}, T, H) = \frac{1}{3}a(T - T_c)\mathcal{S}^2 + \frac{2}{27}B\mathcal{S}^3 + \frac{1}{9}C\mathcal{S}^4 - h\mathcal{S},$$

where $h = \frac{1}{2}\Delta\chi H^2$, and $\Delta\chi$ measures the degree of anisotropy of the molecules. Show that:

(a) The model has a critical point at $T_{cp} = (T_c + B^2/18aC)/2$ and $h_{cp} = -B^3/486C^2$.

(b) Determine the critical exponents α, β, γ and δ of this model.

7

Glassy phenomena in materials with disorder

The terms order and disorder are commonly used to specify the presence or absence of some symmetry or correlation in a system constituted of many components. So far, we have seen that phase transitions involve the formation of a long-range ordered state associated with the long-distance correlation of a given property that usually supposes the breaking of some symmetry in the system. In this chapter, disorder will designate any extrinsic effect that can be described with variables that are *randomly* distributed and causes the breaking of some basic symmetry, either translational, orientational or any other of the clean system considered. As a consequence, it is expected that this class of disorder has some influence on phase transitions that might occur in the clean system [228, 229]. For instance, impurities may produce a rounding of the expected order parameter discontinuity in first-order transitions. In general, annealed and frozen disorder are considered depending on whether they evolve in time or not. In this chapter, we will only consider the case of frozen disorder.

Disorder, from the preceding point of view, is ubiquitous in nature and has a strong impact on many fundamental phenomena and, often, is crucial in determining the properties of many materials. In systems undergoing phase transitions, the most notorious effect is probably to impede the formation of long-range ordered states and to cause glassiness effects. In solids, disorder may arise from many different sources. For instance, the existence of point defects, such as vacancies or impurities, line defects such as dislocations or plane defects such as twin planes or grain boundaries, may be the cause of breaking of the translational crystalline symmetry by, for instance, displacing atoms from their equilibrium positions, which may eventually lead to amorphization. In ferromagnetic systems, the replacement of some magnetic atoms by non-magnetic impurities that are distributed randomly may completely modify the properties by inhibiting the formation of the long-range ferromagnetic order.

Structural glasses form without the influence of disorder when a liquid is rapidly cooled to avoid the transition to the crystalline phase. As a result, the supercooled liquid becomes frozen and a disordered aperiodic metastable or an amorphous solid forms at low temperature, with a very long, almost divergent, relaxation time to the equilibrium crystal state [230]. In spite that this process is often known as a glass transition, it is not a transition in the thermodynamic sense but a kinetic phenomenon where the amorphous solid is dynamically arrested, which means that the process has taken place in a short time and the system has not had enough time to relax to the equilibrium state on experimental timescales [231]. This occurs in a vast majority of liquids if they are cooled fast enough, thanks to the excluded volume effects or, similarly, to the frustrating short-range local interactions, which are the basic physical ingredients that allow the glass transition to take place.

In this chapter, we are mostly interested in glassy behaviour that occurs due to the existence of some extrinsic frozen disorder. Similarly to the structural glasses, at low-enough temperature, a disordered configuration of entities associated with some degrees of freedom becomes frozen due to the influence of disorder on the interactions that are in *conflict* with each other and impede the propagation of correlations. The consequence is that the extrinsic disorder prevents the occurrence of a conventional long-range ordered state. Notwithstanding, in some cases, an unconventional long-range order can be established and thus, some kind of freezing phase transition can still exist.

This behaviour is a common feature observed in ferroic and multiferroic materials, and often originates from the presence of point defects. It has been studied for decades in the case of magnetic systems [232]. In polar materials, it is observed in the so-called dipolar glasses and relaxor ferroelectrics [265]. In these two systems, glassy behaviour is associated with a frozen state of magnetic moments or electric dipoles, respectively. More recently, the concept of strain glass has been introduced in order to designate a similar behaviour in ferroelastic and martensitic systems [234, 235, 236]. However, this has given rise to some controversy since it has been argued that in contrast to the magnetic or polar glasses where the ferroic entities (magnetic moments or spins and electric dipoles) can be identified at a microscopic level, the basic units in a strain glass state are nanoclusters that are mesoscopic entities which can only be properly defined if they have a certain extension. In fact, as already indicated in Chapter 3, the difference already exists in ferroelastics where the spontaneous deformation characterizing a ferroelastic state is not an intrinsic property, but can be defined only with respect to the undeformed paraelastic state in a relatively extended region of the material. In spite of these differences, it has been conjectured that there is a wide variety of ferroic materials that, when doped with point defects or impurities, show a similar glassy phenomenon with associated non-ergodic behaviour originating from the existence of frozen magnetic, polar, or strain clusters of the incoming ferroic phase. Even more, these systems display a similar generic phase diagram as shown schematically in Figure 7.1. In all these systems, glassy features are observed above a given amount of disorder, which is typically of a chemical nature. An interesting aspect is the fact that, usually, the glassy region is preceded by a precursor region where clusters exist but are not frozen and thus the system still shows an ergodic behaviour. Therefore, the concept of a ferroic glass has emerged in recent years, which has given rise to a new paradigm for the study of this class of materials [237]. Within this framework, the existence of a toroidal glass has also been proposed recently [238].

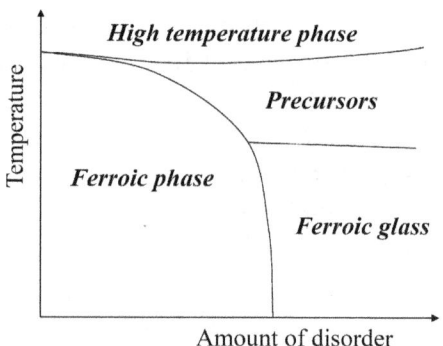

Figure 7.1 Generic phase diagram of ferroic materials with chemical disorder. The glassy region occurs when the amount of disorder is large enough and, usually, it is preceded by a precursor region.

7.1 Magnetic glasses

Disordered magnetic materials can exhibit a large variety of glassy states, all of them associated with specific characteristics of the disorder, magnetic moments and specific features of their interaction. For instance, Hurd [239] distinguished speromagnetic from asperomagnetic glassy states. The former corresponds to a state in which localized moments of a given species are frozen into random orientations, with no net magnetization nor a regular pattern of local ordering. The latter is invoked in systems with randomly placed localized moments of a given species, frozen in various orientations below some ordering temperature, but still with some orientations more likely than others. Here, we will just consider two broad categories, spin glasses and cluster glasses. In spin glasses, the basic entity is the magnetic moment defined at the atomic scale, while cluster glasses occur in systems with magnetic nanoparticles, which determine the length scale at which the magnetic entities that become frozen are defined. Actually, it is worth pointing out that materials that contain magnetic nanoparticles may show a rich variety of magnetism that is, usually, encompassed within the term supermagnetism [240].

7.1.1 Spin glasses

Historically, the spin glass phenomenon was first discovered in intermetallic compounds constituted of noble metals such as Au or Cu alloyed with transition metals such as Fe or Mn. In these alloys, magnetic moments, that are localized to a very good approximation at magnetic transition metal atoms, undergo an indirect exchange coupling via the polarization of the conduction band electrons. This coupling is usually known as the RKKY interaction [241], which stands for Ruderman–Kittel–Kasuya–Yosida. The basis of this indirect exchange was first established by Ruderman and Kittel [242] in the context of nuclear magnetism and was later adapted by Kasuya to d-electron spins interacting through conduction electrons [243] and expanded by Yosida [244]. This polarization oscillates in sign as a function of the distance between spins, which gives rise to an exchange parameter that depends on the distance as

$$J(\boldsymbol{r}) \sim \frac{\cos{(2k_F r + \theta)}}{(k_F r)^3}, \tag{7.1}$$

where k_F is the wave number of the Fermi sphere and θ a constant phase. Due to this interaction, this class of alloys displays a rich variety of magnetic phases depending on the concentration of the magnetic transition metal. As an example, Figure 7.2 shows the phase diagram of the $Au_{1-x}Fe_x$ alloy. This material shows spin glass features at low temperature for low concentration of magnetic atoms. When x increases, the system still shows glassy features but is associated with the existence of magnetic clusters. For $x \geq 16\%$, cluster glass phase and ferromagnetism can coexist at low temperature. For larger x, this alloy is ferromagnetic. At temperature above the region where spin and cluster glass occur, the system displays a paramagnetic behaviour. Nevertheless, for intermediate concentrations, cluster glass behaviour is preceded in temperature by a region where the system shows a superparamagnetic behaviour. In this phase diagram, the amount of disorder increases as x decreases. Therefore, it is clear that this phase diagram is qualitatively similar to the generic diagram depicted in Figure 7.1. It is noteworthy that the superparamagnetic region corresponds to the precursor region in the generic case.

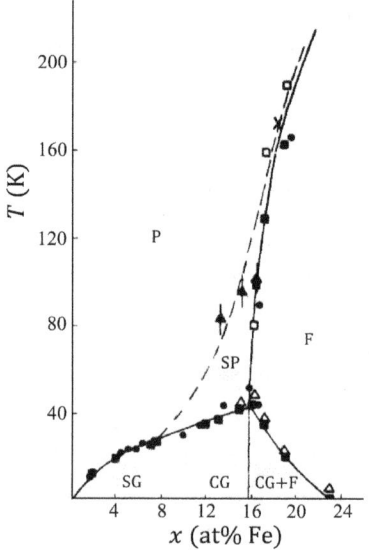

Figure 7.2 Magnetic phase diagram of the Au-Fe alloy. The regions where the different phases occur are indicated. SG: spin glass, CG: cluster glass, SP: superparamgnetic, P: paramagnetic, F: ferromagnetic. Adapted from [245].

Spin glass behaviour occurs at very low concentration of magnetic impurities and is characterized by remarkable memory and relaxation properties. In the low dilution limit, it is expected that due to the oscillatory character of the RKKY interaction, some pairs of spins will be coupled ferromagnetically and others antiferromagnetically, meaning that a conflictive competition between both classes of interactions may occur, which can give rise to frustration effects responsible for the glassy behaviour. Note that, the actual behaviour of a given material may show some dependence on specific treatments as, for instance, quenching or annealing, since this may modify the distribution of disorder. In what follows, we will consider that disorder is randomly distributed and we will not consider such quenching or annealing effects in the present chapter.

It is important to point out that the spin glass behaviour is not limited to this class of materials. Very similar glassy behaviour has been observed in magnetic insulators and amorphous (non-crystalline) alloys, where the nature of the exchange between magnetic moments is completely different from the one in noble metal alloys. Well-known examples are some Eu compounds such as the $Eu_xSr_{1-x}S$ compound. This system belongs to the family of EuX monochalcogenides with the anion X being O, S, Se and Te. $Eu_xSr_{1-x}S$ is ferromagnetic for large enough x [246] and shows canonical spin glass features for $x < 0.5$. Due to the compositional disorder associated with the substitution of Eu by Sr non-magnetic atoms, a competition between ferro- and antiferromagnetism occurs that gives rise to frustration effects. This behaviour is a consequence of an indirect exchange between f-electrons due to an intra-atomic d-f Coulomb interaction [247]. It is useful to note that the dominant exchange is strongly dependent on the anion X. For instance, in $Eu_xSr_{1-x}Te$ a similar spin glass behaviour is observed at low x, but the dominant exchange is antiferromagnetic instead of ferromagnetic [248].

Spin glasses represent the simplest case of a magnetic glass, and their study is interesting as they indicate common conceptual origins and features that can be anticipated in other classes of more complex ferroic glasses. Moreover, the existence of simple models that have been proposed for spin glasses have the advantage of determining likely fundamental origins of glassy phenomena in the broad class of ferroic materials and, thus, permits establishing the basis for the study of these different materials within a common framework.

From a very general viewpoint, a spin glass can be defined as a collection of magnetic moments localized in a given periodic lattice or non-periodic structure that shows a frozen disordered state at low temperature instead of the long-range uniform or periodic states that are commonly found in conventional magnets. The two basic ingredients that are necessary for such a behaviour to occur are randomness originating from disorder and frustration associated with competing interactions.

It is important to analyse in more detail what a frozen disordered state represents. For that purpose, suppose that m_i is the thermal average of a local magnetic moment defined at a given lattice site i, which can be, for instance, associated with an Ising spin variable S_i. This means that $m_i \sim \langle S_i \rangle$. In a frozen state, this local magnetic moment is non-zero, while the mean value over all, N, lattice spins, $m = N^{-1} \sum_i m_i$, as well as any staggered magnetization, $m_{\boldsymbol{q}} = N^{-1} \sum_i m_i \exp(-i\boldsymbol{q} \cdot \boldsymbol{r}_i)$, vanish, for all wave vectors $\boldsymbol{q}$. This means that the system does not show any kind of long-range order. Note that this is essentially different from the behaviour of a paramagnetic phase where not only $m = 0$ and $m_{\boldsymbol{q}} = 0$, but also $m_i = 0$.

The transition from a high temperature paramagnetic phase to a spin glass phase on cooling is indicated by a reduction of susceptibility below a given temperature. If h_i is the magnetic field acting on site i, the single site, diagonal susceptibility can be defined as, $\chi_{ii} = \partial m_i / \partial h_i$. In systems with Ising spins, this susceptibility can be expressed as

$$kT\chi_{ii} = \langle (S_i - \langle S_i \rangle)^2 \rangle = \langle S_i^2 \rangle - \langle S_i \rangle^2 = 1 - m_i^2, \qquad (7.2)$$

where we have taken into account that $S_i^2 = 1$. Averaging over all lattice sites gives

$$\chi_d = \frac{1}{N} \sum_i \chi_{ii} = \frac{1 - N^{-1} \sum_i m_i^2}{kT}, \qquad (7.3)$$

which expresses the reduction of susceptibility when the system becomes frozen, since $m_i = 0$ in the paramagnetic phase and $m_i \neq 0$ in the spin glass phase. However, it must be pointed out that, in fact, χ_d is not the susceptibility measured in experiments, which is given by $\chi = N^{-1} \sum_{ij} \partial m_i / \partial h_j$. Nevertheless, when the interactions between spins are symmetrically distributed, the off-diagonal elements χ_{ij} vanish in the limit of zero applied field. Therefore, in this situation, the experimental susceptibility χ should coincide to a good approximation with χ_d. Therefore, the susceptibility is expected to display a sharp cusp at a well-defined temperature so that the difference between the measured susceptibility and the extrapolation at low temperature of the high temperature susceptibility represents a good measure of the degree of freezing. In fact, experimental results confirm that the susceptibility of spin glass materials shows a cusp-like behaviour as illustrated in Figure 7.3 for a number of dilute alloys. Note, however, that due to the fact that an applied field is necessary to measure the susceptibility, when the scale is strongly expanded, it is observed that the cusp is not strictly sharp but instead shows some rounding. The temperature of the maximum of the susceptibility represents a suitable estimate of the freezing temperature T_f.

In conventional continuous transitions taking place from a disordered to a long-range ordered phase, we know that the approach to the critical temperature is accompanied by a dramatic increase in the range of the correlations, which is related to the divergence of susceptibility. Similarly, in spin glasses as T_f is approached on cooling, the range of correlations in space also increases. Nevertheless, in a spin glass system, the two-point correlation

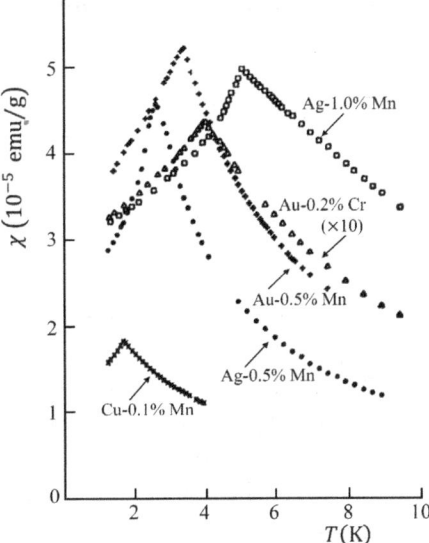

Figure 7.3 Behaviour of the susceptibility of a number of dilute magnetic alloys in the region of the spin glass transition. Adapted from [249].

function $G(r_{ij}) \sim \langle S_i S_j \rangle$ is not the relevant correlation function that reflects the occurrence of long-range order since it is expected to randomly change from positive to negative due to the random distribution of the exchange interaction between spins and, therefore, should not extend beyond a few lattice parameters. Actually, in a spin glass the relevant quantity that diverges at the spin glass transition is the square of this correlation, $\sim \langle S_i S_j \rangle^2$, which is related to a second-order susceptibility defined as

$$N\chi^{(2)} = \sum_{i,j} \chi_{ij}^2 = \frac{1}{k^2 T^2} \sum_{i,j} \langle S_i S_j \rangle^2. \tag{7.4}$$

This quantity can be experimentally obtained from the deviations from linearity of the dependence of magnetization m on an applied field h. This deviation is related to the non-linear susceptibility, χ_{nl}, which is usually defined as the coefficient of the h^3 term in the expansion of magnetization in powers of the external field, $m = \chi h - \chi_{nl} h^3 + ...$ The non-linear susceptibility is expected to diverge at a critical temperature as, $\chi_{nl} \sim |t|^{-\gamma_2}$, where $t = (T - T_f)/T_f$. The order parameter q for this transition is related to an average value over disorder of the square of the magnetization. Close to the freezing region, it is expected to show power-law critical behaviour. That is, $q \sim |t|^\beta$ at zero field and $q \sim h^{1/\delta}$ at $T = T_f$. More generally, in the critical region, it has been proven that q satisfies the following scaling behaviour [251]

$$q(T, h) = |t|^\beta h G\left[\frac{h^2}{|t|^{\beta + \gamma_2}}\right],\qquad (7.5)$$

where G is a scaling function. This equation represents the analogue to scaling Eq. 1.150 for a conventional phase transition. Then, it can be obtained that the non-linear susceptibility satisfies

$$\chi_{nl}(T, h) = h^{2/\delta}\Psi\left[\frac{t}{h^{2/\varphi}}\right],\qquad (7.6)$$

where Ψ is a scaling function that must satisfy that it is a constant for $x \to 0$ and that $\Psi(x) \sim x^{-\gamma_2}$ for $x \to \infty$. Here φ is a critical exponent that satisfies, $\gamma_2 = \varphi(1 - 1/\delta)$. This behaviour has been experimentally verified [252], which supports the idea that the freezing in spin glasses occurs through an unconventional phase transition.

Usually, the susceptibility, as that shown in Figure 7.3, is measured by first cooling the material down to a temperature far below T_f without an applied magnetic field, and then, a weak constant magnetic field (in the range of 10 Oe) is used to determine the susceptibility during small heating steps. Nevertheless, if instead, the material is cooled in the presence of a weak applied field, the obtained result is very different. In that case, below the temperature T_f, the measured susceptibility is larger than the one measured with the zero-field cooling protocol and remains roughly independent of temperature up to T_f. This larger susceptibility is related to the remanent magnetization that results when the system is cooled down below T_f in the presence of an applied field. In reality, associated with this behaviour $m(h)$ curves display hysteresis. At high temperature, above T_f, the susceptibility shows a behaviour independent of the measurement protocol. The two measurement protocols are usually denoted as zero-field cooled (ZFC) and field cooled (FC) protocols. These two protocols are illustrated in Figure 7.4. It is interesting to note that FC measurements are quite reproducible in the sense that the same susceptibility evolution is obtained in successive measurements performed up and down within a given temperature interval. Instead, ZFC measurements show a certain dependence on the actual value of the applied field used to measure the susceptibility. In practice, any splitting of the measured susceptibilities using these two protocols is used as a test to detect the existence of glassy behaviour in a given material.

The fact that the susceptibility shows a significant dependence on the measurement protocol is indeed a consequence of the glassy nature of the low temperature phase and proves the system shows a non-ergodic behaviour. In

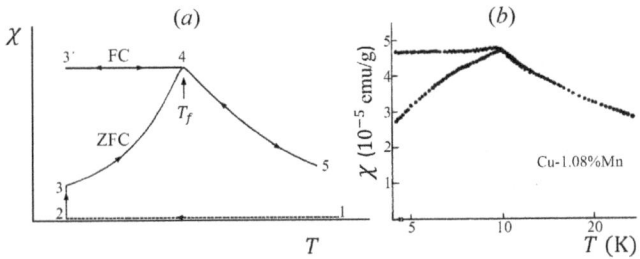

Figure 7.4 (a) Susceptibility measurement protocols. ZFC: 1 → 2 (cooling at zero applied field), 2 → 3 → 4 → 5 (susceptibility measurement under the application of a weak applied field. The step 2 → 3 depends on the value of the applied field.) and FC: 5 → 4 → 3' (cooling under an applied field) 3' → 4 → 5 (susceptibility measurement). (b) ZFC and FC susceptibility measurements in a Cu-1%Mn spin glass dilute alloy. Adapted from Ref. [250].

fact, the spin glass phase, and in general glassy phases, are characterized by the existence of many roughly equivalent spin configurations at low temperature. Therefore, the low temperature state reached in a given measurement protocol crucially depends on small details such as history effects or values of the applied field used to perform measurements.

The measurement of the magnetization response to a periodic field of frequency ω provides very important information. In this class of measurements, a complex susceptibility of the form

$$\chi(T, \omega) = \chi'(T, \omega) + i\chi''(T, \omega), \tag{7.7}$$

is obtained. The real part, $\chi'(T, \omega)$ is the in-phase response to the field, while the imaginary part, $\chi''(T, \omega)$, is the out-of-phase response, which is associated with the existence of dissipative effects. An interesting result is the strong dependence of the susceptibility on the field frequency, as shown in Figure 7.5. The freezing temperature T_f is commonly defined as the temperature of the maximum of the real part of the complex susceptibility corresponding to the zero frequency limit.

The strong frequency dependence of the freezing temperature indicates that collective effects near the freezing point give rise to long relaxation times and, thus, to a very slow dynamics, which is characteristic of glassy behaviour. In spin glasses, it seems that characteristic times grow even faster than the relevant correlation length. From a phenomenological point of view, it has been proposed that below T_f the real part of the complex susceptibility shows a logarithmic behaviour such as

$$\chi'(\omega) = \chi - a \ln \omega, \tag{7.8}$$

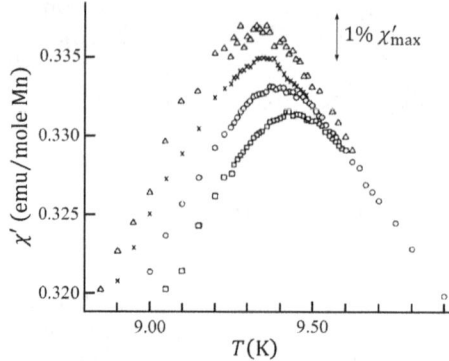

Figure 7.5 Frequency dependence of the real part of the complex susceptibility of a Cu-0.94 at.% Mn dilute alloy in the region of the freezing temperature. The frequency of the field is $\omega = 2.6$ Hz (open triangles), 10.4 Hz (crosses), 234 Hz (open circles), 1.33 kHz (open squares). At the large scale of the figure, the rounding of the peak is clearly defined. In this class of experiments the freezing temperature $T_f \simeq 9.35$ K. Adapted from Ref. [253].

where a is a positive constant. This expression represents a good fit to experimental data in the typical region of frequencies ranging from Hz to kHz where experiments are usually performed. Thus, according to Kramers-Kronig relations that mathematically connect the real and imaginary part of a complex function, the imaginary part should be roughly independent of frequency. This is quite in agreement with experimental results [253].

Other laws have also been proposed, as for instance, a power law with a low value of exponent also provides a good fit to the experimental data. In fact, within the hypothesis of a true phase transition, the time scale $\tau \sim 1/\omega$ should diverge according to the following critical behaviour

$$\tau \sim \left[\frac{T_f(\omega)}{T_f(0)} - 1 \right]^{-\nu z}, \tag{7.9}$$

where ν is the critical exponent that determines the divergence of the correlation length and z is a dynamical exponent [254].

In systems with magnetic clusters, the behaviour of relaxation times is well accounted for by a Vogel-Fulcher empirical law, $\tau = \tau_0 \exp[E/k(T_f(\omega) - T_f(0)]$, that can be justified in terms of the cluster interactions [255].

The fact that glassy behaviour is associated with such long characteristic time scales suggests the presence of many metastable spin configurations with a broad distribution of energy barriers separating them. In the next section, we will see that this is an important point to take into account when looking for solutions of spin glass models.

7.1.2 Models for spin glasses

In this section, we are interested in introducing models that can reproduce the physical behaviour of spin glasses. The fact that spin glass phenomena show some universal behaviour suggests that simple models that incorporate the essential features arising from disorder should be adequate for such a purpose. From this viewpoint, it seems reasonable to consider the Heisenberg model as the reference model adequate to describe the behaviour of clean magnetic systems with the absence of disorder and proceed to introduce disorder in a simple but still realistic way. We can consider, for instance, the particular case of $A_{1-x}M_x$ alloys where M are the magnetic constituents that interact magnetically through an RKKY exchange interaction (Eq. 7.1). Note that this interaction is essential to understand specific features of the phase diagram of the Au-Fe alloy illustrated in Figure 7.3, which belongs to this class of magnetic alloys.

These systems can be modelled by an extension of the Heisenberg model defined on a lattice with an interaction between magnetic atoms that alternates sign and decreases in magnitude as the distance between magnetic atoms increases. Compositional disorder is accounted for by introducing extra variables $\{c_j\}$ that couple to the spin variables $\{S_i\}$, that take values $c_i = 1$ if site i is occupied by a magnetic atom and $c_i = 0$, otherwise. These variables $\{c_j\}$ are assumed to be distributed randomly according to the concentration $x = N^{-1} \sum_i c_i$. Then, the hamiltonian of this model is

$$\mathcal{H}(\{S_i\}, \{c_i\}, h) = -\frac{1}{2} \sum_{i,j=1}^{N} J_{ij} c_i c_j S_i \cdot S_j - h \cdot \sum_{i=1}^{N} c_i S_i. \qquad (7.10)$$

Indeed, the model can also be formulated with Ising or other kind of spin variables. In general, this class of models is known as random-site lattice models. It is also interesting to note that beyond alloys with RKKY exchange, these models are also appropriate to describe magnetic insulators with compositional disorder where the exchange coupling between magnetic moments also alternates its sign with increasing distance. This is, for instance, the case of the $Eu_xSr_{1-x}S$ compounds, which show spin glass behaviour for low concentrations, $x < 0.5$. In this compound, the dependence of the sign and strength of the exchange interaction with distance can be determined from low temperature inelastic neutron scattering experiments.

These experiments confirm that, compared with materials with RKKY exchange, in these insulator materials, the interaction is short range and can be limited to next-nearest neighbours assuming that $J_{nn} = -2J_{nnn} > 0$, which is a reasonable approximation.

Random site models, especially when formulated with Heisenberg variables, are quite realistic but too complex for statistical mechanics treatments. This, together with the fact that spin glass behaviour is observed in a wide variety of materials with very different characteristics of the exchange interaction, suggests that other simpler models might still capture the essential features of spin glasses and, thus, be adequate to describe spin glass properties. With this aim, Edwards and Anderson [257] proposed a lattice model with spin variables where disorder is included in a distribution of exchange coupling parameters. In these models, Ising, Heisenberg, etc., spin variables that lie on the N-sites of a regular lattice are assumed to interact randomly according to a given distribution. Generically, these models are called random bond models and no significant differences in describing a spin glass behaviour are expected when compared with random site models. In terms of Ising spin variables, which supposes an additional simplification, the Edwards-Anderson random bond hamiltonian has the form

$$\mathcal{H}(\{S_i\}, \{J_{ij}\}) = -\frac{1}{2} \sum_{i,j=1}^{N} J_{ij} S_i S_j - h \sum_{i}^{N} S_i, \qquad (7.11)$$

where $\{J_{ij}\}$ are assumed as random variables with a distribution that only depends on the lattice vector separation, $\mathbf{r}_{ij} = \mathbf{r}_j - \mathbf{r}_i$, between sites i and j. A very convenient case corresponds to assuming that the distribution is a symmetric gaussian distribution, centred at zero mean with variance Δ_{ij}. That is, the distribution function $P(J_{ij})$ is of the form

$$P(J_{ij}) = \frac{1}{\sqrt{2\pi\Delta_{ij}}} \exp\left(-\frac{J_{ij}^2}{2\Delta_{ij}}\right), \qquad (7.12)$$

where the range of the interaction can be varied. Non-symmetric models with non-zero mean can also be considered if the existence of long-range order, either ferro- or antiferromagnetic, must be taken into account.

In fact, other distributions can be proposed. Another simple distribution is, for instance, the double-delta function distribution. In this case, it is given by

$$P(J_{ij}) = \frac{1}{2}\delta(J_{ij} - \Delta_{ij}^{1/2}) + \frac{1}{2}\delta(J_{ij} + \Delta_{ij}^{1/2}). \qquad (7.13)$$

In both cases, gaussian or double-delta, the model is specified by the relationship

$$[J_{ij}^2]_c \equiv \Delta_{ij} \equiv \Delta(\mathbf{r}_{ij}), \qquad (7.14)$$

where $[J_{ij}^2]_c$ is an average over the distribution of random variables and represents a measure of the amount of disorder. These averages must be distinguished from thermal averages, which will be indicated by the symbol $\langle \cdot \rangle$. If the interaction is limited to nearest neighbours, only the term $\Delta_{ij} = \Delta_{nn}$ must be considered. On the other hand, the infinite-range case corresponds to the situation in which Δ_{ij} is a constant independent of the separation between spins i and j.

Beyond models for the pure spin glass phase, in some cases, it is adequate to take into account randomness in the field term of the hamiltonian. The Ising hamiltonian with randomness in the field is written as

$$\mathcal{H}(\{S_i\}, \{h_i\}) = -J \sum_{\langle ij \rangle_{nn}} S_i S_j - \sum_i h_i S_i, \qquad (7.15)$$

where an interaction between nearest neighbour spins is considered. The field is assumed to be site dependent, which reflects the existence of randomness. Therefore, $\{h_i\}$ must be considered as local random variables that follow a certain distribution, usually gaussian or double-delta distribution. In the case of a gaussian distribution

$$P(h_i) = \frac{1}{\sqrt{2\pi\Delta_h}} \exp\left(-\frac{h_i^2}{2\Delta_h}\right), \qquad (7.16)$$

which, in fact, assumes that no correlations exist between local fields at different sites. This class of models is usually known as random-field Ising models. The main difference between random field and random bond models arises from the fact that the existence of random fields assumes a local symmetry-breaking effect. On the other hand, random-field Ising models are less realistic than random-bond Ising models since it is considered virtually impossible to realize site-dependent fields that change from site to site in the lattice. Nevertheless, it turns out that dilute antiferromagnets under a uniform applied field or fluids in random porous media in a lattice gas description are examples of systems that can be conveniently modelled by a random-field Ising model [258].

7.1.3 Mean-field treatment

Mean-field treatment of models for spin glasses is useful because this represents a first step in the understanding of collective behaviour in this class of systems. Various effective-field approximation schemes have been proposed [259] that take into account the effects of fluctuations due to random spatial disorder in magnetic spin systems. Sherrington and Kirkpatrick (SK) [260] found an exact solution of a random bond Ising model in which the spins are coupled by infinite-ranged random interactions independently distributed with a gaussian probability density. This is an interesting case since, as discussed in Chapter 1, Section 1.4.1, infinite range interaction is a way of formally defining a mean-field theory, which is exact in this particular situation.

Proceeding in analogy with the mean-field treatment of the Ising model discussed in Chapter 1, the self-consistent equation giving the mean-field solution of the random-bond Ising hamiltonian 7.11 is

$$\langle S_i \rangle = \tanh\left(\frac{1}{k_B T} \sum_j J_{ij} \langle S_j \rangle \right), \tag{7.17}$$

where the sum extends over the full range of the interaction. According to this equation, the onset of ferromagnetic long-range order should occur at a temperature T_c at which

$$\langle S_i \rangle = \frac{1}{k_B T_c} \sum_j J_{ij} \langle S_j \rangle. \tag{7.18}$$

Compared with the clean Ising model, in the present model the exchange parameters J_{ij} are also random variables. Thus, averaging over the bond distribution and ignoring correlations between exchange parameter and local magnetization averages, lead to

$$[\langle S_i \rangle]_c = \frac{1}{k_B T_c} \sum_j [J_{ij}][\langle S_j \rangle]. \tag{7.19}$$

Then, for a symmetric bond distribution, $[J_{ij}] = 0$, no non-trivial solution with $\langle S_i \rangle \neq 0$ can be found. Even more, an analogous consideration eliminates any periodic long-range ordered solution. However, if Eq. 7.18 is first squared and then averaged again ignoring correlations, then it is obtained that below the temperature

$$kT_c = \sqrt{\sum_i [J_{ij}^2]}, \tag{7.20}$$

a frozen-spin state without long-range order occurs characterized by the order parameter $[\langle S_i \rangle^2]$ that takes non-zero values in the low temperature phase.

The preceding simple discussion certainly lacks sufficient rigour. It can be better understood within the framework of the replica theory as introduced by Edwards and Anderson [257], which permits justifying averages over quenched disorder. From a general point of view, in a disordered system, we are interested in statistically representative quantities and, thus, it seems reasonable to look at averages over a given distribution of disorder. Therefore, thermodynamic quantities of interest will be obtained from the corresponding average value of the free energy, $F = -k_B T \ln Q$. This requires averaging over disorder $\ln Q$ instead of the partition function Q. The replica approach starts from the mathematical identity

$$\ln Q = \lim_{n \to \infty} \frac{1}{n}(Q^n - 1). \tag{7.21}$$

In the replica theory, Q^n is interpreted as the partition function of n replicas of the original system, each of them associated with a specific realization of the distribution of disorder. Therefore, for a random-bond Ising model with spins interacting through an infinite range exchange

$$[Q^n] = \int \left(\prod_{i<j} dJ_{ij} P(\{J_{ij}\}) \right) \sum_{\{S_i^\mu\}} \exp \left(\beta \sum_{i<j} J_{ij} \sum_{\mu=1}^{n} S_i^\mu S_j^\mu + \beta h \sum_{i=1}^{N} \sum_{\mu=1}^{n} S_i^\mu \right), \tag{7.22}$$

where the superscript μ is a replica label. Now, instead of a symmetric gaussian distribution of bonds, we will consider a more general distribution with non-zero mean, $J_0 > 0$.[1] That is,

$$P(\{P_{ij}\}) = \sqrt{\frac{N}{2\pi J^2}} \exp \left\{ -\frac{N(J_{ij} - J_0/N)^2}{2J^2} \right\}. \tag{7.23}$$

Then the integral over J_{ij} can be carried out and $[Q^n]$ can be expressed as a function of the moments of the bond distribution. After some straightforward algebra and dropping terms that vanish in the thermodynamic limit, $[Q^n]$ can be written as

$$[Q^n] = e^{N\beta^2 J^2 n/4} \sum_{\{S_i^\mu\}} \exp \beta \left\{ \mathcal{H}_{eff}(\{S_i^\mu\}) + h \sum_i \sum_\mu S_i^\mu \right\}, \tag{7.24}$$

where $\mathcal{H}_{eff}$ plays the role of an effective hamiltonian, which is given by

$$\mathcal{H}_{eff}(\{S_i^\mu\}) = \frac{\beta J^2}{2N} \sum_{\mu<\nu} \left(\sum_i S_i^\mu \right)^2 + \frac{J_0}{2N} \sum_\mu \left(\sum_i S_i^\mu \right)^2. \tag{7.25}$$

---

[1] For $J_0 > 0$, the system is expected to show a ferromagnetic behaviour for low-enough disorder. The case $J_0 < 0$ corresponding to a system with antiferromagnetic behaviour at low disorder could also be considered with an appropriate definition of the order parameter, different from the ferromagnetic order parameter considered in the formalism that will be presented.

By means of the preceding procedure, we have replaced the original model with disorder by one with no disorder described by an effective hamiltonian that includes temperature and involves higher dimensional spins with more complicated interactions than the original hamiltonian.

Now, with the introduction of the effective hamiltonian, the problem can be reduced to a single-particle problem by applying the gaussian integral formula to $(\sum_i S_i^\mu S_i^\nu)^2$ with integral variable $q_{\mu\nu}$, and to $(\sum_i S_i^\mu)^2$ with integral variable m_μ. Then,

$$[Q^n] = e^{N\beta^2 J^2 n/4} \int \prod_{\mu<\nu} dq_{\mu\nu}$$

$$\times \int \prod_\mu dm_\mu \exp\left(-\frac{N\beta^2 J^2}{2}\sum_{\mu<\nu} q_{\mu\nu}^2 - \frac{N\beta J_0}{2}\sum_\mu m_\mu^2\right)$$

$$\times \sum_{S_i^\mu=\pm 1} \exp\left(\beta^2 J^2 \sum_{\mu<\nu} q_{\mu\nu}\sum_i S_i^\mu S_i^\nu + \beta\sum_\mu (J_0 m_\mu + h)\sum_i S_i^\mu\right).$$

$$(7.26)$$

The second sum $\sum_{S_i^\mu}$, can be expressed as

$$\left\{\sum_{S^\mu=\pm 1} \exp \mathcal{L}(S^\mu)\right\}^N = \exp\left\{N\ln\sum_{S^\mu=\pm 1} e^{\mathcal{L}(S^\mu)}\right\}, \qquad (7.27)$$

where $\mathcal{L}(S^\mu) \equiv \beta^2 J^2 \sum_{\mu<\nu} q_{\mu\nu}\sum_i S_i^\mu S_i^\nu + \beta\sum_\mu (J_0 m_\mu + h)\sum_i S_i^\mu$. We can take advantage of the fact that the exponent of the integrand in the expression of $[Q^n]$ is proportional to N, which is large, and, thus, we can evaluate the integral using the saddle-point method. Then it is obtained that

$$[Q^n] \simeq 1 + nNA(q_{\mu\nu}, m_\mu), \qquad (7.28)$$

where

$$A(q_{\mu\nu}, m_\mu) \equiv \left\{-\frac{\beta^2 J^2}{4n}\sum_{\mu<\nu} q_{\mu\nu}^2 - \frac{\beta J_0}{2n}\sum_\mu m_\mu^2 + \frac{1}{n}\ln\sum_{S^\mu=\pm 1} e^{\mathcal{L}(S^\mu)} + \frac{1}{4}\beta^2 J^2\right\}.$$

$$(7.29)$$

Now, taking the limit $n \to 0$ and keeping N large, but finite, the free energy f per spin is obtained as

$$-\beta f = \lim_{n\to 0} \frac{[Q^n]-1}{nN} = \lim_{n\to 0} A(q_{\mu\nu}, m_\mu), \qquad (7.30)$$

where $q_{\mu\nu}$ and m_μ are the saddle-point values that maximize $[Q^n]$, which are solutions of

$$q_{\mu\nu} = \langle S^\mu S^\nu \rangle_{\mathcal{L}} = \frac{1}{\beta^2 J^2} \frac{\partial}{\partial q_{\mu\nu}} \ln \sum_{S^\mu = \pm 1} e^{\mathcal{L}(S^\mu)}, \qquad (7.31)$$

$$m_\mu = \langle S^\mu \rangle_{\mathcal{L}} = \frac{1}{\beta J_0} \frac{\partial}{\partial m_\mu} \ln \sum_{S^\mu = \pm 1} e^{\mathcal{L}(S^\mu)}. \qquad (7.32)$$

The preceding equations represent a self-consistent solution of the problem that would be straightforward to solve for finite n. However, with the limiting procedure $n \to 0$, the problem becomes much more difficult and requires an appropriate analytical continuation. This is, however, quite easy within the replica-symmetry *ansatz* that assumes that $q_{\mu\nu}$ and m_μ must be independent of the replica. Therefore, it is assumed that

$$q_{\mu\nu} = q \quad \forall \mu \neq \nu,$$
$$m_\mu = m \quad \forall \mu.$$

Then, within this hypothesis, before performing the limit $n \to 0$, we estimate the term $\ln \sum e^{\mathcal{L}}$ that appears in the expression of $A(q_{\mu\nu}, m_\mu)$ using the gaussian integral method. Taking into account Eq. 7.27 that defines $\mathcal{L}$, it is obtained that

$$\ln \sum_{S^\mu = \pm 1} e^{\mathcal{L}} = \ln \sum_{S^\mu = \pm 1} \sqrt{\frac{\beta^2 J^2 q}{2\pi}}$$

$$\times \int dz \exp \left\{ -\frac{\beta^2 J^2 q}{2} \frac{z^2 + n}{2} + \beta(\beta J^2 q z + J_0 m + h) \sum_\mu S^\mu \right\}$$

$$= \int Dz \exp \left\{ n \ln 2 \cosh(\beta J \sqrt{q} z + \beta J_0 m + \beta h) - \frac{n}{2} \beta^2 J^2 q \right\}, \quad (7.33)$$

where $Dz = dz \exp(-z^2/2)/\sqrt{2\pi}$. Then, defining

$$\tilde{h}(z) \equiv J \sqrt{q} z + J_0 m + h, \qquad (7.34)$$

the preceding Eq. 7.33 can be rewritten in the form,

$$\ln \sum_{S^\mu = \pm 1} e^{\mathcal{L}} = \ln \left\{ 1 + n \int d\tilde{h} P(\tilde{h}) \ln 2 \cosh \beta \tilde{h} - \frac{n}{2} \beta^2 J^2 q + \mathcal{O}(n^2) \right\},$$

$$(7.35)$$

where $P(\tilde{h}) = \exp\{-(\tilde{h} - J_0 m - h)^2\}/2J^2 q)/\sqrt{2\pi J^2 q}$. It is interesting to note that $\tilde{h}$ plays the role of a mean-field, which is gaussian distributed about the mean-field $(J_0 m + h)$ with a variance $J^2 q$. Note that in the preceding expression 7.33, the dependence on all S^μ variables is linear with the same coefficient which represents the simplest analytical continuation that allows the calculation of the free energy. Now, taking the limit $n \to 0$ the following free energy is obtained:

$$f = \frac{[F]}{N} = -\frac{\beta^2 J^2}{4}(1-q)^2 + \frac{1}{2}\beta J_0 m^2 - \int d\tilde{h} P(\tilde{h}) \ln 2 \cosh \beta \tilde{h}, \quad (7.36)$$

which is the replica-symmetric solution of the SK model. Here q and m are the order parameters of the model that are obtained self-consistently from the equations,

$$q = \int d\tilde{h} P(\tilde{h})(\tanh \beta \tilde{h})^2, \quad (7.37)$$

$$m = \int d\tilde{h} P(\tilde{h}) \tanh \beta \tilde{h}. \quad (7.38)$$

Therefore, these order parameters can also be identified as $q = [\langle S_i \rangle^2]$ and $m = [\langle S_i \rangle]$. The case $q \neq 0$ corresponds to a cooperative frozen magnetic state, while $m \neq 0$ supposes that the frozen state has a ferromagnetic component. Thus, the model can describe the following three different phases:

- Paramagnetic phase (P) with $m = 0$ and $q = 0$.
- Ferromagnetic phase (F) with $m \neq 0$ and $q \neq 0$.
- Spin glass (SG) phase with $m = 0$ and $q \neq 0$.

The spin glass phase is expected to occur for $h = 0$ and low-enough values of J_0. The transition temperature[2] T_g from the paramagnetic to spin glass phase can be obtained by expanding Eq. 7.36 in powers of q. With $h = 0$, to the lowest order, the result is

$$f = -\frac{1}{4}\beta J^2 - \ln 2 + \frac{\beta J^2}{4}(1 - \beta^2 J^2)q^2 + \ldots, \quad (7.39)$$

which indicates that the model shows a critical temperature at $\beta^2 J^2 = 1$ that must be identified with the spin glass transition temperature given by $T_g = J/k_B$. Alternatively, we can proceed by expanding Eq. 7.37. In the case $h = 0$, this gives

$$q = \beta^2 J^2 q + \beta^2 J_0 m^2 + \ldots, \quad (7.40)$$

which, indeed, is consistent with the same spin glass transition temperature $T_g = J/k$ while $m = 0$. This requires that $J_0 \leq J$. For larger J_0, a transition from the paramagnetic to ferromagnetic phase can exist at a temperature T_c. This temperature can be obtained by expanding Eq. 7.38 about $m = 0$

---

[2] This transition temperature should be identified with the experimental freezing temperature T_f.

with $h = 0$. This leads to $m = \beta J_0 m + ...$, which indicates that $T_c = J_0/k$. The boundary between spin glass and ferromagnetic phases must be estimated by numerically solving Eqs. 7.37 and 7.38. The phase diagram of the model in the case $h = 0$ is depicted in Figure 7.6. When the effect of the magnetic atom concentration x is properly taken into account on the effective interaction parameters J and J_0, the obtained phase diagram shows a good qualitative agreement with experimental phase diagrams as, for instance, that of the RKKY Au-Fe alloy shown in Figure 7.2. Actually, in this class of spin glasses, the parameter J can be related to the distribution of magnetic atoms and it can be shown that $J \sim x$ to a good approximation. This, indeed, leads to a spin glass temperature that increases with x consistently with experiments. On the other hand, the boundary line, in the SK-phase diagram that separates the spin glass and ferromagnetic phases is not correct. For J_0/J in the range from 1 to 1.25 (see Figure 7.6), as temperature decreases the system transforms first from a paramagnetic to a ferromagnetic phase and on further cooling to a spin glass phase. This peculiar reentrant behaviour is in fact an artefact of the replica-symmetry hypothesis. This behaviour disappears when the replica symmetry breakdown is properly taken into account [262], which requires an infinite number of order parameters. Within the improved treatment, the boundary line is slightly modified and, more importantly, a low temperature region below this line is shown to have features of both ferromagnetic and spin glass phases in the sense that the ferromagnetic order parameter is finite and corresponds to complex spin states. In this mixed region, instead of spin glass, the frozen phase often corresponds to a cluster glass phase.

Finally, note as well that when the distribution of disorder is strictly symmetric, which means that $J_0 = 0$, no solutions with $m \neq 0$ are found and thus, there is no ferromagnetic phase in the phase diagram.

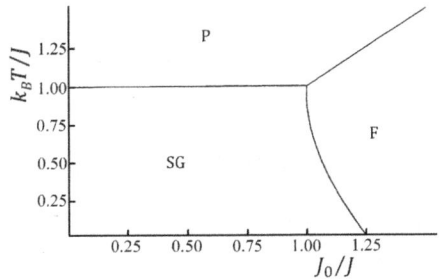

Figure 7.6 Phase diagram of the SK-model in the case $h = 0$. Adapted from Ref. [260].

It is important to point out that the case of a gaussian random-bond Ising model with interactions limited to z nearest neighbours can also be treated within the same mean-field approximation. In that case, the same solution is obtained after replacing J_0 and J by zJ_0 and $\sqrt{z}J$, respectively. However, this solution is no longer an exact solution as in the infinite range case but instead an approximate mean-field solution of the model.

As the spin glass transition is approached from above, the behaviour of the order parameter q can be obtained from an expansion of the free energy given by Eq. 7.39. The result is $q = |t| + t^2/3 + ...$, where $t = (T - T_g)/T_g$. The susceptibility can then be computed by differentiating Eq. 7.38 with respect to h and taking the limit $h \to 0$. In the case $J_0 = 0$, it is found that

$$\chi_0 = \frac{1}{T_g}(1 - |t| + ...) \quad \text{for } T > T_g, \tag{7.41}$$

$$\chi_0 = \frac{1}{T_g}\left(1 - \frac{1}{3}t^2 + ...\right) \quad \text{for } T < T_g, \tag{7.42}$$

which shows a cusp at T_g, in agreement with experiments. In the general case, when $J_0 \neq 0$, the susceptibility can be expressed in terms of χ_0 as

$$\chi = \frac{\chi_0}{(1 - J_0\chi_0)}, \tag{7.43}$$

which also shows a cusp at the transition temperature. The behaviour of the susceptibility is depicted in Figure 7.7. It is important to point out that in the presence of an applied magnetic field, the cusp of the susceptibility is rounded as found experimentally.

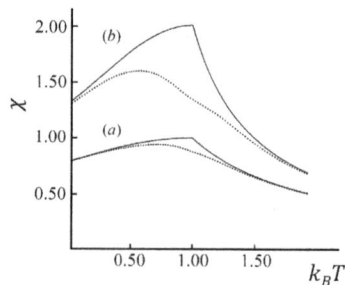

Figure 7.7 Susceptibility of the SK-model without an external applied field (continuous lines) and with an applied field $h = 0.1J$ (dashed lines), in the cases, (a) $J_0/J = 0$, and (b) $J_0/J = 0.5$. Note that the effect of an applied external field is to round the cusp that occurs at T_g. Adapted from Ref. [260].

It can be shown that the heat capacity also shows, in the absence of an applied field, a cusp behaviour similar to the one shown by the susceptibility. This, in fact, as we have already seen in Section 7.1.1, is in contrast with the behaviour of response functions found at a critical point in continuous phase transitions. At the spin glass transition, it is the non-linear susceptibility that is expected to diverge at the freezing point T_g. Within the SK mean-field treatment, a non-linear susceptibility can be obtained as $\chi_{SK} = \partial q^2/\partial \bar{h}$, where $\bar{h}$ is a mean value of a distribution of random fields $\{h_i\}$, which can be shown to be linearly related with the non-linear susceptibility χ_{nl} defined from the expansion, $m = \chi h + \chi_{nl} h^2 + ...$. Then the non-linear susceptibility is found to diverge as, $\chi_{nl} \sim t^{-1}$.

It has been already mentioned that the replica-symmetry *ansatz* assumed in the SK model leads to a striking but in fact incorrect boundary line that separates the spin glass and ferromagnetic phases. Actually, the replica symmetry solution obtained by Sherrington and Kirkpatrick that was based on the steepest-descent method does not provide a stable solution in both the spin glass and ferromagnetic phases. Almeida and Thouless showed that when obtaining Eqs. 7.37 and 7.38 using the saddle-point method, the resulting free energy is not a minimum everywhere in both the spin glass and ferromagnetic phases [261], which means that in these regions the obtained solutions are not thermodynamically stable. This, for instance, leads to an unphysical negative zero-temperature entropy of the spin glass phase. The problem can be clearly revealed by analysing the normal modes of the fluctuations of $q_{\mu\nu}$ and m^μ in the replica space about their replica-symmetry values. This is accomplished by defining fluctuations as $\delta q_{\mu\nu} \equiv q_{\mu\nu} - q$ and $\delta m_\mu \equiv m_\mu - m$ and expanding the free energy 7.29 to second order about the replica-symmetry order parameters. Then, from the study of the normal-mode spectrum in the limit $n \to 0$, it is found that the states beneath a given surface in the space $(h/J, J_0/J, k_BT/J)$, which include the spin glass phase, are unstable against replica-symmetry modes. This surface is shown in Figure 7.8. The projection of this surface on the space $(h/J, k_BT/J)$ defines the Almeida-Thouless line. Beneath this line, a more sophisticated treatment, as the one proposed by Parisi [262] is needed to find mean-field stable solutions, which explicitly take into account specific features of the highly degenerate ground state in this class of systems.

To briefly introduce the Parisi solution[3], we must remind that the SK model discussed above is characterized by a spontaneous symmetry breaking of the replica symmetry, which shows up in a non-trivial form of the matrix

---

[3] Giorgio Parisi was awarded the Nobel prize in physics for this contribution in 2021.

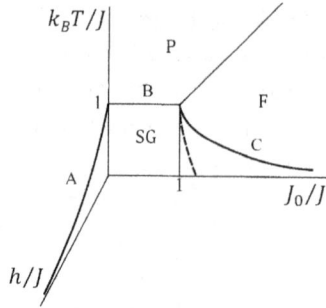

Figure 7.8 Almeida-Thouless surface defined by the lines A, B and C. Beneath this plane replica-symmetric states are unstable. Line A is usually denoted as the Almeida-Thouless line. Adapted from Ref. [249].

$q_{\mu\nu}$. The stability problem of this solution is reflected by the fact that it yields a negative zero temperature entropy. To solve this problem, it must be taken into account that besides this solution many others exist due to the symmetry of the replica free energy function with respect to replica permutations. The solution to the problem proposed by Parisi can then be viewed in terms of a sequence of *ansatze* that progressively approximate the true solution in a better way. This scheme introduces order parameters q_i, m_i at each step i. The important issue is that in the limit of the infinite sequence, a continuous function $q(x)$ is obtained such that the free energy can be expressed as a functional of this order parameter function. Finally, the problem reduces to extremizing this functional and the solution can be obtained numerically with a high degree of approximation.

7.2 Glassy behaviour in polar materials: dipolar glasses and relaxors

Beyond magnetic materials, glassy behaviour is observed in polar materials as well. In some cases, even a similar ideal spin glass behaviour has been reported in this class of systems. Nevertheless, in general, polar materials display more complex glassy features that, in any case, can be understood on the basis of the spin glass paradigm as discussed in the preceding section. Here we will discuss two interesting classes, dipolar glasses and relaxor ferroelectrics.

Dipole glasses

Dipole glasses are a class of glassy materials which have a strong relationship with ferroelectric materials. A prototypical example is the mixed solid solutions $(KDP)_{1-x}(ADP)_x$ consisting of the ferroelectric KH_2PO_4 (KDP) and the antiferroelectric $NH_4H_2PO_4$ (ADP). At low temperature, KDP-type systems show configurations comprising 180° domains with dipole moments along $+c$ and $-c$ axes of the (a, b, c) tetragonal unit cell. Instead, ADP shows four 90° antiferroelectric domains with dipole moments along $+a$, $-a$ and $+b$, $-b$ axes, respectively. Therefore, in these mixed solid solutions, competition between ferroelectric and antiferroelectric interactions exists that together with random electric field effects causes frustration and may lead to a glassy state. This glassy state can be understood in terms of quenched random interactions between pseudospin degrees of freedom, which represent the two equilibrium positions of protons corresponding to a double-well potential characteristic of $O-H\cdots O$ bonds. In fact, the simplest microscopic description of this interaction is based on a tunnelling model that can be expressed as a random bond version of the Ising model with a transverse field that represents the tunneling frequency of the protons. Moreover, in some circumstances, the fact that NH_4 groups can be positioned nonsymmetrically with respect to the surrounding cations, tilting the proton double-well potential in a random manner, induces an effective local random field. This behaviour occurs, for instance, in the mixed hydrogen ferro- and antiferroelectric dipolar glass $Rb_{1-x}(NH_4)_xH_2PO_4$ (RADP).

Consequently, in contrast with spin glasses in magnetism, a general model for dipolar glasses must include the effect of disorder in the form of both random bonds and random fields. The difference essentially arises from the fact that while magnetic spins can be usually assumed uncoupled from the lattice to a good approximation, $O-H$ bonds are intrinsically a part of the lattice and, thus, strongly affected by substitutional disorder. Therefore, at first glance, it is expected that due to the symmetry-breaking effect associated with the random fields, which will be reflected in the fact that the corresponding variance will act as an effective ordering field, the Edwards-Anderson order parameter q is non-zero at all temperatures and no phase transition to a dipolar glassy phase might exist.

The phase diagram of the $Rb_{1-x}(ND_4)_xD_2PO_4$ (DRADP) mixed solution has been studied in detail in Ref. [263]. This mixed solid solution is similar to the RADP system but instead of a proton (H) glass it is a deuteron (D) glass. At intermediate concentrations, in a range $x_{FE} \leq x \leq x_{AFE}$, a pseudospin-like dipolar glassy phase with no long-range order associated with macroscopic symmetry breaking occurs at any temperature. It is a consequence of the competition between ferroelectric and antiferroelectric interactions, which gives rise to random frustration effects. At low temperature, along a given concentration line, the randomly frozen dipoles undergo a transition from an ergodic to a non-ergodic phase, which can be experimentally detected from the splitting of the susceptibility in ZFC-FC measurements. This line can be identified with an Almeida-Thouless line that extends between the two extreme regions of compositions $0 \leq x \leq x_{FE}$ and $x_{AFE} \leq x \leq 1$, respectively, where inhomogeneous ferroelectric and antiferroelectric regions coexist with glassy domains.

Within the pseudospin description, the behaviour of this class of glassy polar materials can be described by a random-bond-random-field version of the SK model

$$\mathcal{H} = -\frac{1}{2} \sum_{i,j} J_{ij} S_i S_j - \sum_i f_i S_i - E \sum_i S_i, \qquad (7.44)$$

where the exchange coupling parameters $\{J_{ij}\}$ are gaussian distributed with mean J_0/N and variance J^2/N. The random fields $\{f_i\}$ are assumed to have an independent gaussian distribution with zero mean and variance Δ_f. The third term in the hamiltonian represents the coupling of the pseudospins with an applied external electric field, E. Using the replica formalism discussed in the preceding section, it is obtained that above a temperature T_f the order parameter q is given by a self-consistent equation equivalent to Eq. 7.37 but now in the definition of $\tilde{h}$, given by Eq. 7.34, $J\sqrt{q}z$ must be replaced by $J\sqrt{q + \tilde{\Delta}_f}z$, where $\tilde{\Delta}_f = \Delta_f/J^2$. The freezing temperature must then be obtained from the study of the stability of the replica-symmetry solution [264]. The crossing of this line is connected with a phase transition that, in fact, persists even in the presence of random fields.

Relaxor ferroelectrics

The other important family of polar glassy materials is that of relaxor ferroelectrics or simply relaxors. This is a class of disordered materials that show peculiar structure and physical properties. At high temperature, they exist in a non-polar paraelectric phase, which is similar in many respects to the paraelectric phase of normal ferroelectrics. Upon cooling, precursor regions of nanometre scale with randomly distributed directions of dipole moments appear. This state is often qualified as the ergodic relaxor state and occurs below the so-called Burns temperature, T_B. Usually, this temperature is hundreds of degrees above the expected ferroelectric transition temperature. While the transition to the ferroelectric phase is associated with a crystallographic symmetry change, the occurrence of the relaxor state is not accompanied by any change of crystal symmetry at the macroscopic scale. Nevertheless, the existence of polar nanoregions (PNRs) strongly affects crystallographic features, which gives rise to unique physical properties.

For this reason, the state of crystal at $T < T_B$ with precursors of the low temperature phase is often considered as a new phase different from the paraelectric phase.

At temperatures close to T_B, the PNRs are mobile but, on cooling, their dynamics slows down and, in some cases, instead of the expected transition to a ferroelectric phase, at a temperature T_f the PNRs become frozen into a non-ergodic state but, in any case, the average symmetry of the crystal does not change. This behaviour shows some similarities with the freezing transition that takes place in dipolar glasses. Nevertheless, the non-ergodic relaxor state that occurs below T_f can be irreversibly transformed into a ferroelectric state by the application of a strong enough external electric field. This is an important property of relaxors that distinguishes them from typical dipolar glasses. Upon heating, the obtained ferroelectric transforms to an ergodic paraelectric state at the expected ferroelectric transition temperature, which is close to T_f.

In general, the relaxor state has been characterized by the frustration of the local polarization, which prevents the development of global long-range ferroelectric order [265]. Cross [266] proposed to denote relaxor ferroelectrics in the materials that fulfil the following three properties: (*i*) The temperature-dependent dielectric permittivity exhibits a broad and smeared maximum. (*ii*) The temperature for the maximum dielectric permittivity is frequency dependent, which indicates the presence of dielectric relaxation. (*iii*) There is no macroscopic symmetry breaking as a function of temperature. All these features are characteristics of systems that can exhibit glassy behaviour.

A representative example of a relaxor material is $PbMg_{1/3}Nb_{2/3}O_3$ (PMN). This oxide crystallizes in the cubic perovskite structure (see Figure 3.9) with the Mg ion, that contributes charge 2+, and the Nb ion, that contributes charge 5+, randomly distributed over the B sites of the perovskite structure. The formation of polar nanoclusters is due to compositionally or frustration-induced charge disorder [267]. Charge fluctuations are then the source of quenched random electric fields in a way similar to the case of dilute antiferromagnets [268].

At high temperature, clusters comprise reorientable regions of typically 2–3 nm size embedded in a polarizable neutral matrix characterized by a high dielectric constant. In this class of materials, the properties of the matrix determine whether nanoclusters will grow in size to percolate the whole sample at a lower temperature to constitute a long-range ferroelectric ordered state or whether they will remain small enough at all lower temperatures to form a non-ergodic relaxor state.

The existence of a non-ergodic relaxor state is usually detected by the splitting of FC and ZFC measurements of the electric susceptibility or dielectric constant. An interesting signature that corroborates the existence of polar nanoclusters is the presence of elastic diffuse scattering, that is temperature dependent, and that can persist down to very low temperatures in non-ergodic relaxors.

Relaxors, such as PMN, can be described within the general framework of the family of random-bond-random-field models, which bring this class of materials close to the category of dipolar glasses. However, there are important differences between the two classes of systems. In relaxors, the pseudospin variables are associated with the effective dipole moment of polar nanoclusters, which must be treated as discrete vectors with variable magnitude that depends on the size of the clusters and can point along a large but finite number of orientations in space. Actually, the dipole moment of a given nanocluster i, $\boldsymbol{P}_i^{nc}$, is given as the sum of the elemental dipolar moments of the unit cells that constitute the cluster. That is, if $\boldsymbol{P}_i^l$ is the dipolar moment of the unit cell l in cluster i, $\boldsymbol{P}_i^{nc} = \sum_l \boldsymbol{P}_i^l$. In PMN, since Mg and Pb ions have the same charge, the contribution of Mg-type cells to $\boldsymbol{P}_i^{nc}$ should be small and, thus, only Nb-cells are relevant in order to determine $\boldsymbol{P}_i^{nc}$. If n_i is the number of Nb atoms in cell i, then

$$\boldsymbol{P}_i^{nc} \cong n_i \boldsymbol{P}_0(i), \tag{7.45}$$

where it is assumed that all Nb-type cells of cluster i contribute with the same dipole moments $\boldsymbol{P}_0(i)$. A pseudospin vector variable, $\boldsymbol{S}_i$, associated with each cell can then be defined as

$$\boldsymbol{S}_i = \left(\frac{3}{\langle n^2 \rangle}\right)^{1/2} \frac{\boldsymbol{P}_i^{nc}}{\boldsymbol{P}_0(i)}, \tag{7.46}$$

where $\langle n^2 \rangle = (\sum_i n_i^2)/N$ is the average of n_i^2 over the total number N of nanoclusters. With this definition, the pseudospin variables satisfy the closure relation, $\sum_i (\boldsymbol{S}_i)^2 = 3N$.

Pirc and Blinc [269] proposed a simplified version where each allowed direction of pseudospin components $\boldsymbol{S}_i$ can fluctuate over the entire space constrained to satisfy the previously established closure condition. This leads to the spherical vector random-field-random-bond model that can be solved with the replica formalism. The solution predicts the existence of either a ferroelectric phase or a spherical glass phase that does not show long-range order. A phase transition between these two phases is only expected to exist in the absence of random fields. Nevertheless, for weak enough random fields the third-order susceptibility is still predicted to show a sharp peak at a temperature close to the freezing temperature of the spherical model without random fields. This behaviour is in agreement with the experimental behaviour reported in the $Pb_{1-x}La_x(Zr_yTi_{1-y})_{1-x/4}O_3$ (PLZT) relaxor ceramic [270].

7.3 Strain glasses

Strain glasses are disordered frozen states of local lattice strains. These states can be considered conjugates of long-range ordered ferroelastic states in the same sense that cluster glasses or relaxors are conjugates of ferromagnetic or ferroelectric states, respectively. In ferroelastic materials, nanoscale regions in the form of islands with the symmetry of the approaching low temperature phase form at high temperature within the paraelastic phase due to the presence of quenched-in disorder usually originating from point defects that give rise to composition fluctuations. In any case, other defects such as dislocations or precipitates can be at the origin of these nanoscale precursors. Usually, the existence of precursors is detected as contrast anomalies in transmission electron microscopy and their presence is confirmed by a significant elastic diffuse scattering in a broad range of temperatures.

In the precursor state, the symmetry of the high temperature phase is preserved at the macroscopic scale and the system displays ergodic behaviour. Depending on specific features of these precursors, a strain glass phase may occur associated with the frozen-in domains below a certain temperature. In the strain glass state, the system shows non-ergodic behaviour, which is confirmed by the splitting of ZFC and FC strain versus temperature measurements.[4] A representative family of materials that displays strain glass behaviour is that of TiNi-based shape memory alloys. For instance, the binary $Ti_{1/2-x}Ni_{1/2+x}$ alloy shows a cubic structure at high temperature and undergoes a ferroelastic/martensitic transition at room temperature towards a monoclinic phase. However, when the alloy is doped with an excess of Ni exceeding $x = 0.13$, the ferroelastic transition is inhibited and instead a strain glass phase forms [271]. Similar behaviour has been reported in TiNi alloys doped with small amounts of Fe, Co, Cr, or Mn.

An interesting class of precursors is that of tweed textures. These are spatially inhomogeneous states comprising cross-hatched strain modulations. In fact, they consist of a multi-phase pattern of coexisting regions with properties varying over nanometre distances. Tweed is observed well above the actual phase transition temperature as diffuse striations parallel to the traces

---

[4] The applied field in this case is a stress conjugate to the measured strain.

of specific planes of the high temperature crystallographic phase. It is not strictly periodic but can be characterized by two characteristic length scales. The longitudinal extent of the long diagonal striations and their transversal width are typically on the scale of ten to hundreds of lattice spacings. Materials with the presence of tweed textures show intense anisotropic diffuse scattering along given crystallographic directions associated with low energy phonon branches that, eventually, may show dips at specific q-positions which are indicative of the selected structure of the ground state. Interestingly enough, systems showing this class of tweed precursors do not become frozen in a non-ergodic glass phase but instead, below a given temperature, tweed regions grow and percolate into a ferroelastic phase. In contrast, strain glass occurs in materials where no signatures of tweed contrast are detected. These results confirm that tweed shows strong anisotropic properties and indicate that specific features of precursors in this class of materials are controlled by a combined effect of disorder and elastic anisotropy that determines the strength of long-range interactions, which are essential for a ferroelastic phase to form. Actually, the presence of disorder gives rise to a distribution of energy barriers that, above a certain critical amount that should depend on the elastic anisotropy, may cause a screening of the long-range potential thus inducing a cutoff of correlations, which can lead to the suppression of the transition to the ferroelastic twinned structure. From this point of view, glassy behaviour in ferroelastic materials occurs because while a local phase transition takes place at short time scales through the system, coalescence into a global phase takes much longer times which tend to diverge as temperature is lowered. Therefore, as the strength of long-range elastic interaction weakens, more and more time is required for the system to reach for paths in phase space leading to the final state, which eventually leads to a glassy state due to a kinetic arrest process.

Models aimed at elucidating the nature of strain glass phase have been proposed taking into consideration the interplay between elastic long-range interactions and disorder. In general, these studies are based on phase-field Landau models for ferroelastic systems as the model discussed in Chapter 3. Disorder is included as an additional factor that may perturb the establishment of a long-range ordered state and promote the formation of a glassy phase. As the strain glass behaviour has often been reported in alloys, disorder is usually associated with composition fluctuations or chemical inhomogeneities, which are almost unavoidable in this class of metallic materials. Since the transition temperature to the ferroelastic/martensitic phase in alloys such as Ti-Ni is known to be very sensitive to very small changes in composition, it is then assumed that disorder gives rise to a distribution in local transition temperatures. Then, formally, it is introduced in the harmonic term of the Landau free energy, which is related to the relevant elastic modulus[5]. For instance, in the case of a material with a high temperature cubic phase and a low temperature tetragonal phase, the relevant modulus is $C' = C_{11} - C_{12}$, which is then supposed to be of the form

$$C'(T, \eta) = a[T - T_c - \eta(\boldsymbol{r})], \tag{7.47}$$

where T_c is the low stability limit of the high temperature phase in the absence of disorder, and η determines the disorder field, which is often taken to be spatially correlated and gaussian distributed with zero mean, $g(\eta) = (\sqrt{2\pi}\zeta)^{-1} \exp(-\eta^2/2\zeta^2)$. This choice ensures that local transition temperatures vary smoothly in space. By this method, regions with different degrees of metastability separated by finite free-energy barriers are introduced. Anisotropy is taken into account in the long-range compatibility term of the free energy. This term is essential to establish long-range correlations in specific lattice directions that may give rise to a ferroelastic phase with long-range order. These correlations are strongly smoothed out for low values of the elastic anisotropy that in systems with cubic symmetry is measured by the ratio, $A = C_{44}/C'$, between the deviatoric and the shear elastic constants.

Results obtained from numerical simulations [272] show that when a system is cooled down from high temperature, for a low amount of disorder and high enough elastic anisotropy, tweed first occurs at temperatures well above the phase transition and, below a certain temperature, a twinned ferroelastic phase is obtained. In contrast, for high disorder and low values of the elastic anisotropy, instead of tweed, droplets of the low-temperature phase appear at high temperature and, on further cooling, no phase transition to a ferroelastic phase occurs. Instead, below a certain temperature, the system shows typical features of a strain glass phase. Examples of patterns obtained from numerical simulation for increasing values of the amount of disorder and anisotropy are shown in Figure 7.9. Strain glass can be detected from the splitting of strain versus temperature curves below a certain temperature in ZFC and FC simulation results. These results indicate that the disorder threshold above which strain glass occurs depends on the elastic anisotropy as $\zeta \sim \sqrt{A}$, where ζ is the standard deviation of the disorder distribution, which is expected to provide a reasonably good measure of the excess of

---

[5] The relevant elastic modulus is the one that is associated with the shear strain component that plays the role of order parameter for the transition considered and is the inverse of the elastic susceptibility associated with the order parameter.

doping at which the ferroelastic transition is suppressed in a given material and strain glass appears.

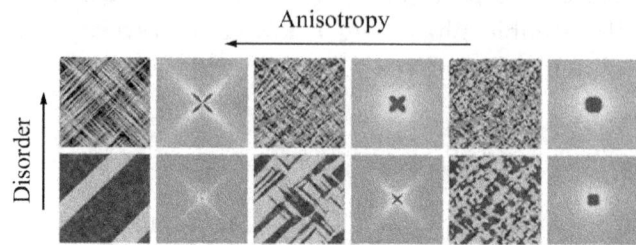

Figure 7.9 Snapshots of structural patterns formed at low temperature for decreasing values of elastic anisotropy and increasing amount of disorder, as indicated by the arrows. In the upper row, a typical tweed pattern is shown at the left corner, while the snapshot at the right corner illustrates a system with nanodroplet precursors. The corresponding patterns in the lower row correspond to a twinned ferroelastic phase and a strain glass phase, respectively. Results are obtained from numerical simulations of a $2d$ model for a square-rectangle transition that mimics a cubic-tetragonal ferroelastic transition. Regions with different tonality indicate rectangular domains with the long axis in the x and y direction, respectively. Each snapshot is accompanied by its corresponding Fourier transform, which clearly shows that by decreasing anisotropy and increasing the amount of disorder patterns lose directionality along the diagonals of the square. For a square-rectangular transition, diagonal directions are the directions along which correlations propagate to form the long-range ordered ferroelastic phase. Adapted from Ref. [273].

An alternative to continuum Landau models for ferroelastics and strain glasses based on a pseudo-spin model has been proposed, which is obtained from appropriate discretization of the continuum model [274]. This has, in principle, the advantage of permitting the applications of the same statistical mechanics techniques that are typically used to describe spin glasses, dipolar glasses, and relaxors, in the case of strain glasses.

For a ferroelastic system with a g-degenerate ground sate, where g are the differently oriented domains that can grow in the ferroelastic phase, the idea is to consider a lattice obtained from coarse graining and define the local state of each lattice grain by a spin variable that can take $g + 1$ values that represent the high temperature phase and the low temperature domains. For instance, for the simple case of a $2d$ square to rectangle transition, the continuum to discrete limit is achieved by replacing the strain order parameter e_2 (see Chapter 3) by $|e_2|S$, where the pseudo-spin variable S can take the values $S = 0, \pm 1$, which correspond to the minima of the free energy representing the high temperature phase, and the two ferroelastic domains, respectively. Then, taking into account that all even powers of the order parameter in the free energy expansion are 0 or 1, the homogeneous terms of the Landau free energy can be written as $w(T)S^2$, where $w(T)$ is a function of temperature. Similarly, using the finite difference definition of a gradient, the term, $(\nabla e_2)^2$, should be expressed as products of S values of neighbouring grains, $S_i S_j$. Then, the following hamiltonian for square-rectangle transition is obtained

$$\mathcal{H}_{S-R} = \sum_{\langle ij \rangle_{nn}} J_{ij} S_i S_j + w(T) \sum_i S_i^2 + A \sum_{i,j} U_{ij} S_i S_j, \qquad (7.48)$$

where A is the strength of the long-range anisotropic interaction, U_{ij}, which depends on the elastic anisotropy. The resulting model is a particular version of the Blume–Emery–Griffiths spin-1 model introduced in Section 1.3.1.[6]

The model offers different options to include disorder. Keeping the analogy with the Landau model from which the spin hamiltonian 7.48 has been derived, disorder should be included in the term $w(T)$. Note that this is different from what is usually done in the case of spin glasses where disorder is included in the exchange parameter that determines the interaction between magnetic moments. In Ref. [274] the effect of assuming that the parameters J_{ij} and $w(T)$ are gaussian distributed has been studied from the replica-symmetry approach. The interesting result is that the existence of strain glass may be reproduced regardless of particular features of disorder.

---

[6] This simplified version of the Blume–Emery–Griffiths model is also denoted as the Blume–Capel model.

Exercises

7.1 Consider the random-field Ising model defined on a lattice with z nearest neighbours with hamiltonian

$$\mathcal{H} = -J \sum_{\langle ij \rangle_{nn}} S_i S_j - \sum_{i=1}^{N} h_i S_i,$$

where $J > 0$ and the random fields $\{h_i\}$ follow the double-delta distribution

$$P(h_i) = \frac{1}{2}\delta(h_i - h_0) + \frac{1}{2}\delta(h_i + h_0). \tag{7.49}$$

(a) Defining the magnetization m as the configurational average of the mean spin per site, $m = [\langle S_i \rangle]_c = \frac{1}{N}\sum_{i=1}^{N}\langle S_i \rangle$, show that a mean-field hamiltonian can be written in the form

$$\mathcal{H}_{MF} = -\frac{1}{2}NzJ\left(\frac{2m}{N}\sum_{i=1}^{N} S_i - m^2\right) - \sum_{i=1}^{N} h_i S_i.$$

(b) The equation of state reads

$$m = \beta Jz[1 - \tanh^2(\beta h_0)]m$$
$$- \frac{1}{3}(\beta Jz)^3[1 - 4\tanh^2(\beta h_0) + 3\tanh^4(\beta h_0)]m^3.$$

(c) Show that a line of critical points $T_c(h_0)$ exists given by the condition:

$$T_c = \frac{zJ}{k}[1 - \tanh^2(\beta_c h_0)].$$

(d) Show that the critical line ends at a tricritical point at $\tanh^2(\beta_{tc}h_0) = 1/3$ and that for higher fields the transition is first order.

(e) Find the equation of state assuming that the random fields are gaussian distributed.

7.2 The lower critical dimension d_l is the least space dimension for which a phase transition does not occur for a given universality class. For instance, in the Ising model $d_l = 1$. Consider a random-field Ising model with a ferromagnetic exchange J and a random field of zero mean and variance, $[h_i h_j]_c = h^2 \delta_{ij}$. Using qualitative arguments *à la Peierls* show that the lower critical dimension of the random-field Ising model is $d_l = 2$. Should this dimension be different for systems with spins that can rotate continuously in space?

7.3 Consider that the probability density of energy configurations of a disordered model defined by a hamiltonian $\mathcal{H}(\{S\})$ is given as the configurational average:

$$P(E) = [\delta(E - \mathcal{H}(\{S\}))]_c.$$

Suppose that this probability density is gaussian and has the form:

$$P(E) = \frac{1}{\sqrt{N\pi}J} \exp\left[-\frac{E^2}{NJ^2}\right],$$

where J is a measure of the ferromagnetic strength. This is, in fact, the probability density of energy configurations of the Sherrington-Kirkpatrick model, where $J = [J_{ij}]_c$.

Using the microcanonical formalism, show that there is a critical freezing temperature

$$T_f = \frac{J}{2k_B\sqrt{\ln 2}}, \tag{7.50}$$

such that, temperatures below T_f correspond formally to being in the range of E where there are no states. This means that the system must freeze into the last available state, $|E| = E_0$ at these temperatures. Note that this result supposes that the low temperature phase is thermodynamically trivial and has zero entropy.

8

Non-equilibrium phase transitions

This chapter deals with phase transitions in externally driven materials. In Chapter 2, we have already seen that the response to an external field of cooperatively interacting many body systems essentially depends on the relation between three time scales associated with the driving, τ_h, relaxation τ_d, and thermal fluctuations, τ_{tf}, where the last one depends not only on temperature but also on specific features of the system considered, which characterize the energy barriers that separate high and low symmetry phases. Now we are interested in discussing the following two important athermal situations. In the former, we consider the rate-dependent case dominated by the competition between the two first time scales, $\tau_h/\tau_d \sim 1$, while the latter refers to the quasistatic limit $\tau_h/\tau_d \gg 1$ in systems containing quenched disorder. The athermal condition supposes that in the two cases thermal fluctuations are considered as irrelevant. We will see that both classes of systems may undergo phase transitions that occur out of equilibrium driven by the driving frequency and disorder, respectively.

In the first case, suppose that the system is subjected to an external field that oscillates in time. Then the response of the system, for instance, the magnetization in a magnetic material, will also oscillate and, in general, it will lag behind the applied field due to the effect of the relaxational delay. This will give rise to hysteresis. When the time period of the driving field becomes much shorter than the typical relaxation time of the thermodynamic system, the hysteresis loop becomes asymmetric around the origin, which can be understood as related to a new phase that occurs via a dynamic phase transition. This transition is indeed a non-equilibrium phase transition, which arises spontaneously from a dynamically broken symmetry due to the competition of time scales [275].

Close to the quasistatic athermal limit, when $\tau_h/\tau_d \gg 1$, we have seen that a transition is expected to occur abruptly when the energy barrier that separates the two phases involved in the transition disappears, which happens at (or close to) a spinodal point. However, due to quenched disorder, a wide coexistence region may exist characterized by a complex energy landscape with multiple minima separated by energy barriers. Then, within a certain range of rates, when driving the material, it responds intermittently by successive relaxations from one metastable state to another within the energy landscape. These local transformations define avalanches. The interesting situation occurs when these avalanches appear with the absence of characteristic time and size scales, a situation that defines the so-called *crackling noise* [276], which is a sort of non-equilibrium criticality also denoted as *avalanche criticality*. We will discuss this behaviour in detail in Section 8.2.

An important point to be remarked is that both problems can be treated within mean-field approaches that provide formally analogous solutions to the mean-field solution corresponding to equilibrium thermal transitions. Pushing the analogy beyond this approximation, one might explain the fact that equilibrium solutions do so well when applied to the study of phase transitions in real materials, where non-equilibrium effects seem, in principle, unavoidable.

8.1 Dynamic phase transitions

Dynamic phase transitions may occur in any material that may undergo a phase transition driven by a time-varying field and manifest as a change in a material's dynamical behaviour at a given value of the characteristic frequency of the driving field. A symmetry-breaking effect should be reflected in the hysteresis features of the driven transition. Magnetic materials are indeed prototypical materials where such a behaviour is expected. In any case, a broad class of physical systems may display this kind of dynamic phase transition. Interesting examples include dynamics of vortices in superconducting materials [277, 278], charge density waves [280], and laser-driven bistable optical cavities [281]. For instance, the last example shows optical hysteresis that is maximized at a given laser frequency due to effective photon–photon interactions. Actually, most of these systems are invariant under reversal of the order parameter and can, thus, be described by a two-state microscopic model. Therefore, from a microscopic point of view, the system may be conveniently represented by a dynamic Ising model. This is the case of ferromagnetic films with uniaxial magnetization, which can be consistently described as an Ising system that displays a frequency-induced dynamic phase transition when subjected to an oscillatory external magnetic field [279].

Consequently, let us consider a prototypical system to study dynamic phase transitions in a simple uniaxial magnetic material subjected to a time-dependent external field $h(t)$. For this system, the order parameter is the magnetization m, which is a non-conserved quantity. Therefore, the dynamics will be conveniently described by a model A dynamics given by Eq. 2.33, with a free energy functional suitable to describe the system considered. Assuming that magnetization is homogeneous, in a mean-field approximation, we can use the free energy function given by Eq. 1.70 corresponding to a ferromagnetic Ising system. Then the following dynamic equation is found:

$$\tau \frac{dm}{dt} = -m + \tanh\left(\frac{zJm + h(t)}{kT}\right), \tag{8.1}$$

where m is the magnetization per particle and τ is a relaxation time (related to the coefficient Γ_{nc}^{-1} in Eq. 2.33). This model represents the simplest model relevant to the study of a dynamic phase transition. It was introduced in Ref. [282] as the minimal extension of the mean-field approach to the equilibrium Ising model adequate to study this class of transitions. Note that in equilibrium, $dm/dt = 0$ and the magnetization is given as a solution of Eq. 1.71, as expected. It is insightful to analyse this dynamical equation when the external field is a sinusoidal field given by

$$h(t) = h_0 \sin \Omega t + h_b, \tag{8.2}$$

which permits to study the dependence of the response of the system on the frequency Ω. In the preceding expression, h_b is a bias field. It is clear that τ and Ω^{-1} are proportional to the two time scales that must be associated with relaxation and driving rate, respectively. The response of the system is determined by the competition of these two time scales and is thus controlled by the product $\tau\Omega$. It is convenient to define reduced variables such as $\tilde{t} \equiv t/\tau$, $\tilde{h} \equiv h/zJ$, $\omega = \tau\Omega$, and $\tilde{T} \equiv kT/zJ = T/T_c$, where T_c is the critical temperature of the equilibrium model. That is, $T_c = zJ/k$. Note that the field period $P = 2\pi/\Omega$ is then given as $P = \tilde{P}\tau$, where $\tilde{P} = 2\pi/\omega$. Thus, the equation of motion in the new dimensionless variables reads

$$\frac{dm}{d\tilde{t}} = -m + \tanh\left(\frac{m + \tilde{h}_0 \sin \omega\tilde{t} + \tilde{h}_b}{\tilde{T}}\right). \tag{8.3}$$

It is expected that for given initial conditions, after a certain transient time, the magnetization obtained by solving the preceding equation is a periodic function (not necessarily sinusoidal) of time with the same period as the external applied field but lagging behind the field due to the existence of a relaxation time τ. Indeed, the amount of lag between the magnetization and the field will depend on the product $\tau\Omega = \omega$. Consequently, when magnetization is represented as a function of the field, a loop will result with non-zero area associated with the existence of dynamic hysteresis.[1] In the reduced units, the loop area A will be given as

$$A = - \oint md\tilde{h}. \tag{8.4}$$

For small frequencies ($\omega \ll 1$), the area A should approach zero since the magnetization will be able to follow the external field. For increasing values of ω, hysteresis is expected to increase since the magnetization will need quite a long time to relax and thus will not be able to follow field oscillations. In this regime, for $h_b = 0$, the resulting hysteresis loops should be symmetric and, thus, centred at the origin of the m-h plane. Nevertheless, above a certain frequency threshold, which is expected to depend on the amplitude of the field and temperature, the loop area should start decreasing and, at the same time, becoming asymmetric, which means that it would be no longer centred at the origin of the m-h plane. In fact, it should occur either in the upper half-plane or in the lower half-plane. This striking change of behaviour reflects the existence of symmetry breaking induced by dynamical effects, which suggests the existence of a phase transition between a dynamical paramagnetic (DP) phase showing symmetric hysteresis loops and a dynamical ferromagnetic (DF) phase characterized by asymmetric hysteresis loops. Indeed, this transition is athermal, induced by the frequency of the external driving field. Numerical solution of Eq. 8.1 confirms this behaviour and results corresponding to $h_b = 0$ are depicted in Figure 8.1, where the symmetric and asymmetric time dependence of the magnetization are shown for low and high frequency, respectively. The corresponding evolution of the hysteresis loops with frequency is illustrated in Figure 8.2. Note that within this deterministic model, the occurrence of the loop in the upper or in the lower half-plane will depend on initial conditions.

It is important to compare the behaviour of the hysteresis shown in Figure 8.2 with experimental results. Figure 8.3 shows the frequency dependence of magnetic hysteresis loops measured in ultrathin ferromagnetic Fe/Au(001) samples[2] reported in Ref. [284]. Loops were obtained by

---

[1] We will only consider the hysteresis associated with the competition between relaxation and driving time scales with respect to equilibrium and will not take into account any extra term that may arise from the fact that thermal fluctuations are not operative.

[2] Films are constituted of few monolayers of single-crystalline Fe deposited on a bulk-terminated (1×1) structure of Au(100) (see Ref. [283]).

applying an oscillating magnetic field generated by a time-varying current, and magnetization was estimated by using the surface magneto-optic Kerr technique.[3] In Figure 8.4, the intensity of the Kerr signal (proportional to magnetization) is given as a function of the field frequency for a given amplitude of the field. It is clear that the evolution of the loops with increasing frequency shows the same qualitative behaviour as obtained from the dynamic model. In particular, results confirm that symmetry breaking occurs above a certain critical frequency, which in the experiments seems to be located slightly above 22.5 Hz. As in the model, the loop area first increases as frequency increases, reaches a maximum, which is close to the critical frequency, and then gently decreases to very low values. It is important to remark that for low values of the frequency and amplitude of the magnetic field, the loop area scales as

$$A \sim h_0^{\alpha_{hys}} \omega^{\beta_{hys}}, \qquad (8.5)$$

where α_{hys} and β_{hys} are critical exponents, which are temperature independent. Numerical studies indicate that in dimension $d > 2$, $\alpha_{hys} \cong 2/3$ and $\beta_{hys} \cong 1/3$, which are in agreement with values reported from experiments performed on Fe/Au(001) thin films in Ref. [279]. However, different values have been reported by other authors. For instance, in Ref. [285], much lower values, $\alpha_{hys} \cong 0.3$ and $\beta_{hys} \cong 0.06$, were determined. These discrepancies may suggest [275] that, in the experiments, the observed hysteresis is not entirely dynamic in origin, as confirmed by the fact that A does not go strictly to zero in the limit $\omega \to 0$.

Analytical mean-field treatment

A convenient order parameter Q to study this dynamical transition can be defined as the mean magnetization in a full field oscillation cycle, that is

$$Q \equiv \frac{1}{2\pi} \int_{\eta}^{\eta+2\pi} m(\eta')d\eta', \qquad (8.6)$$

where $\eta = \omega\tilde{t} = \Omega t$. Note that this order parameter vanishes in the dynamic paramagnetic phase while it is non-zero in the dynamic ferromagnetic phase. For very high frequencies, $|Q| \to 1$.

It is expected that the dynamical phase transition is continuous for high enough temperatures and, presumably, that it can become first order at low

---

[3] The magneto-optic Kerr effect (MOKE) is a microscopic technique that makes use of the changes of the light reflected from a magnetized surface to investigate the magnetization evolution of materials. This technique is especially useful for studying the magnetization in thin films.

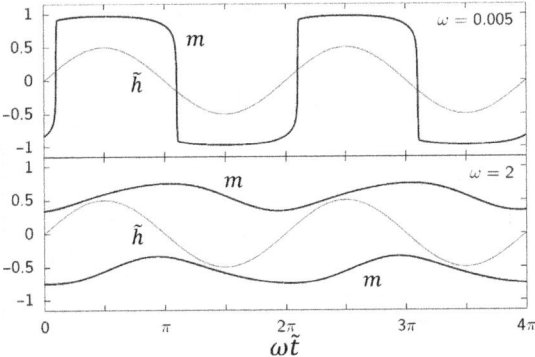

Figure 8.1 Time evolution of the magnetization *vs.* time for two values of the reduced field frequency $\omega = 0.005$ and 2, with $\tilde{h}_0 = 0.5$, $\tilde{T} = 0.7\, T_c$, and $h_b = 0$. In the lower panel, corresponding to $\omega = 2$, the two magnetization curves correspond to results obtained with initial conditions, $m = 1$ and $m = -1$, respectively.

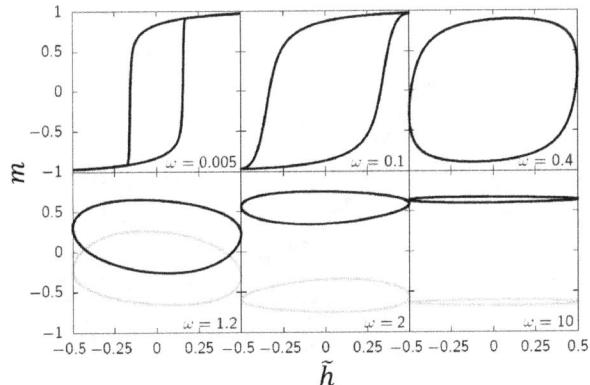

Figure 8.2 Hysteresis loops obtained numerically for increasing values of the frequency ω from 0.005 to 10 with $h_0 = 0.5$, $\tilde{T} = 0.7$, and $h_b = 0$. In the lower panels, corresponding to frequencies $\omega = 1.2, 2$ and 10, upper and lower cycles are obtained with initial conditions, $m = 1$ and $m = -1$, respectively.

temperatures, thus showing a discontinuous jump in the order parameter. When the transition is continuous, below the critical period, $\tilde{P} < \tilde{P}_c$, the magnetization should oscillate about a given finite value so that the order parameter Q is non-zero. As this finite value must tend to zero as $P \to P_c$, it is reasonable to assume that $m(\eta)$ has the following general form:

$$m(\eta) = Q + \xi(P, \eta), \tag{8.7}$$

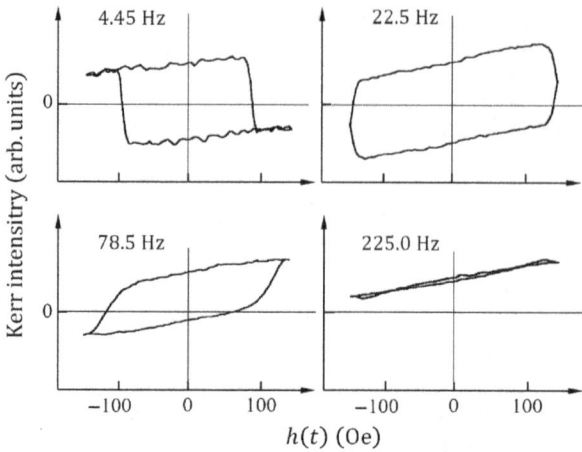

Figure 8.3 Hysteresis loops measured at selected frequencies in a thin film of Fe grown on a gold surface using the magneto-optical Kerr method. The magnetic field amplitude is $h_0 = 145$ Oe. Adapted from Ref. [284].

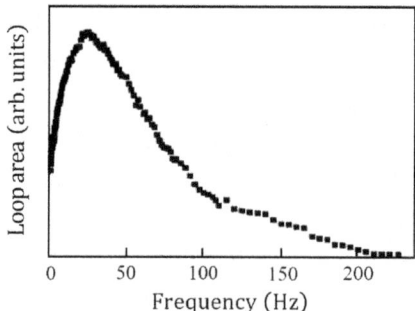

Figure 8.4 Area of the hysteresis loops as a function of frequency measured at room temperature using the magneto-optical Kerr effect method in a thin film of Fe grown on a gold surface. The magnetic field amplitude is $h_0 = 145$ Oe. Adapted from Ref. [284].

where $\xi(P, \eta)$ represents the oscillatory contribution to the magnetization that must have a vanishing time average at the critical point. Furthermore, in the dynamic paramagnetic phase, ξ must oscillate symmetrically about zero thus satisfying $\xi(P > P_c, \eta + \pi) = -\xi(P > P_c, \eta)$. Following the procedure given in Ref. [286], Eq. 8.7 is inserted in the dynamic Eq. 8.3 and, taking into account that the order parameter is small close to a critical point, the resulting equation is expanded in a power series of Q, which leads to

$$\omega \frac{d\xi(\eta)}{d\eta} = -Q + A_0 + A_1 \frac{Q}{\tilde{T}} + A_2 \frac{Q^2}{\tilde{T}^2} + A_3 \frac{Q^3}{\tilde{T}^3} + \dots \tag{8.8}$$

The coefficients A_i are functions of η, $\tilde{T}$, $\tilde{h}_b$, and $X(\eta) \equiv (\xi + \tilde{h}_0 \cos \eta)/\tilde{T}$. Then, integration of the preceding Eq. 8.8 yields the expansion

$$\tilde{h}_b = a_2 Q + a_4 Q^3 + a_6 Q^5 + \dots, \tag{8.9}$$

where the coefficients are given as

$$a_2 \equiv \left[\frac{2\pi\tilde{T} - I_1(P)}{I_1(P)}, \right], \tag{8.10}$$

$$a_4 \equiv \frac{1}{\tilde{T}^2} \left[\frac{I_2(P)}{I_1 P)} \left(\frac{2\pi\tilde{T} - I_1(P)}{I_1(P)} + \frac{1}{3} \right) \right], \tag{8.11}$$

$$a_6 \equiv \frac{1}{\tilde{T}^4} \left[\frac{I_3(P)}{60 I_1(P)} + \frac{1}{3} \left(\frac{I_2(P)}{I_1(P)} \right)^2 \right]. \tag{8.12}$$

In the preceding expression $I_1(P)$, $I_2(P)$, $I_3(P)$, ..., are the following integrals

$$I_1 = \int_\eta^{\eta+2\pi} \text{sech}^2[X(\eta')]d\eta',$$

$$I_2 = \int_\eta^{\eta+2\pi} \{2 - \cosh[2\,X(\eta')]\}\text{sech}^4[X(\eta')]d\eta'.$$

$$I_3 = \int_\eta^{\eta+2\pi} \{26\cosh[2X(P,\eta')] - \cosh[4X(P,\eta')] - 33\}\sec^6[X(\eta')]d\eta'.$$

Equation 8.9 plays the role of the equation of state of the dynamical system. While the coefficient a_6 is positive, the expansion can be limited to fifth-order in Q. In this case, the equation of state can be assumed to derive from the following Landau-type pseudo-free energy function:

$$F_d = \frac{1}{2} \left[\frac{2\pi\tilde{T} - I_1(P)}{I_1(P)} \right] Q^2 + \frac{1}{4} \left[\frac{I_2(P)}{3\tilde{T}^2 I_1(P)} \right] Q^4 + \frac{1}{6}Q^6 - \tilde{h}_b Q. \tag{8.13}$$

For a zero applied bias field, taking into account the general Landau theory for first-order transitions discussed in Section 1.5.2, while the coefficient of the Q^4 term is positive, this free energy confirms the existence of a critical point at $a_2 = 0$, which determines the critical period, P_c. Nevertheless, if the coefficient $a_4 < 0$, a first-order transition can occur. Therefore, a tricritical

point is expected to be present characterized by the simultaneous fulfilment
of the conditions, $a_2 = a_4 = 0$, which lead to

$$I_1(P_t) = 2\pi \tilde{T}, \tag{8.14}$$

$$I_2(P_t) = 0, \tag{8.15}$$

with $I_3(P_t) > 0$. In these expressions, P_t denotes the tricritical period. In
Figure 8.5, the phase diagram in a frequency-temperature plane for two
different values of the magnetic field amplitude is depicted. In each case, the
corresponding location of the tricritical point is indicated.

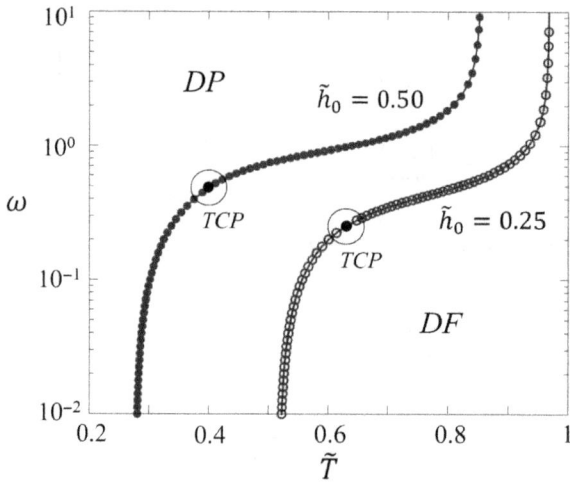

Figure 8.5 Phase diagram in a frequency-temperature space corresponding
to $\tilde{h}_0 = 0.25$ and $\tilde{h}_0 = 0.5$. Note that the frequency axis is given on a
logarithmic scale. The tricritical point (*TCP*) is indicated on both curves.
For frequencies above the tricritical frequency, the transition from the dynamical paramagnetic (*DP*) to the dynamical ferromagnetic (*DF*) phase is
continuous, while it is first-order for lower frequencies.

The behaviour described by the preceding mean-field analysis has been confirmed by the numerical solution of Eq. 8.8. It must be pointed out that
both, continuous and first-order transitions, have been also reported from
Monte Carlo simulations of the full dynamical model in Ref. [275]. It is worth
mentioning that the existence of a first-order transition has been later put
into question by some authors who claim that this is merely an artefact arising from finite size effects in the numerical simulations. In any case, what
cannot be denied is the fact that the dynamic mean-field model shows a
tricritical point. In favour of these results, it must be indicated that some

evidence of the existence of a first-order dynamical transition in an Fe ferro-magnetic thin film has been reported to occur in the low frequency regime in Ref. [285].

8.2 Disorder-induced transitions

In the quasi-static athermal limit, when $\tau_d \ll \tau_h \ll \tau_{tf}$, thermal fluctu-ations play a very minor role on the phase transition dynamics and, as already discussed, when externally driven, this class of systems remains in a metastable minimum until the limit of metastability or unstable spinodal is reached. Whereas this problem is well understood in *clean* systems, the presence of quenched disorder drastically changes the physical behaviour. Actually, in this situation, the study of the relaxation of a metastable state becomes much more interesting. This phenomenon has been reported for a wide variety of non-equilibrium systems, where it has been shown that extreme events such as macroscopic avalanches or rupture phenomena can occur [287, 288]. A prototypical example where such a behaviour has been re-ported is disordered magnets. We have discussed in Chapter 7 that in these materials, disorder has the effect of drastically modifying the free energy landscape by introducing multiple metastable states separated by large free energy barriers. When driven by an externally applied magnetic field, only when the system becomes locally unstable at a local spinodal, relaxation occurs toward another metastable state. This is schematically represented in Figure 8.6. The corresponding process is a collective phenomenon that may involve many local regions and gives rise to an avalanche. It must be pointed out that, strictly speaking, a local spinodal cannot exist at finite temperature since thermal fluctuations should destroy metastability before reaching the spinodal instability through nucleation of the new phase. Nev-ertheless, in many real systems, at low enough but still finite temperatures, the description based on such ideas is very reasonable since fluctuations are known to play a very irrelevant role and, thus, can be neglected.

The physics in this class of systems is dominated by avalanches, which is reflected in the shape of the hysteresis cycle associated with the switching of magnetization induced by the magnetic field. At a temperature below the Curie temperature, as the amount of quenched disorder increases, the hys-teresis loop associated with the switching of the magnetization is expected to change its shape from *sharp* to *smooth* at a certain amount of disorder at which an infinite avalanche should occur. In fact, this is expected to happen at a non-equilibrium critical point, at which avalanches of all sizes should occur characterized by a power-law distribution reflecting emergent scale in-variance. In magnetic materials, avalanches can be detected as Barkhausen

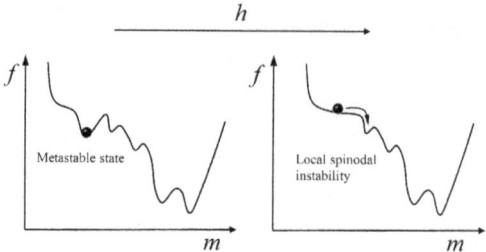

Figure 8.6 Schematic representation of the complex energy landscape of a system with quenched disorder. The relaxation from a local metastable state to another occurs when a local spinodal is reached as the external field is increased.

noise pulses, which originate from the flipping of magnetic domains induced by the applied magnetic field. These pulses can be picked up as voltage pulses measured in a coil surrounding the magnetic material [289]. An example of Barkhausen noise recorded during a magnetization switching in a $Fe_{85}B_{15}$ amorphous magnetic ribbon is shown in Figure 8.7.

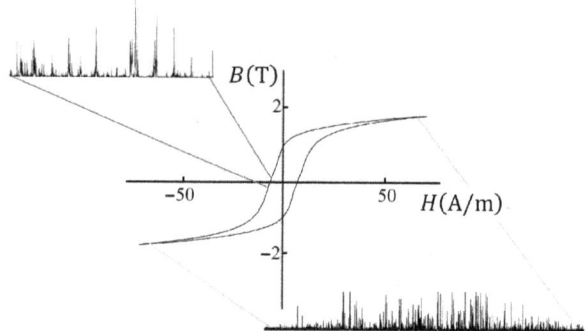

Figure 8.7 Example of Barkhausen noise detected during a magnetization switching loop in a $Fe_{85}B_{15}$ amorphous ribbon. The hysteresis loop is also shown in a magnetic induction *vs.* magnetic field diagram. The bottom-right noisy signal is the Barkhausen noise recorded during the increasing magnetic field path and the top-left signal is an enlargement of the noise detected around the coercive field. In the experiment the sample was subjected to a moderate tensile stress in order to enhance the signal-to-noise ratio. Adapted from Ref. [290].

The Ising model with quenched disorder is the simpler model adequate to describe this class of non-equilibrium transitions. For the study of avalanches, the random field Ising model (RFIM) with ferromagnetic nearest-neighbor interaction of the spins, as introduced in Eq. 7.15, is the most popular version. In this model, quenched disorder is taken into account by a distribution of random fields $\{h_i\}$, which are defined on each lattice site i with a given distribution $\rho(h_i)$, which is often assumed to be gaussian, that is

$$\rho(h_i) = \frac{1}{\sqrt{2\pi}R} \exp\left(-\frac{h_i^2}{2R^2}\right), \tag{8.16}$$

where the width R of the distribution, or disorder parameter, provides a measure of the amount of disorder. For a given value h of the external applied field, the local field F_i acting on site i will be given by

$$F_i = -\frac{\partial \mathcal{H}}{\partial S_i} = J\sum_j S_j + h + h_i, \tag{8.17}$$

where the sum extends over the z nearest neighbours of the spin i. We will assume that a local spinodal is reached at site i when $F_i = 0$. Taking into account this local instability condition, the following deterministic field-driven dynamics can then be introduced. Consider, for instance, that the system is subjected to a very intense negative magnetic field so that all spins are pointing down ($S_i = -1, \forall i$). As the field is quasi-statically increased, at some value of the external field, a local spinodal at site i will be reached at which the effective local field will vanish. Then, the spin at site i will flip and, due to the interaction between spins, neighbouring spins can become unstable. Consequently, the flipping of one spin can induce the flipping of neighbouring spins, which in turn might push other neighbours to flip, and so on until all the spins in the system become stable. The process can thus give rise to an avalanche. The quasistatic change of the field supposes that the field changes in time slowly so that it remains constant during the avalanche process. In this sense, this zero temperature dynamics is also denoted as *adiabatic*.

The described model must reproduce the expected change of shape of the hysteresis loop induced by disorder. Indeed, for a large amount of disorder, the distribution of random fields is wide and only small avalanches are expected. Thus, the magnetization *vs.* magnetic field curves defining the hysteresis loop will be smooth in the thermodynamic limit. On the other hand, the distribution of random fields will be narrow for low disorder and large avalanches should occur. At the critical point, corresponding to an amount of disorder R_c and external field h_c, in the thermodynamic limit, an infinite avalanche will occur for the first time, and the magnetization will show a discontinuity. Therefore, as the amount of disorder is decreased, at this point, the hysteresis loop will change from smooth to a sharp one. Close to this critical point, scale invariance associated with avalanches of all sizes will occur. Therefore, criticality is expected to manifest in the behaviour of the hysteresis loop and in the distribution of avalanche sizes.

The dynamical behaviour previously described allows for the occurrence of athermal nucleation at local spinodals and gives rise to a critical point defined by a critical amount of disorder and a critical field. It is important to note that this behaviour must be distinguished from that associated with the propagation of an interface in a disordered medium. In this case, as the field approaches a critical value, larger and larger Barkhausen avalanches occur that originate from depinning processes, which may develop a power law size distribution. Above a given critical field, the front will be completely depinned thus moving smoothly with an inhomogeneous velocity [291]. This problem will not be discussed in this chapter.

Mean-field solution

A mean-field solution can be obtained [287, 292] that reproduces the critical point at given values of disorder and external magnetic field. In this approximation, we replace the local field F_i in Eq. 8.17 by an effective field $F_i^{eff} = Jm + h_i + h$. This assumes that each spin is coupled to all other spins through the same effective exchange interaction J/N.[4] Therefore, the effective field acting on each spin depends linearly on the magnetization per spin, $m = \sum_i S_i/N$, instead of the actual local configuration of neighbouring spins. Thus, within this framework, the spin configuration of the system is given by the rule

$$S_i = -1 \quad \text{if} \quad F_i^{eff} < 0 \Leftrightarrow \quad h_i < -Jm - h, \tag{8.18}$$

while

$$S_i = +1 \quad \text{if} \quad F_i^{eff} > 0 \Leftrightarrow \quad h_i > -Jm - h. \tag{8.19}$$

Let us now analyse how the magnetization evolves when the external applied field increases from $h = -\infty$ to $h = \infty$. In the initial state, all the spins are pointing down and, thus, the magnetization takes the saturation value $m = -1$. As the field takes a given finite value h, according to Eq. 8.18, the field dependence of the magnetization will be a solution of the following integral expression:

$$m(h) = 1 - 2N_-(h), \tag{8.20}$$

---

[4] If only the nearest-neighbour spin exchange coupling is considered, J should be replaced by zJ, where z is the lattice coordination number.

where

$$N_-(h) = \int_{-\infty}^{-(h+Jm)} \rho(h_i)dh_i. \tag{8.21}$$

The preceding Eq. 8.20 plays the role of the mean-field equation of state of the system. Note that one can proceed starting from a saturated state with all the spins pointing up and decreasing the external field from $h = \infty$ to $h = -\infty$. In this case, the corresponding mean-field equation of state reads

$$m(h) = -1 + 2N_+(h), \tag{8.22}$$

where $N_+(h) = \int_{-(h+Jm)}^{\infty} \rho(h_i)dh_i$. If the distribution $\rho(h_i)$ is assumed gaussian as given by Eq. 8.16, then it is straightforward to see that the equation of state can be expressed as

$$m(h) = \mathrm{erf}\left[\frac{(h + Jm)}{\sqrt{2}R}\right], \tag{8.23}$$

where $\mathrm{erf}(x) = \frac{2}{\sqrt{\pi}}\int_0^z e^{-t^2}dt$, is the error function.[5] Taking into account that the magnetization curves must be antisymmetric since the magnetization must change $m \to -m$ when $h \to -h$, the same equation describes the increasing and decreasing magnetic field processes starting from the corresponding saturated states.

It is worth noting that the obtained mean-field equation of state has the same form as the equilibrium equation of state of the *clean* Ising model (see Eq. 1.71), after replacing the error function with the hyperbolic tangent function. Therefore, the disorder parameter R plays the role of temperature in the equilibrium case. This analogy with the clean equilibrium case indicates that a critical point must exist at a critical amount of disorder $R_c = \sqrt{2/\pi}J$ and $h = 0$. Equation 8.23 has a single-valued solution for $R > R_c$. In this case, the m-h diagram does not display hysteresis. At $R = R_c$ hysteresis and infinite avalanche behaviour begin. For $R < R_c$, the magnetization jump Δm associated with the infinite avalanche occurs at the coercive fields, $\pm h_{coer} \neq 0$.

Expanding the error function in Eq. 8.23 near $h = m = 0$, one obtains,

$$h = \sqrt{\frac{\pi}{2}}(R - R_c)m + \frac{\sqrt{2\pi^3}}{24}R_c m^3 + \dots . \tag{8.24}$$

---

[5] In order to express the equation of state in terms of the error function, it is useful to remember that

$$\mathrm{erf}(x) = 1 - \mathrm{erfc}(x) = 1 - \frac{2}{\sqrt{\pi}}\int_{-x}^{\infty} e^{-t^2}dt,$$

where $\mathrm{erfc}(x)$ is the complementary error function.

From this equation, it is straightforward to see that close to the critical point, the magnetization jump scales as

$$\Delta m \sim |r|^{\beta}, \tag{8.25}$$

where $r = (R - R_c)/R_c$ and $\beta = 1/2$. The susceptibility is found to diverge as, $\chi = dm/dh \sim |r|^{-\gamma}$, with $\gamma = 1$, and, at $r = 0$, $h \sim m^{\delta}$, with $\delta = 3$. In this case, close to the critical point, the susceptibility scales with the field as, $\chi(r = 0, h \to 0) \sim h^{-2/3}$. The obtained exponents are, as expected, mean-field exponents.

It is worth mentioning that the specific nature of the mean-field solution of the problem depends strongly on the actual distribution of disorder [292]. For instance, no critical point exists for a uniform distribution of disorder. This is not a surprising result since the uniform distribution may be viewed as an infinitely broad gaussian distribution. For a bimodal distribution of disorder, $\rho(h_i) = [\delta(h_i + R) + \delta(h_i - R)]/2$, the problem can be reduced to a two-body problem. Half of the spins with $h_i = R$ flip upwards at the coercive field $h = h_{co_1} = J - R$. This further triggers the flipping of the remaining spins if their updated local field $h_{co_1} - R$ is positive, which supposes that $R < J/2$. This gives rise to a global switch from $m = -1$ to $m = +1$ at h_{co_1}. In the high-disorder case, when $R > J/2$, the magnetization remains zero until the local field changes sign at $h = h_{co_2} = R$. In contrast with the gaussian case, in this case the highly disordered phase exhibits hysteresis.

In general, when disorder-driven hysteresis criticality occurs, in mean-field and beyond, it is expected that the magnetization as a function of r and h satisfies the following scaling equation:

$$m(r, h) \sim |r|^{\beta} \mathcal{M}_{\pm} \left[\frac{h}{|r|^{(\beta + \gamma)}} \right], \tag{8.26}$$

where $\pm$ refers to the sign of r. It must be noted that the scaling relation $\beta + \gamma = \beta\delta$ must be fulfilled. The preceding scaling equation is analogous to the corresponding equilibrium scaling equation given by Eq. 1.150. In fact, based on symmetry arguments, it has been argued that hysteresis criticality should be in the same universality class as the corresponding equilibrium criticality [293].

There has been some interest in looking for experimental systems displaying the above-discussed scaling behaviour. From an experimental point of view, this is not an easy task due to the difficulty in controlling the amount of disorder. Anyway, a couple of interesting examples have been reported. The first is Co/CoO bilayers [294], and the second is a Cu-Al-Mn alloy in its spin glass phase [295].

In Co/CoO bilayers, Co is ferromagnetic while its native oxide is antiferromagnetic with a Néel temperature T_N lower than the Curie temperature of Co. Due to the coupling between the two layers, disorder on the ferromagnetic layer increases when temperature is decreased below the Néel point of the oxide, which is a consequence of the appearance of antiphase domains.[6] This is schematically represented in Figure 8.8. Consistently, ferromagnetic hysteresis loops change from sharp to smooth when temperature is decreased. Therefore, the amount of disorder can be controlled by changing the temperature from above to below the Néel point. Consequently, it is assumed that $r \sim (T - T_c)/T_c$, where T_c is the temperature at which, when temperature is decreased, the ferromagnetic hysteresis cycle changes from sharp (low disorder) to smooth (high disorder). The reduced field is defined as, $h_r = (h - h_{co})/h_{co}$, where h_{co} is the coercive field at a given temperature. A scaling plot is shown in Figure 8.8 from which exponents $\beta = 0.022 \pm 0.006$ and $\beta\delta = 0.30 \pm 0.03$ are found.

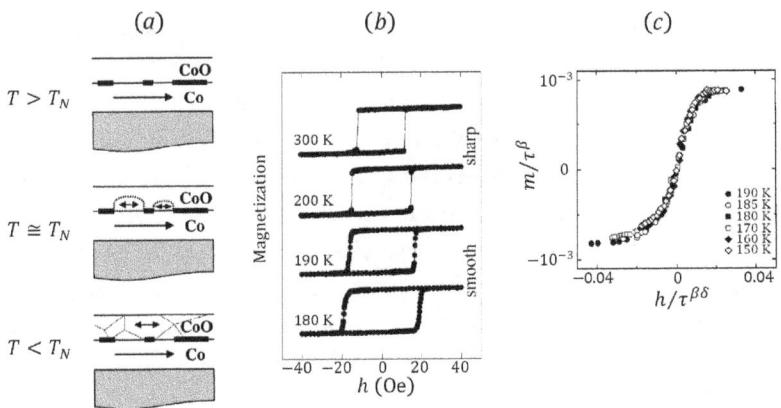

Figure 8.8 (a) Schematic illustration of the magnetic state in Co/CoO-bilayers for $T > T_N$, $T \cong T_N$, and $T < T_N$, where T_N is the Néel temperature of the antiferromagnetic CoO layer. Antiferromagnetic domains appear in this layer below T_N. (b) Ferromagnetic hysteresis loops measured at the temperatures indicated in the figure. By decreasing the temperature, close to $T = 200$ K, the loops change from sharp to smooth. (c) Scaling plot where m/τ^β vs. $h/\tau^{\beta\delta}$ is represented using data measured at the indicated temperatures, which correspond to the region of high disorder. Note that $\tau = (T - T_c)/T_c$ is a measure of the amount of disorder, r. Adapted from Ref. [294].

---

[6] The oxide is homogeneous above the Néel temperature T_N, while domains appear below the antiferromagnetic phase transition.

Similarly, in the case of the Cu-Al-Mn alloy, in the glassy phase, below the freezing temperature, hysteretic magnetic loops occur. Close to the freezing temperature the magnetic loops are smooth but below a given temperature the magnetization shows a discontinuity at a coercive field. This change of behaviour has also been associated with a temperature-controlled disorder-induced transition. In this case, disorder has the origin in the coexistence of ferromagnetism and antiferromagnetism, which originates from the existence of misplaced Mn atoms.[7] In this case, scaling behaviour has also been confirmed and the corresponding critical exponents are $\beta = 0.03 \pm 0.005$ and $\beta\delta = 0.4 \pm 0.1$. Within the error bars, these exponents are consistent with those reported for Co/CoO bilayers, but different from the predicted mean-field exponents. This indicates that, in general, disorder-induced transitions are non-mean-field transitions.

8.2.1 Beyond mean-field

The mean-field approach discussed previously in the case of RFIM with a gaussian distribution of disorder (random fields) predicts no hysteresis for $R \geq R_c$. Only for $R < R_c$, the model gives rise to hysteresis associated with the infinite avalanche, which causes the magnetization discontinuity (or jump) that occurs at the coercive field. This is not a realistic behaviour that seems to be an artefact of the mean-field approach. On the other hand, in addition to this fact, the approach to the infinite avalanche by varying the external field for $R < R_c$ is continuous, while avalanches, which have been reported to occur in many different materials undergoing a phase transition, should be expected.

Avalanche criticality

In general, the study of avalanches associated with disorder-driven transitions must be performed numerically. In the case of the RFIM with Gaussian disorder, the numerical procedure is implemented aimed at simulating the deterministic zero temperature (athermal) adiabatic dynamics described above. Therefore, one starts by assigning a random field to each spin located at the sites of a given lattice that usually is a hypercubic lattice in dimension d. Initially the field has, for instance, a very large negative value and all the spins point down. Then, the field is increased in very small steps until a new site is found at which the local field $F_i \leq 0$. Subsequently, the spin at this

---

[7] In Cu-Al-Mn magnetic moments are located at Mn sites. Below a certain temperature and for a high enough applied magnetic field, the spins are aligned and an incipient ferromagnetism occurs.

site is flipped and the stability condition at neighboring sites is checked. If one or more of these spins become unstable they are flipped,[8] and the process continues until all the spins are stable. Next, the field is increased again and the process starts all over again. The total number of spins that flip at each step defines the avalanche size, s. A numerical algorithm that exactly follows this procedure can be easily implemented and examples of avalanches obtained in this way are shown in Figure 8.9. However, algorithms of this kind are very time and memory consuming and more sophisticated methods have been proposed to efficiently study avalanche dynamics [300].

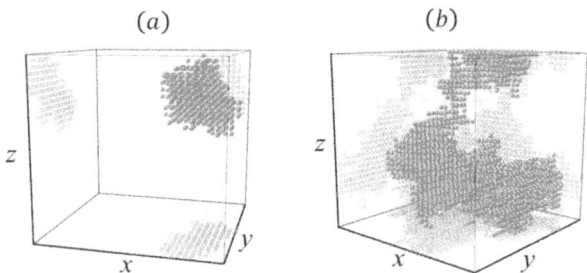

Figure 8.9 Examples of (a) a non-spanning avalanche and (b) an avalanche that spans along the three space dimensions in an $L \times L \times L$ $3d$ simple cubic lattice ($L = 32$, in units of the lattice parameter). Dots represent the spins that have flipped at the represented avalanche. Adapted from Ref. [296] with permission from the author.

Probability distributions of avalanches, sizes integrated over all fields, $D(s)$, obtained from numerical simulation results of a RFIM defined on a $3d$ cubic $L \times L \times L$ lattice with gaussian disorder are shown in Figure 8.10. Results confirm the existence of a critical amount of disorder, $R = R_c$, for which the distribution is power law, characterized by a critical exponent τ. That is, in this case, the probability distribution behaves as

$$D(s) \sim s^{-\tau}. \tag{8.27}$$

In $3d$, these simulations lead to $R_c \simeq 2J$ and $\tau \simeq 2$.

For low disorder, $R < R_c$, the distribution is subcritical and shows a peak about $s/L \sim 1$, which indicates the existence of spanning avalanches. These avalanches propagate along the full system in, at least, one dimension. Note that only in the thermodynamic limit, an infinite avalanche may occur. For large disorder, the distribution is supercritical characterized by a curvature (in a log-log representation) associated with a deficiency of large avalanches.

---

[8] Usually, all the unstable spins are flipped simultaneously. In this case, the dynamics is called *synchronous*.

It is also possible to analyse the distributions of durations[9] and energies of the avalanches that also exhibit a behaviour similar to that of the avalanche sizes, but characterized by different critical exponents at $R = R_c$. The power law behaviour of these distributions indicates that the avalanche process occurs with the absence of size and time scales. Therefore, critical exponents must satisfy exponent equalities [301]. It is insightful to remark that the exponent τ can be expressed as $\tau = \tau' + \sigma\beta\delta$, where τ' is the critical exponent of the avalanche size distribution in the neighbourhood of the coercive field (critical field) and σ^{-1} is an exponent associated with the cutoff of the integrated avalanche size distribution. Values reported from numerical simulations indicate that both τ' and τ do not significantly depend on the space dimension. In this sense, note that τ' takes values[10] about 1.5 and τ about 2 in $3d$, while the corresponding mean-field values are $3/2$ and $9/4$, respectively [293].

Avalanche criticality has been reported for a number of materials, usually associated with a ferroic phase transition, which displays athermal behaviour to a good approximation. In ferromagnetic materials, we have already mentioned that avalanches can be detected from Barkhausen noise measurements. Old results reported in Ref. [297] already showed that Barkhausen pulse areas integrated over the hysteresis loop in a $Ni_{80}Fe_{20}$ compound display a power-law distribution with an exponent τ in the range from 1.7 to 2, which is in quite good agreement with predictions based on numerical simulations of the athermal RFIM. Power-law behaviour of the distribution of avalanches has been reported more recently for many different soft ferromagnetic compounds. Most of these results are summarized in Ref. [298]. It must be taken into account that often, long-range interaction effects are relevant and the Barkhausen noise is dominated by domain wall propagation rather than nucleation. In these cases, the reported exponents must be lower than those theoretically predicted from the simulations of RFIM.

In addition to Barkhausen results, experiments aimed at imaging avalanches have also been performed by means of magneto-optic Kerr effect (MOKE) and magnetic force microscopy (MFM). In Ref. [299], visualization of the Barkhausen effect has been reported in granular thin films with perpendicular anisotropy.[11] The distribution of avalanche sizes estimated in this way yields critical exponents consistent with those obtained from Barkhausen measurements.

In ferroelectrics, such as lead zirconate titanate, PZT, $BaTiO_3$, or some relaxors, polar avalanches can be detected from voltage pulses during

---

[9] When the dynamics is synchronous, the duration of the avalanches can be assumed to be proportional to the number of steps that take place during an avalanche.

[10] Note that for RFIM, the upper critical dimension is $d = 6$.

[11] The studied sample was an 80-nm thick $La_{0.7}Sr_{0.3}MnO_3$ film, epitaxially grown on a $LaAlO_3(001)$ substrate by pulsed laser deposition.

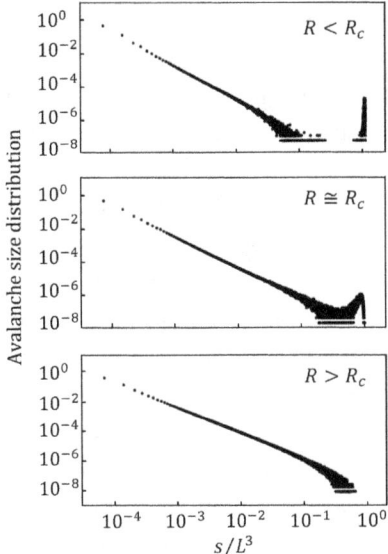

Figure 8.10 Probability distribution of avalanche sizes in log-log representation obtained by numerical simulation of an RFIM with gaussian distribution of the random fields and deterministic zero temperature adiabatic dynamics in an $L \times L \times L$ $3d$ simple cubic lattice ($L = 24$, in units of the lattice parameter) for $R < R_c$, $R \cong R_c$ and $R > R_c$. Data has been obtained after averaging over 10^5 realizations of disorder. Power-law behaviour is obtained at the critical amount of disorder $R = R_c \simeq 2$. The peak in the distribution centred at $s/L^3 \simeq 1$ corresponds to the distribution of spanning avalanches. Adapted from Ref. [296] with permission from the author.

electric field-induced switching of the polarization [302]. Results are similar to those reported for magnetic materials. In ferroelectrics, polar pulses are accompanied by acoustic emission signals generated by local changes of the deformation field that occur during polarization reversal. Thus, avalanches can also be detected and characterized with the acoustic emission pulses. Then, the obtained exponents are consistent with those obtained from electric Barkhausen detection of avalanches [303].

The measurement of the acoustic emission is the most commonly used experimental technique adequate to detect avalanches in ferroelastic and martensitic transitions. The transition in this class of materials shows usually an excellent athermal behaviour. In general, avalanches are detected by thermally inducing the transition and their distribution often occurs with the absence of characteristic scales. Nevertheless, this is not always the case and it has been shown that in the case of a single-crystalline material that is prepared with a very low density of defects, the transition occurs very sharply in the first transition cycle with the acoustic emission located in a very narrow temperature range, with a subcritical distribution of avalanche sizes. As the material is cycled back-and-forth across the transition, the acoustic emission occurs within a much broad temperature range and, after a certain number of cycles, it becomes reproducible from cycle to cycle to a very good approximation. It is at this stage when the avalanche size distribution displays a power law behaviour [304]. This result has been interpreted in the sense that during cycling disorder effectively increases due to the generation of dislocations and the system is able to find an optimal path with lower energy barriers separating metastable states, at which avalanche criticality occurs. This means that disorder cannot be considered as strictly quenched. The evolution of disorder induces a training effect where the system evolves from a subcritical behaviour to a critical steady state situation during the cycling process. In Ref. [305], avalanche criticality models have been generalized to account for this training process that self-drives the system to the critical state.

It is important to note that both experiments and numerical simulations indicate that in ferroelastic/martensitic transitions, avalanche critical exponents seem to depend on the symmetry reduction at the phase transition, that is, on the variant multiplicity, which is given by the ratio of symmetry operations in the parent and ferroelastic phases [306]. This suggests that by increasing the variant multiplicity, at each transformation step within the two-phase coexistence region, the system is able to find more pathways connecting the high- and the low-symmetry phases and, consequently, the probability that large avalanches occur should decrease with respect to the probability of small avalanches, which should be reflected in the critical exponents. The situation is somewat analogous to the equilibrium case where universality classes are established by space dimension and symmetry properties of the order parameter.

Figure 8.11 shows examples of the distribution of avalanches detected in ferromagnetic, ferroelectric, and ferroelastic materials. In the first case, Barkhausen avalanches were detected during field-induced magnetization switching in an Fe 7.8%wt Si polycrystalline alloy. For the ferroelectric case, polar avalanches were recorded during the temperature-induced ferroelectric transition in $0.7Pb(Mg_{2/3}Nb_{1/3})O_3$-$0.3PbTiO_3$ crystals. In the ferroelastic Fe-Pd alloys, acoustic emission experiments were performed in the region of its ferroelastic transition. In all cases, a good power law behaviour over several decades is depicted, which confirms the occurrence of avalanche criticality associated with the corresponding transitions in the three different classes of ferroic materials.

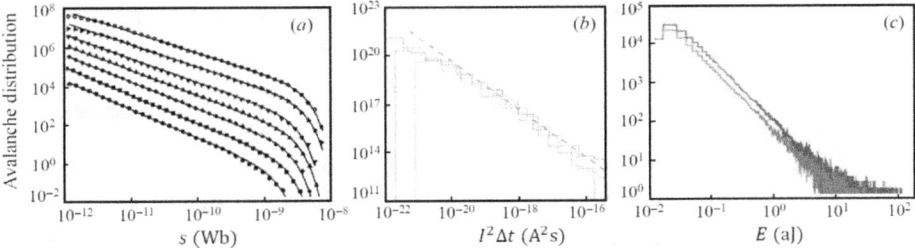

Figure 8.11 Histograms showing the avalanche distribution detected in ferromagnetic (*a*), ferroelectric (*b*) and ferroelastic (*c*) materials in log-log representations. In (*a*) the plot represents the size distribution of Barkhausen avalanches during magnetic field-induced magnetization switching in an Fe;7.8%wt Si polycrystalline alloy. The curves from top to bottom were obtained at magnetic field sweeping frequencies of 5, 10, 20, 30, 40, 50, and 60 mHz. The slope of the linear part of the curves (critical exponent) shows some dependence (approximately linear) on the field frequency, which is probably associated with the effect of avalanche overlapping with increasing frequency. Adapted from Ref. [298]. In (*b*) the histogram determines the energy distribution of polar avalanches ($I^2\Delta t$ is proportional to the energy of the avalanches). Avalanches were detected from thermally stimulated depolarization current measurements during the temperature induced ferroelectric transition in PMN-PT ($0.7Pb(Mg_{2/3}Nb_{1/3})O_3$-$0.3PbTiO_3$) crystals. The three curves were obtained in samples with different crystallographic orientations. Adapted from Ref. [307]. In (*c*) the histograms represent the distribution of the energy of acoustic emission avalanches recorded during the temperature induced ferroelastic/martensitic transition in a $Fe_{68.8}Pd_{31.2}$ single-crystal alloy. The top curve was obtained during the transition on cooling and the bottom curve during the reverse transition on heating. Adapted from Ref. [308].

Hysteresis properties and memory effects

In contrast to the mean-field solution discussed previously, numerical results of magnetization *vs.* magnetic field show hysteresis for high, critical, and low disorder. Hysteresis properties often display an interesting behaviour associated with the so-called *return point memory* property [309]. This property refers to the fact that every time that the field rate is reversed and partial loop is completed, the magnetization returns to its original value when the larger loop is rejoined. Therefore, no gap or crossing occurs in the hysteresis paths. More generally, this means that if in an increasing (decreasing) field branch, the field undertakes a quasi-static excursion $h(t)$, not necessarily monotonic, in the interval 0 to $\bar{t}$, bounded in the range $h(0) \leq h(\bar{t})$ [$h(0) \geq h(\bar{t})$], then the final state depends only on $h(\bar{t})$ and is independent of

$\bar{t}$ or history $h(t)$. In particular, a system coming back to a previous extremal field will return to exactly the same state, provided that the field remains within the bounds. Return point memory property has been reported to occur to a quite good approximation in a number of magnetic materials [310]. It can be reproduced by the RFIM with zero temperature adiabatic dynamics. In this case, not only the magnetization is recovered but the entire state of the system is identical to the state it had when the field was reversed in the larger loop. This is illustrated in Figure 8.12.

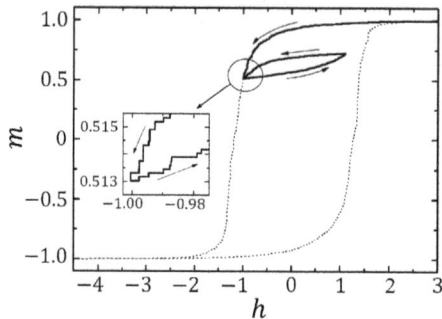

Figure 8.12 Illustration of the return point memory property. Results are obtained from numerical simulation of the RFIM with zero temperature adiabatic synchronous dynamics. The external loop (discontinuous line) is the saturation hysteresis cycle obtained by varying the field from $h = +\infty$ to $h = -\infty$ and back again to $h = +\infty$. The internal loop is obtained by reversing the sign of the field rate at a given state and completing a partial loop at which the field returns to the same field when the field was reversed. The inset is a zoom of the region where the field is first reversed and returns back to the same value. The result clearly demonstrates that the model reproduces the return point memory property. Adapted from Ref. [296] with permission from the author.

It has been rigorously demonstrated in Ref. [287] that the return point memory property must be accomplished by any system that satisfies the following three conditions:

(i) *A partial ordering of states.* This supposes that a given microstate A is *ahead* of another microstate B, if every spin in A is greater than or equal to the corresponding spin in B.

(ii) *No passing rule.* That is, the dynamics preserves the partial ordering.

(iii) *Strict quasi-static (or adiabatic) dynamics.* The external driving field changes very slowly so that the system does not lag behind.

Note that when thermal fluctuations play a significant role or the system is driven fast enough, these conditions will not be fulfilled and the system will not display the return point memory property.

The disorder-induced hysteresis criticality discussed so far refers to the saturation cycle. In the RFIM with athermal dynamics, the condition of saturation is not required and the critical point occurs for subloops at the critical amount of disorder. This is a very important result since, in experiments, it is often difficult to take the studied ferroic material (a magnet, for instance) to the saturation state due to the unavailable high field required.

Other kinds of disorder: random bonds, random anisotropy, dilution, etc.

Beyond a distribution of random fields, disorder can be introduced in many other different ways. For instance, in the form of random bonds, random anisotropy, or site dilution, or, even, a combination of them. On the other hand, models such as the Blume-Emery-Griffiths model, in which the spins take three, ± 1, 0, or more values, with similar forms of disorder, must be considered. In all these cases, when these systems are driven by a zero-temperature adiabatic dynamics, similar disorder-induced hysteresis and avalanche criticality are found. In fact, symmetry arguments indicate that critical exponents should be the same in all these models, which has been corroborated in some cases by means of numerical simulations. Nevertheless, the return point memory property is not reproduced by some of these models. This is, for instance, the case of the random bond Ising model (RBIM). This is due to the fact that reverse spin flips may occur during an avalanche. These reverse flips are a consequence of antiferromagnetic bonds that may be present in the system. This destroys the dynamical partial ordering of the metastable states necessary for the return point memory property to be fulfilled. In any case, for a low amount of disorder, the number of such reverse flips in a complete loop is expected to be very small compared to the total number of flips, and the return point memory will be approximately satisfied to a good approximation. Note that reverse spin-flips can never happen in the RFIM at zero temperature in which the return point memory property is strictly satisfied.

8.2.2 Avalanches in other systems: breakdown phenomena in systems with disorder

Apart from phase transitions in systems with disorder, many other systems crackle when externally driven. That is, all these systems respond intermittently with discrete events or avalanches of a very large variety of sizes and durations. The Earth is perhaps the prototypical example of a system that responds with discrete events known as earthquakes when tectonic plates collide with each other in subduction zones. As a matter of fact, it is well known from a long time ago that the earthquake magnitudes and energies span many orders of magnitude and distribute as a power law with no characteristic scale. This behaviour is described by the Gutenberg-Richter law [311], which basically indicates that the distribution of energies of the earthquakes is a power law.

Other examples include plasticity and breakdown phenomena in disordered materials under external forces. Specific examples are the breakdown of a disordered conductor under applied current or voltage drop, the fracture of some porous materials under compression, and fibre bundles under tension. For instance, the failure of a porous material subjected to compressional forces is often heralded by a significant crackling activity. In this regime, the response of the system to the applied compressive stress is not smooth and continuous as classically expected for elastoplastic materials but, instead, occurs as a sequence of avalanches that stem from sudden changes of the internal strain field and can be detected by means of acoustic emission measurements. Their statistical characteristics share many similarities with Earth seismicity. In particular, not only the distribution of energies is also a power law with an exponent similar to the exponent of the Gutenberg-Richter law, but even the statistics of aftershocks and waiting times share a very similar behaviour with earthquakes [312].

In fibre bundles, when the tension stress increases, a fibre can break when the stress it undergoes exceeds a certain threshold, which distributes among the fibres according to a given probability distribution. The problem is clearly similar to the RFIM previously discussed. In the fibre bundle case, the external tension plays the role of the external magnetic field in the RFIM and disorder is associated with the threshold distribution of fibre breaking, that has the same effect as the random distribution of local fields. Actually, a treatment similar to the one discussed for the RFIM is possible to describe this class of systems. Then, critical amount of disorder is found at which the failure changes from occurring sharply to occurring smoothly. The difference between the magnetization switching phenomena in disordered magnets and failure phenomena in fibre bundles is due to the fact that while rupture is a completely irreversible effect that cannot be undone, magnetization switching can, indeed, be reversed with a certain hysteresis. Then, a question arises: can all these breakdown phenomena be associated with a phase transition as is the case of magnetization switching in magnets? While this is a controversial point, it is worth pointing out that some authors have classified such phenomena within the group of non-equilibrium phase transitions [313].

Exercises

8.1 Consider a temporal sequence of avalanche-type events, which assumes that the random occurrence of an event during a time interval $(t, t+dt)$ is determined by an intensity function $\mu(t)$ that accounts for the rate of events at t.

(a) Show that in the case $\mu(t) = \lambda$ is a constant, the sequence of events is a Poisson process with independent occurrence of events.

(b) In the *Epidemic Type After Shock* model (usually known as ETAS model),[12] the intensity $\mu(t)$ is supposed to increase additively after the occurrence of an event k of energy E_k that occurs at time t_k by an amount $\phi_k(t)$ given by:

$$\phi_k(t) = K\frac{(E_k/E_0)^{2\alpha/3}}{(c-t-t_k)^{1+\theta}} \quad \text{for } t > t_k,$$

where the energies of the events are assumed randomly and independently distributed according to a power-law normalized between E_0 and ∞ as:

$$P(E)dE = (\varepsilon - 1)\frac{E_0^{\varepsilon-1}}{E^\varepsilon}dE.$$

Find the intensity $\mu(t)$ at a given time and the average number of events generated by an individual main event with energy E_k that has taken place at t_k.

8.2 In a Barkhausen avalanche process, near the critical point, (R_c, h_{co}), the voltage of avalanche pulses, $V(\mathcal{T}, t)$, detected as a function of time, t, scales as:

$$V(\mathcal{T}, t) = \mathcal{T}^{1/(v-1)}f_{shape}(t/\mathcal{T}),$$

where the exponent $1/v$ relates the avalanche size, s, and avalanche pulse duration, $\mathcal{T}$, as $s \sim \mathcal{T}^{1/v}$, and $f_{shape}(t/\mathcal{T})$ is a universal shape function. Then, the probability $P(V|s)$ that a voltage V occurs at some point per avalanche of size s should be of the form:

$$P(V|s) = V^{-x}f_{voltage}(Vs^{-y}),$$

where $f_{voltage}$ is also a universal function.

---

[12] The Epidermic Type After Shock model was introduced by Ogata, Y., *Statistical Models for Earthquake Occurrences and Residual Analysis for Point Processes*, J. Am. Stat. Assoc., **83**, 9–27 (1988).

(*a*) From the normalization condition of $P(V|s)$ and the average voltage in an avalanche, show that $x = 1$ and $y = 1 - v$.

(*b*) Show that the energy of avalanche pulses, $E(s)$, given as the average squared voltage times the average avalanche duration, must scale as:

$$E(s) \sim s^{2-v}.$$

9

Superconductors

A material is called a superconductor when it behaves approximately as an ideal electric conductor and an ideal diamagnet. This striking behaviour emerges in a number of materials below a given temperature through a phase transition that occurs from a high-temperature normal conductor to a low-temperature superconductor. The microscopic mechanism that gives rise to this behaviour can only be understood within the framework of quantum mechanics and, therefore, superconductivity is a quantum phenomenon that shows up at a macroscopic scale. In spite of that, the transition from a normal conductor phase to a superconductor phase is still driven by thermal fluctuations and is, thus, a thermal phase transition.

The phenomenon of superconductivity was discovered in 1911 by Kamerlingh Onnes just after he was the first person to liquefy helium, which opened entirely new perspectives in low-temperature physics [314]. With liquid He available, Onnes was interested in knowing how low the electric resistance of a pure metal can be at low temperature. The motivation for these experiments was the fact that the finite resistance of normal conductors was known to be a consequence of the deviations from perfect periodicity that occur in real crystals, which mainly originate from vibrations (phonons) and lattice defects that break translational symmetry. Therefore, according to Drude's model [315], it was expected that in a perfect crystal, the resistance must go to zero at the absolute zero and that any residual resistance should be attributed to the existence of defects. Nevertheless, the result of the experiments was highly unexpected. Kamerlingh Onnes found that at a finite temperature of about 4.2 K, the resistance of mercury falls to an unmeasurable low value at a quite fast rate [316]. This was the discovery of a new state of matter which he called the superconducting state that occurs through a phase transition. Soon after many other materials were found to show similar superconducting behaviour.

Importantly enough, the superconducting state is not only defined by the zero resistance to an electric current, but the material must also show perfect diamagnetism. This property supposes that, in the presence of a magnetic field, materials in a superconducting state expel the magnetic flux so that the magnetic induction is zero inside the material. This second feature that is usually denoted as the Meissner-Ochsenfeld effect[1] was discovered by Meissner and Ochsenfeld [317] and is essential to ensure that the superconducting state is an actual thermodynamic equilibrium state. It is easy to see that the low-temperature state of an ideal conductor that does not satisfy the condition of ideal diamagnetism would be a history-dependent state so that under zero-field cooling and field cooling protocols different states should be reached. In this respect, two classes of materials are usually distinguished: type I and type II superconductors. In the first case, the magnetic induction vanishes inside the superconductor for fields below a given field H_c, usually denoted as a critical field, at which suddenly superconductivity is destroyed. Instead, in type II superconductors, two critical fields can be defined, the lower critical field H_{c1} and the upper critical field H_{c2}. While the behaviour of these materials is analogous to type I superconductors for fields $H \leq H_{c1}$, for $H \geq H_{c1}$ magnetic flux can start to penetrate inside the superconductor and upon increasing the field, the magnetic flux density increases and, finally, at H_{c2} superconductivity is destroyed. Therefore in the range $H_{c1} \leq H \leq H_{c2}$, the superconductor is in a mixed state. Abrikosov showed that in this state, the magnetic flux can enter the superconductor in the form of vortices consisting of regions of circulating supercurrents around a small central core, which behaves as a normal conductor.

From a microscopic viewpoint, superconductivity cannot be understood within a one electron theory. In fact, pairing of electrons originates from a phonon-mediated interaction which is the essential mechanism that permits understanding the superconducting state as a sort of condensation of electron pairs with bosonic character. This is the starting point of the Bardeen-Cooper-Schrieffer (BCS) theory that successfully explains the properties of superconductors. This theory represents a mean-field approach to the superconducting transition. Within this framework, it was acknowledged for many years that the maximum superconducting temperatures were limited to a temperature below 30 K [318]. Nevertheless, in 1986, Bednorz and Müller [319] discovered a new class of materials with a superconducting temperature slightly above 30 K. This breakthrough opened the new era of high-T_c

---

[1] Usually, it is simply denoted as the Meissner effect.

superconductors. Figure 9.1 shows the evolution of the superconductivity transition temperature records during the twentieth century.

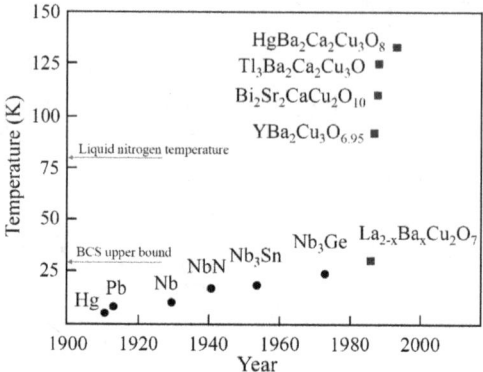

Figure 9.1 Superconductivity temperature record through the years. Solid circles stand for old superconductors and solid squares for high-T_c superconductors. In most of these high-T_c materials, superconducting temperatures are above the BCS upper bound and above the temperature of liquefaction of nitrogen. This last feature is important because liquid nitrogen could then be used as a refrigerant. This is quite useful for applications since nitrogen is very abundant and liquid nitrogen can be produced cheaply.

According to the BCS theory, the transition from normal to the superconducting state is continuous in the absence of an applied magnetic field while it is first-order at the so-called critical field in type I superconductors but still second order in type II superconductors. Experiments show that when the transition is continuous, it is characterized by mean-field exponents to a high precision. The reason for the agreement with the BCS theory arises from the fact that the Ginzburg temperature interval is extremely narrow in these systems and, thus the influence of fluctuations is so limited that most experiments are not sensitive enough to probe the *true* critical behaviour [320]. It must be noted that the situation is however different in high-T_c superconductors, where the Ginzburg interval is large, up to several kelvin and experiments clearly show a non-mean-field critical behaviour.

9.1 Phenomenology and thermodynamics

Two main classes of superconductors, type I and type II, are considered depending on whether a finite range of magnetic fields exists where superconducting and non-superconducting regions can coexist. In the superconducting regions, the magnetic induction vanishes and thus, magnetization and magnetic field intensity are related as, $M = -H$, which expresses the perfect diamagnetism of superconductors. According to this equation, when, at a give temperature below the critical temperature T_c, the material is subjected to an increasing magnetic field, the magnetization first increases linearly up to a given field. In type I superconductors, the magnetization falls suddenly to zero at this field, at which the superconducting and normal phases coexist in equilibrium. Usually, this field is denoted as the critical field H_c in spite that the transition is first-order. Therefore, in this class of materials, the line $H_c(T)$ is a line of first-order transitions that ends at the critical point, $T = T_c$ and $H = 0$. In contrast, in type II superconductors, the magnetization smoothly goes to zero in the range between this field, which denotes the lower critical field H_{c_1}, and an upper critical field, H_{c_2}. This range defines the mixed state. Phase diagrams of type I and type II superconductors are schematically depicted in Figure 9.2.

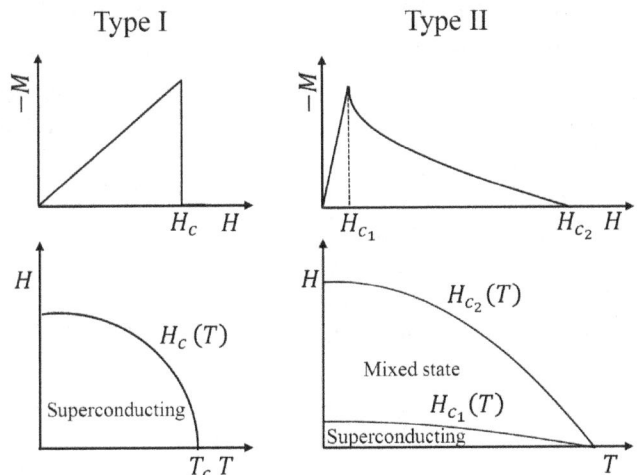

Figure 9.2 Schematic phase diagrams in the *M-H* and *H-T* spaces of type I and type II superconductors.

A number of pure metals such as Al, Hg, Pb, Nb or Sn are superconductors at a quite low temperature. Among them, Nb has the highest critical temperature T_c of about 9.2 K. All these elements belong to the group of type I superconductors. In these superconductors, the critical field at 0 K is quite low of the order of few 10^{-2} T. Most superconductors, including high-T_c superconductors, belong to the type II group.[2] From a practical point of

---

[2] Most type I superconducors are pure metals. Only a very small number of alloys seem to belong to this group. Among them, TaSi$_2$ is the one that shows more clearly type I behaviour [321].

view, the advantage of type II superconductors is due to the fact that they can remain superconductors even at very large applied magnetic fields since the upper critical field can often reach tens of Teslas.

London and London were the first to propose a phenomenological theory of superconductivity that takes into account the Meissner effect [322]. The theory is based on a modification of Maxwell equations suited to incorporate the two essential features that define superconductivity. That is, the fact that electric carriers can move without dissipation and that magnetic induction vanishes in the presence of a magnetic field. Therefore, within this approach, when subjected to an electric field $\boldsymbol{E}$, superconducting carriers must move according to the equation

$$m\frac{d\boldsymbol{v}}{dt} = -2e\boldsymbol{E}, \tag{9.1}$$

where m is the mass of carriers, $\boldsymbol{v}$ the velocity and we have taken into account that carriers are pairs of electrons of charge $q = 2e$, with e being the electron charge. If the density of superconducting carriers is n_s then the current density will be, $\boldsymbol{j}_s = -qn_s\boldsymbol{v}$, which leads to the first London equation,

$$\frac{d\boldsymbol{j}_s}{dt} = \frac{n_s q^2}{m}\boldsymbol{E}. \tag{9.2}$$

This equation, together with the Maxwell equation, $\nabla \times \boldsymbol{E} = -\frac{\partial \boldsymbol{B}}{\partial t}$ yields

$$\frac{\partial}{\partial t}\left(\frac{m}{n_s q^2}\nabla \times \boldsymbol{j}_s + \boldsymbol{B}\right) = 0. \tag{9.3}$$

Integration of the preceding equation over a conducting loop must keep the magnetic flux constant. Meissner effect is then taken into account by imposing that the integration constant is zero. That is,

$$\nabla \times \boldsymbol{j}_s = -\frac{n_s q^2}{m}\boldsymbol{B}, \tag{9.4}$$

which is the second London equation. In some cases, it is convenient to express London equations in terms of the vector potential $\boldsymbol{A}$. Within the Coulomb or London gauge, $\nabla \cdot \boldsymbol{A} = 0$, ensures that the density of superconducting carriers is conserved as required from the continuity equation. Then, $\boldsymbol{B} = \nabla \times \boldsymbol{A}$, and from Eq. 9.4, it follows that $\boldsymbol{j}_s = -\frac{n_s q^2}{m}\boldsymbol{A}$.

When the two London Eqs. 9.2 and 9.4 are combined with the Maxwell equation, $\nabla \times \boldsymbol{B} = \mu_0 \boldsymbol{j}_s$, the following equations are obtained:[3]

$$\nabla^2 \boldsymbol{B} - \frac{1}{\lambda_L^2} \boldsymbol{B} = 0, \tag{9.5}$$

$$\nabla^2 \boldsymbol{j}_s - \frac{1}{\lambda_L^2} \boldsymbol{j}_s = 0, \tag{9.6}$$

where λ_L is a characteristic length defined as

$$\lambda_L \equiv \sqrt{\frac{m}{\mu_0 n_s q^2}}. \tag{9.7}$$

Application of the preceding Eqs. 9.5 and 9.6 to study a semi-infinite superconductor in the region $z > 0$ in contact with a semi-infinite vacuum at $z < 0$ under an applied field, $\boldsymbol{B} = (B_x, 0, 0)$ leads to

$$B_x = B_x^0 \exp(-z/\lambda_L), \tag{9.8}$$

$$j_{sy} = j_{sy}^0 \exp(-z/\lambda_L), \tag{9.9}$$

which shows that the field penetrates the superconductor but decays exponentially from the surface over a distance λ_L, which is then denoted as the London penetration depth. Consistently it is found that the superconducting currents, that shield the interior of the superconductor against the external field, also decay exponentially with distance inside the solid. An estimation of the London penetration length gives small values that range between 50 nm and 500 nm.

In the mixed state of type II superconductors, the field enters the superconductor in the form of vortices, where each vortex consists of a region of circulating supercurrents around a central core that essentially behaves as a normal metal [328]. Therefore, the circulating currents serve to eliminate the magnetic flux outside the cores of the vortices along which the field can cross the material. Figure 9.3 shows a schematic representation of the vortex in a type II superconductor. It is interesting to note that due to the quantization of the magnetic flux, each vortex carries a fixed unit of flux, $\phi_0 = h/q$, which means that if the number of vortices per unit area of the material is n_v, then the average density of magnetic induction inside the material is $B = n_v h/q$.

Vortices in type II superconductors can be modelled as very long cylindrical tubes embedded inside a superconducting material. It can be assumed

---

[3] The calculation must take into account that, for a given vector $\boldsymbol{V}$, $\nabla \times \nabla \boldsymbol{V} = \nabla(\nabla \cdot \boldsymbol{V}) - \nabla^2 \boldsymbol{V}$. Since $\boldsymbol{B}$ and $\boldsymbol{j}_s$ satisfy that $\nabla \cdot \boldsymbol{B} = 0$ and $\nabla \cdot \boldsymbol{j}_s = 0$, then $\nabla \times \nabla \boldsymbol{B} = -\nabla^2 \boldsymbol{B}$ and $\nabla \times \nabla \boldsymbol{j}_s = -\nabla^2 \boldsymbol{j}_s$.

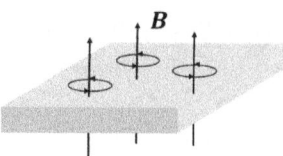

Figure 9.3 Schematic representation of vortices in a type II superconductor. Field lines inside the superconductor are indicated by arrows and the superconducting currents by circular loops where the arrow indicates the direction of the screening currents.

that the field inside the core, $\boldsymbol{B} = (0, 0, B_z)$, points along the direction of the long axes of cylinders, which are parallel to the z-direction. London equations, expressed in cylindrical coordinates, (r, θ, z), can be used to calculate the field and the supercurrents. In cylindrical coordinates, Eq. 9.5 for the field reads as

$$\frac{d^2 B_z}{dr^2} + \frac{1}{r}\frac{dB_z}{dr} - \frac{B_z}{\lambda_L^2} = 0. \tag{9.10}$$

This equation is a modified Bessel equation with solutions being modified Bessel functions of the second kind, $K_\nu(z)$ [324]. In the present case, $\nu = 0$ and the solution is given as

$$B_z(r) = \frac{\phi}{2\pi\lambda_L^2} K_0\left(\frac{r}{\lambda_L}\right), \tag{9.11}$$

where ϕ is the total magnetic flux through the cross-section of a vortex. For $r/\lambda_L \ll 1$, it can be expressed as $B_z(r) = \frac{\phi}{2\pi\lambda_L^2}\ln\left(\frac{\lambda_L}{r}\right)$, where it must be noted that the divergence at the core centre, $r = 0$, is not physical since superconductivity should be suppressed for sufficiently small r where the material is expected to behave as a normal material, and the field should be constant in the region of the core centre. For $r/\lambda_L \gg 1$, the solution can be written in the form, $B_z(r) = \frac{\phi}{2\pi\lambda_L^2}\sqrt{\frac{\pi\lambda_L}{2r}}\exp\left(-\frac{r}{\lambda_L}\right)$, that shows that, similarly to the case of the semi-infinite superconductor, the field penetrates the superconductor region but decays exponentially with the same characteristic length λ_L.

Thermodynamics of superconductors

From a thermodynamic point of view, the relevant generalized displacement to describe a superconductor is the magnetization and, thus, the corresponding fundamental thermodynamic equation reads

$$dU = TdS + \mu_0 \Omega H dM, \tag{9.12}$$

where isotropy is assumed and, thus, M is the magnetization along the direction of the applied external field $\boldsymbol{H}$. In the preceding equation, Ω is the volume of the body, which is assumed to be constant, independent of the magnetic state. In terms of the magnetic Gibbs free energy, $G = U - TS - \mu_0 \Omega H M$, the fundamental equation can be written in the form

$$dG = -SdT - \mu_0 \Omega M dH, \tag{9.13}$$

where now T and H are the independent variables that can be externally controlled. In the case of a type I superconductor, we are interested in determining the gain in the magnetic free energy, $G_s(T,0) - G_n(T,0)$, of the superconducting state with respect to the normal state at a temperature T below T_c. For such a purpose, it is convenient to compute first the change of magnetic free energy of the superconducting phase at T induced by an increase of the magnetic field from 0 to H_c. It is given by

$$G_s(T, H_c) - G_s(T, 0) = -\mu_0 \Omega \int_0^{H_c} M dH = \frac{1}{2} \mu_0 \Omega H_c^2, \tag{9.14}$$

where the Meissner effect, expressed as, $M = -H$, that holds in the superconducting phase has been taken into account. The preceding equation indicates that as the applied field increases the energy increases as H^2. At the coexistence point, $H_c(T)$, $G_s(T, H_c) = G_n(T, H_c)$ and, therefore

$$G_s(T, 0) - G_n(T, 0) = -\frac{1}{2} \mu_0 \Omega H_c^2, \tag{9.15}$$

where it has been taken into account that $G_n(T, H_c) - G_n(T, 0) = 0$, since $M = 0$ in the normal state. Here $\frac{1}{2} \mu_0 \Omega H_c^2$ is the so-called condensation energy.

The change of entropy between superconducting and normal phases can be obtained by taking into account that on the coexistence line, the change of magnetization between superconducting and normal phases is $\Delta M = -H_c$ and, then, using the Clausius–Clapeyron equation, it follows that the entropy change, $\Delta S = S_s(T, H_c) - S_n(T, H_c)$, is given as

$$\Delta S = \mu_0 \Omega H_c \frac{dH_c}{dT}. \tag{9.16}$$

Therefore, the transition latent heat, $\ell = T\Delta S$, vanishes at the critical point where $H_c = 0$.

The difference in heat capacities between the two phases, $\Delta C = C_s - C_n$, can be obtained by taking into account that $\frac{\Delta C}{T} = \frac{\partial \Delta S}{\partial T}$. Therefore, at critical point, this leads to

$$\Delta C(T_c) = \mu_0 \Omega T_c \left(\frac{dH_c}{dT} \right)^2_{T=T_c}. \tag{9.17}$$

Since $(dH_c/dT)^2$ does not vanish at the critical point, this equation predicts that in type I superconductors, the heat capacity should show a discontinuity at this point. This is in agreement with experiments that confirm this behaviour. Figure 9.4 shows measurements of the heat capacity of Al about the region of its superconducting transition.

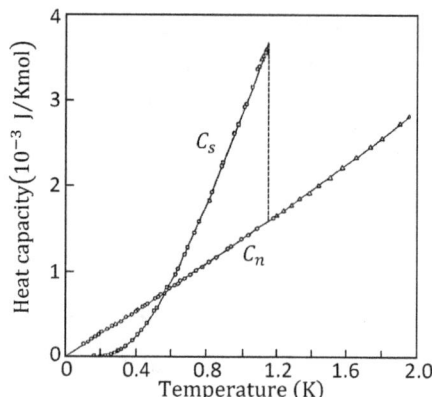

Figure 9.4 Heat capacity of pure Al in the normal (C_n) and superconducting (C_s) phases. The critical temperature is at $T_c \simeq 1.2$ K where the heat capacity shows a discontinuity. Below the critical temperature, C_n was measured by applying a weak magnetic field of 3 10^{-2} T which is larger than the critical field and permits keeping the system within the normal phase. Adapted from Ref. [325].

In the case of a type II superconductor, the difference in magnetic free energies, $G(T, H_{c_2}) - G(T, 0) = \mu_0 \Omega \int_0^{H_{c_2}} M dH$, where the integral of the right-hand side is the area limited by the curve $M(H)$ in Figure 9.2 for a type II superconductor. Therefore, an effective critical field can be defined as $\frac{1}{2} H_c^{eff} \equiv \int_0^{H_{c_2}} M dH$, and all the thermodynamic expressions obtained for a type I superconductor can be used for a type II superconductor replacing H_c by H_c^{eff}.

9.2 The Ginzburg–Landau theory of superconductors

Ginzburg and Landau proposed a phenomenological model based on a generalization of the Landau theory of phase transitions motivated by the possibility of addressing the surface tension associated with the boundary between the normal and the superconducting phases and correctly describing the destruction of superconductivity by a magnetic field or a current. This, indeed, required considering inhomogeneous states, and with this aim, they proposed to include gradient terms of the order parameter in the expansion of the free energy [326]. From an historic point of view, this is the first time that a *Ginzburg–Landau* model was proposed to deal with a phase transition. A detailed introduction of the theory in modern physics language can be found in Ref. [327].

Ginzburg and Landau postulated that a suitable order parameter for a superconducting transition is the macroscopic wave function that describes the superconducting state. Therefore, they chose a complex function ψ as the order parameter. This function must obviously satisfy that $\psi = 0$ in the normal state and $\psi \neq 0$ in the superconducting state. The fact that ψ is a complex function means that all the observable quantities must depend on ψ and its conjugate ψ^* in such a way that they must remain unchanged when ψ is multiplied by a phase constant. Consistently, $|\psi|^2$ must be identified with the concentration of superconducting carriers.

Consider first the case of a uniform superconductor, which is an adequate situation in the absence of (or for a low enough) applied magnetic field. In this case, the singular part of the free energy density should be of the form

$$f_s(T, \psi) = f_n(T) + a(T)|\psi|^2 + \frac{1}{2}b(T)|\psi|^4 + \dots . \tag{9.18}$$

Since the superconducting transition is continuous, the expansion can be limited to fourth order and taking, as usual, $a(T) = a_0(T - T_c)$ and b constant, the following solution is found

$$|\psi| = \begin{cases} 0 & \text{for } T > T_c, \\ \sqrt{\frac{a_0(T_c - T)}{b}} & \text{for } T \leq T_c. \end{cases} \tag{9.19}$$

An important feature to point out is the fact that there are an infinite number of low-temperature solutions of this kind since the order parameter is a complex function of the form $\psi = |\psi|e^{i\theta}$. This infinite number of minima of the free energy density correspond to all possible values of the complex phase θ, since all these values lead to the same free energy. Nevertheless, at the transition, the system chooses one of these possible values. This fact reveals the spontaneous symmetry breaking that takes place at the superconducting transition.

The thermodynamic free energy density can be obtained by replacing the solution for the order parameter in the free energy density function. It is then obtained that in the superconducting phase, $f_s(T) - f_n(T) = -\frac{a_0^2(T-T_c)^2}{2b}$. Taking into account that $f_s(T) - f_n(T) = \frac{1}{\Omega}[G_s(T,0) - G_n(T,0)]$, from Eq. 9.15, it is obtained that for a type I superconductor, the critical field as a function of temperature, that can be obtained as a function of model parameters, is

$$H_c = \frac{a_0}{\sqrt{\mu_0 b}}(T_c - T). \tag{9.20}$$

On the other hand, differentiating $f_s(T) - f_n(T)$ with respect to temperature leads to

$$s_s(T) - s_n(T) = -\frac{a_0^2}{b}(T_c - T), \tag{9.21}$$

where $s = S/\Omega$ is an entropy density. A further differentiation permits obtaining the heat capacity. The discontinuity of the heat capacity at the critical temperature is given as

$$c_s(T_c) - c_n(T_c) = \frac{a_0^2}{b}T_c. \tag{9.22}$$

Therefore, from Eq. 9.17, it results that $\frac{a_0^2}{b} = \mu_0 \left(\frac{dH_c}{dT}\right)^2_{T=T_c}$.

Another useful way to identify free energy parameters with measurable quantities is by taking advantage of the fact that $|\psi|^2 \propto n_s$, where n_s is the concentration of superconducting carriers. From Eq. 9.7, it results that $n_s \propto \lambda_L^2$ and, thus,

$$\frac{\lambda_L^2(0)}{\lambda_L^2(T)} = \frac{|\psi(T)|^2}{|\psi(0)|^2} = |\psi(T)|^2 = \frac{a_0(T_c - T)}{b}, \tag{9.23}$$

where it has been taken into account that $|\psi(0)|^2 = 1$. Then, using Equation 9.20. it is obtained that

$$b = \mu_0 H_c^2 \frac{\lambda_L^4(T)}{\lambda_L^4(0)}. \tag{9.24}$$

To account for non-uniform states, gradient terms of the order parameter ψ must be included in the free energy density expansion. To the lowest order, which is adequate for smoothly varying inhomogeneities in space, the free energy density at a point r should be of the form

$$f_s(T, (r)) = f_n(T) + a(T)|\psi(r)|^2 + \frac{1}{2}b(T)|\psi(r|^4 + \frac{\hbar^2}{2m^*}|\nabla\psi(r)|^2. \tag{9.25}$$

As it is usual in the Ginzburg–Landau theory, the gradient term determines the energy cost associated with inhomogeneities of the order parameter.

Since ψ can be interpreted as the macroscopic wave function of the super-conducting state, the coefficient of the gradient term has been expressed as $\hbar^2/2m^*$, which highlights that it has the form of a density of kinetic energy. Therefore, the coefficient m^* can be understood as an effective mass of the superconducting carriers. Taking into account that carriers are pairs of electrons m^* can be identified with twice the mass of the electron, $2m$.

Minimization of the Ginzburg–Landau functional, $\mathcal{F}_s(T) = \int f_s(T, \psi(\boldsymbol{r}))d\boldsymbol{r}$, leads to

$$-\frac{\hbar^2}{2m^*}\nabla^2\psi + (a + b|\psi^2|)\psi = 0, \tag{9.26}$$

which is the differential equation that permits determining the equilibrium order parameter function under given boundary conditions. Interestingly, this equation has the form of a non-linear Schrödinger equation for the wave function of the superconducting phase. This equation can be used to compute the interface energy σ between a normal and a superconducting material. A simple case is that of an interface lying in the y-z plane which separates a normal material in the $x < 0$ region and a superconducting material in the $x > 0$ region. In this case, for symmetry reasons, ψ should only vary along the x-axis and using the results obtained in *Example 1*, Section 2.2.1, it is obtained that the wave function or order parameter in the superconducting region will be given as

$$\psi(x) = \psi_0 \tanh\left[\frac{x}{\sqrt{2}\xi(T)}\right], \tag{9.27}$$

where it has been assumed that $\psi(0) = 0$ and that ψ_0 is the order parameter in the bulk of the superconducting material, far from the interface. Here $\xi(T)$ is the coherence length that provides a measure of the distance from the interface to the region where the order parameter has already reached its bulk equilibrium value. Indeed, this length coincides with the correlation length in the superconducting phase. In terms of model parameters, it is given as $\xi(T) = \sqrt{\frac{\hbar^2}{2m^*a_0}}(T_c - T)^{-1/2}$.

The interface energy can be obtained as

$$\sigma = \int_0^\infty \{f_s[T, \psi(x)] - f_s(T, \psi_0)\}dx = \frac{\hbar^2}{m^*}\int_0^\infty \left(\frac{d\psi}{dx}\right)^2 dx, \tag{9.28}$$

where $f_s(T, \psi_0) = f_n(T) - a(T)^2/2b$ is the bulk free energy density of the superconducting phase. Then, using the expression of $\psi(x)$ given by Eq. 9.27, the surface energy can be given as[4]

$$\sigma = \frac{2}{3}\sqrt{2}\mu_0 H_c^2(T)\xi(T). \tag{9.30}$$

It is important to note that this surface energy goes to zero at T_c as $\sigma \sim (T_c - T)^{3/2}$, which is the expected mean-field behaviour.

9.2.1 The Ginzburg–Landau theory in the presence of an applied magnetic field

The main impact of the Ginzgurg–Landau theory shows up when the effect of a magnetic field is considered and the Meissner effect is taken into account. In this case, terms associated with the field energy and with the appearance of field-induced spatial inhomogeneities in the order parameter must be added to the free energy density. The first term, which reads as $|B|^2/2\mu_0$, is constant for a given applied field and since it does not affect the state of the superconductor it is usually neglected.[5] The second term can be introduced with the usual replacement in quantum mechanics within the Coulomb gauge of the canonical momentum operator $p = \frac{\hbar}{i}\nabla$ by $\frac{\hbar}{i}\nabla - qA$ that applies when the wave function ψ describes charged particles subjected to a time-independent magnetic field. In this replacement, A is the vector potential. Therefore, in the presence of a magnetic field, the Ginzburg–Landau free energy functional reads

$$\mathcal{F}_s(T) = \mathcal{F}_n(T) + \int_\Omega \left[a|\psi|^2 + \frac{b}{2}|\psi|^4 + \frac{\hbar^2}{2m^*}\left|\left(\frac{\hbar}{i}\nabla + 2eA\right)\psi\right|^2 \right] d\,r$$

$$+ \frac{1}{2\mu_0}\int_\Omega |B(r)|^2 d\,r, \tag{9.31}$$

where now it has been taken explicitly into account that $q = -2e$. Minimization of the preceding functional leads to the general Ginzburg–Landau equation, which also has the form of a non-linear Schrödinger equation

$$-\frac{\hbar^2}{2m^*}\left(\nabla^2 + \frac{i2e}{\hbar}A\right)^2 \psi + (a + b|\psi^2|)\psi = 0. \tag{9.32}$$

---

[4] Note that,

$$\int_0^\infty \left(\frac{d\psi}{dx}\right)^2 dx = \frac{\psi_0^2}{\sqrt{2}\xi(T)}\int_0^\infty \text{sech}^4 y\, dy = \frac{2}{3\sqrt{2}}\frac{\psi_0^2}{\xi(T)}. \tag{9.29}$$

[5] This term is present even in the absence of a superconducting material.

Magnetic field-induced supercurrents, j_s, can be obtained from the functional derivative of the free energy functional with respect to the vector potential. The result is

$$j_s = \frac{\delta F_s}{\delta A} = -\frac{i2e\hbar}{2m^*}(\psi^*\nabla\psi - \psi\nabla\psi^*) - \frac{(2e)^2}{m^*}|\psi|^2 A. \qquad (9.33)$$

It must be noted that the vector potential should be obtained from the magnetic field B by taking into account the Maxwell equation, $\nabla \times B = (\mu_0(j_s + j_{ext})$, where j_{ext} are, for instance, external currents in the device that create the applied magnetic field.

At this point, it is useful to discuss in depth the relevance of the gauge transformation used to introduce the vector potential. Under a gauge transformation, $A \to A + \nabla\varphi(r)$, the wave function must undergo a change of the phase as, $\psi(r) \to \psi(r)e^{i\theta(r)}$. Therefore, since the application of momentum operator to the transformed order parameter yields

$$\mathbf{p}\, \psi(r)e^{i\theta(r)} = \left\{ \frac{\hbar}{i}\nabla + 2e\left[A + \frac{\hbar}{2e}\nabla\theta(r) \right] \right\} \psi(r)e^{i\theta(r)}, \qquad (9.34)$$

it is concluded that the free energy must remain unchanged under the two changes, $\psi(r) \to \psi(r)e^{i\theta(r)}$ and $A(r) \to A(r) + \frac{\hbar}{2e}\nabla\theta(r)$. This means that both the vector potential and the phase of the order parameter depend on the choice of the gauge in such a way that all the physical observables must remain unchanged. It is important to note that according to Eq. 9.34, the change of phase of the order parameter supposes an excess of free energy, $\Delta\mathcal{F}_{phase}$, with respect to the uniform ground state corresponding to constant θ and $A = 0$, given by

$$\Delta\mathcal{F}_{phase} = \frac{\hbar^2}{2m^*}\int_\Omega \left(\nabla\theta + \frac{2e}{\hbar}A \right)^2 d\,r. \qquad (9.35)$$

This result assumes that for a given gauge, as for instance the Coulomb gauge, the ground state corresponds to a state in which the gradients of the phase are minimized as much as possible. In the absence of an applied magnetic field, taking $A = 0$, the minimum is a state with a uniform phase throughout the whole system. This is an interesting result that corroborates the spontaneous symmetry breaking that occurs at the superconducting transition is related to the phase of the order parameter.

Flux lattice in type II superconductors

As an application of the Ginzburg–Landau theory, it is useful to see that in the presence of an applied magnetic field, a solution can be found that correctly describes the mixed state in type II superconductors close to H_{c_2}

where the transition to the normal state can be considered as continuous. This solution was first derived by Abrikosov for the case of a bulk superconductor [328].

Consider a superconductor at a temperature $T < T_c$ subjected to an applied magnetic field. Due to the continuous character of the transition at H_{c2}, the order parameter should be small close to this field at which it should vanish. Therefore, it is a good approximation to assume that $B = \mu_0 H$, where H is the field intensity created by an external device. Indeed, this supposes that $M \cong 0$ in the region considered. Furthermore, in this region, spatial variations of B can be neglected. Therefore, assuming that a constant magnetic field is applied along the z-axis, the magnetic induction will be $B = (0, 0, B)$, with B being constant and, within the Coulomb gauge, the corresponding vector potential will be of the form, $A = (0, xB, 0)$. The Ginzburg–Landau equation will thus read as

$$\left(-\frac{\hbar^2}{2m^*}\nabla^2 + -i\hbar\omega_c x \frac{\partial}{\partial y} - \frac{m^*\omega_c^2}{2} \right) \psi(\boldsymbol{r}) = -a(T)\psi(\boldsymbol{r}), \qquad (9.36)$$

where only linear terms in ψ have been retained and the cyclotron frequency, $\omega_c \equiv 2eB/m^*$, has been introduced. The obtained differential equation is formally equivalent to the Schrödinger equation of a charged particle in the presence of an applied magnetic field. A general solution can be assumed to have the form of a plane wave in the y- and z-directions combined with a function $\varphi(x)$.[6] That is,

$$\psi(\boldsymbol{r}) = \varphi(x)e^{i(k_y y + k_z z)}. \qquad (9.37)$$

Substituting this solution in Eq. 9.36 leads, after some straightforward calculation, to the following differential equation for the function φ

$$-\frac{\hbar^2}{2m^*}\frac{d^2\varphi}{dx^2} + \frac{m^*\omega_c^2}{2}(x - x_0)^2\varphi = -\left[a(T) - \frac{\hbar^2 k_z^2}{2m^*} \right]\varphi. \qquad (9.38)$$

This is the equation of a simple harmonic oscillator of natural frequency ω_c with the origin of coordinates located at $x = x_0$, which is given by, $x_0 = -\frac{\hbar k_y}{m^*\omega_c}$. Therefore,

$$-\left[a(T) - \frac{\hbar^2 k_z^2}{2m^*} \right] = \left(n + \frac{1}{2} \right)\hbar\omega_c, \qquad (9.39)$$

where $n = 0, 1, 2, \ldots$. It is concluded that for each n the functions φ are the wave functions of the simple harmonic oscillator shifted by x_0. Suppose now that the material is at a temperature lower than T_c, close to H_{c2} the

---

[6] The solution of this problem is often denoted as the Landau-level solution.

solution of Eq. 9.38 should be the minimum or the ground-state solution corresponding to $n = 0$ and $k_z = 0$, which leads to, $\hbar e B / m^* = a_0(T_c - T)$. Then, introducing the coherence length $\xi(T)$ and the flux quantum $\phi_0 = 2\pi\hbar/2e$, it is obtained that

$$\mu_0 H_{c2} = \frac{\phi_0}{2\pi\xi(T)^2} = \frac{\phi_0}{2\pi\xi(0)^2}\frac{T_c - T}{T_c}, \tag{9.40}$$

where $\xi(0) = \sqrt{\frac{\hbar^2}{2m^* a_0 T_c}}$ is the coherence length at $T = 0$. It is worth comparing this field H_{c2} with the critical field H_c in type I superconductors given by Eq. 9.20. Introducing also the coherence length and flux quantum in the expression of H_c, the result is,

$$H_c = \frac{\phi_0}{2\pi\sqrt{2}\xi(T)\lambda_L(T)} = \frac{H_{c2}}{\sqrt{2}\kappa}, \tag{9.41}$$

where $\kappa \equiv \lambda_L(T)/\xi(T)$ is the ratio of the two relevant length scales. The preceding equation points out that when $\kappa > 1/\sqrt{2}$, $H_{c2} > H_c$ and, on cooling down from high temperature the transition will take place at H_{c2} showing a continuous character. Instead, when $\kappa < 1/\sqrt{2}$ a first-order transition will take place at H_c. From these results, it can be concluded that for $\kappa < 1/\sqrt{2}$ the superconductor will be of type I, while for $\kappa > 1/\sqrt{2}$ it will be of type II.

To find the structure of vortex that forms at H_{c2} in a type II superconductor, the non-linear Ginzburg–Landau equation should, in principle, be solved. While this can be done numerically, Abrikosov [328] was able to figure out analytically a solution, which is essentially exact. He argued that the solution of the linearized Ginzburg–Landau Eq. 9.36 corresponding to $n = 0$ and $k_z = 0$ is the only one that must be significant. Therefore, he suggested that the required solution should be of the form

$$\psi(\boldsymbol{r}) \propto e^{ik_y y}e^{-(x-x_0)^2/[\xi(T)]^2}, \tag{9.42}$$

where $e^{-(x-x_0)^2/[\xi(T)]^2}$ is the ground-state solution of Eq. 9.38. Then the idea of Abrikosov was to look for a periodic solution in y defined on a lattice, which supposes that k_y must be restricted to $k_y = 2\pi m/d_y$, with d_y being the corresponding period and m is an integer. Then, $x_0 = -(2\pi\hbar/m^*\omega_c a)m = -(\phi_0/d_y B)m$. Therefore, Abrikosov tried a periodic solution of the form

$$\psi(\boldsymbol{r}) = \sum_{m=-\infty}^{m=\infty} C_m e^{i(2\pi m y/d_y)}e^{-\left(x+m\frac{\phi_0}{d_y B}\right)^2/[\xi(T)]^2}, \tag{9.43}$$

where the coefficients C_m can be considered as variational parameters that must result from the minimization of the Ginzburg–Landau functional. To ensure that the solution is not only periodic in y but also in x these coefficients must satisfy that $C_{m+\nu} = C_m$ for some integer ν. The period in the x direction should then be, $d_x = (\phi_0/d_y B)\nu$. The case $\nu = 1$ corresponds to a square lattice, while $\nu = 2$ corresponds to a triangular lattice, which is the structure of minimum energy. In both cases, $\phi(\mathbf{r})$ vanishes at one point of the corresponding unit cell and each vortex is crossed by a flux quantum, ϕ_0. Thus, the solution represents a lattice of vortices.

The hexagonal symmetry of the minimum energy flux lattice associated with the triangular unit cell has been confirmed to a quite good approximation in many type II superconductors using a variety of experimental techniques, including neutron scattering, transmission electron microscopy, or scanning tunnelling microscopy.[7] Figure 9.5 shows the amplitude of the order parameter for a triangular lattice and an image of the flux lattice in the superconductor NbSe$_2$ induced by an applied magnetic field of 1 T.

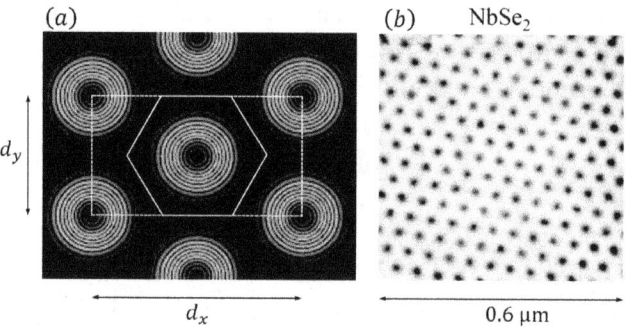

Figure 9.5 Abrikosov flux lattice. (*a*) Amplitude of the order parameter, $|\psi(\mathbf{r})|^2$, for the triangular lattice. The periodicities along x and y directions, d_x and d_y, are indicated. (*b*) Scanning tunnelling microscopy image of an Abrikosov vortex lattice in NbSe$_2$ produced by an applied field of 1 T. The triangular unit cell is clearly observed. Adapted from Ref. [329].

The obtained flux lattice solution is correct just below H_{c_2}. In this region, the vortices are separated by distances on the order of the coherence length, $\xi(T)$, which means that they are very close to each other. In contrast, near H_{c_1}, the number of vortices is expected to be low and, thus, well separated from each other. Then, the field H_{c_1} can be estimated from the energy balance between the energy of a single vortex and the work associated with the field penetration due to the presence of the vortex in the mixed phase.

---

[7] In a few cases, a square flux lattice has been reported [330]. This anomalous behaviour is commonly attributed to the unconventional character of these superconductors, which seems to be associated with the symmetry of Cooper pairs.

The energy of a single vortex can be estimated from the kinetic energy of the rotating supercurrents. Its density is given as, $\frac{1}{2}n_s mv^2 = \frac{1}{2}\mu_0\lambda_L^2 j_s^2$, where it has been taken into account that $v = -j_s/qn_s$. Then, the vortex energy per unit length is given as

$$\varepsilon_V = \frac{\pi}{2}\int_\Omega \mu_0\lambda_L^2 |j_s|^2 rdr = \int_\Omega \lambda_L \left[\frac{\partial B_z(r)}{\partial r}\right]^2 rdr, \qquad (9.44)$$

where the Maxwell equation $\mu_0 j_s = \nabla \times B$ with $B = (0,0,B_z)$ has been taken into account to write the second equality. The magnetic field $B_z(r)$ embedded within a vortex core is given by Eq. 9.11 with ϕ replaced by the flux quantum, ϕ_0. For the case of interest, $\xi \leq r \ll \lambda_L$, which corresponds to $\kappa \gg 1$,[8] the order parameter and the magnetic field in the vortex region are schematically represented in Figure 9.6. In this limit, it is found that

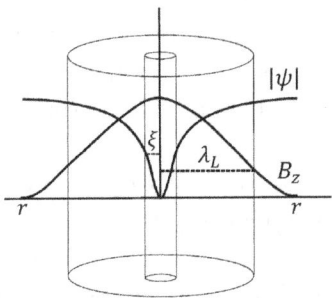

Figure 9.6 Schematic representation of the order parameter, $|\psi|$, and the magnetic field, B_z, in the region of a vortex core.

the energy per unit length is given by

$$\varepsilon_V = \frac{\phi_0^2}{4\pi\mu_0\lambda_L^2}\ln\left(\frac{\lambda_L}{\xi}\right). \qquad (9.45)$$

Suppose now that N_A is the number of flux lines per unit area of the system. The energy per unit volume of these vortices is $N_A\varepsilon_V$ and the corresponding magnetic induction that crosses the system is $B = N_A\phi_0$. Then, at constant applied field H, the work performed to create this flux is $HN_A\phi_0$. It is concluded that when $\varepsilon_V < H\phi_0$, it will be energetically favourable that the field enters the system, which provides the condition defining the lower critical field, H_{c1}. It is given by

$$H_{c1} = \frac{\phi_0}{4\pi\mu_0\lambda_L}\ln\left(\frac{\lambda_L}{\xi}\right). \qquad (9.46)$$

---

[8] Note that in this limit, B_z is constant over a distance ξ from the centre of the vortex and shows a logarithmic decay, $\ln(\lambda_l/r)$ for larger distances.

It is important to note that this expression is only adequate when $\kappa = \lambda_L/\xi \gg 1/\sqrt{2}$.

Vortex avalanches in type II superconductors

Flux lattice has been reported to be periodic over quite long scales in many type II superconductors. Nevertheless, defects analogous to those occurring in crystalline solids such as point defects or dislocations, often occur and break the translational symmetry. These defects concentrate near crystalline defects of the underlying crystal lattice and, in particular, in grain boundaries and impurities. In some cases, due to a high concentration of lattice defects, a competition between vortex interactions and quenched disorder arising from these defects occurs, which may give rise to a complex avalanche dynamics of flux lines. This happens when a type II superconductor is externally driven by, for instance, increasing an applied magnetic field and vortices are forced to move in a landscape with pinning sites that can temporarily trap the vortices. Avalanche dynamics is then associated with local instabilities at which relaxation from one metastable to another metastable minimum of the landscape occurs [331]. In this case, as the driving field increases at a very low rate above the lower critical field H_{c_1}, within the mixed state, the flux that penetrates the superconductor evolves through a sequence of flux jumps consisting of individual vortices. The number of vortices in each jump determines the avalanche size, which can change from one avalanche to another by orders of magnitude. Actually, vortex avalanches in superconductors show avalanche criticality with scaling features similar to those reported in externally driven ferroelastic or ferromagnetic materials discussed in Chapter 8. In particular, it has been shown that avalanche sizes often show power-law distribution [332]. Furthermore, it is important to indicate that in contrast with clean type II superconductors where the flux enters the superconductor in the form of a regular triangular lattice of quantized magnetic flux lines, in systems displaying avalanche dynamics the spatial evolution of the flux may be much more complex and, in some cases, even dendritic patterns can form which are quite common in thin-film superconductors [331].

9.2.2 *The analogy between type II superconductors and smectic A liquid crystals*

To end this section, it is useful to note that there is a strong formal analogy between type II superconductors and smectic-A liquid crystals. This analogy, that was first pointed out by de Gennes [333], arises from the fact that in both cases the order parameter is a complex function and the corresponding phase plays a similar role.

The analogy shows up in an astonishing one-to-one correspondence between properties of superconductors and of smectic-A liquid crystals that can be revealed by comparing Ginzburg–Landau free energies for both the smectic–nematic and superconducting phase transitions. To be specific, consider the free energy discussed in Section 6.3.2, focusing on the smectic A–nematic transition. The smectic-A phase is assumed to be constituted of layers separated by a mean distance d with elongated molecules showing perfect orientational order with the director, $\boldsymbol{n}$, pointing along the z-direction. In the absence of any effect that induces some orientational distortion, the free energy density given by Eq. 6.52 must be of the simpler form

$$f_{n-s} - f = \frac{1}{2}A_s|\psi|^2 + \frac{1}{4}C_s|\psi|^4 + \frac{1}{2}\Lambda_1(\nabla\psi)^2, \qquad (9.47)$$

where now $\psi = \psi_0 \exp(-i\phi_0)$ is the complex order parameter for the nematic–smectic transition that determines the average location of the layers. Similar to the superconductors, higher-order gradient terms have not been considered. Then, this free energy density has the same form as the free energy density of superconductors given in Eq. 9.31 with $\boldsymbol{B} = 0$ (and $\boldsymbol{A} = 0$). The important point is to note that this free energy density must be invariant by simultaneously rotating the director and the layers. Due to the complex scalar character of ψ, this imposes that

$$\frac{1}{2}\Lambda_1(\nabla\psi)^2 = \frac{1}{2}\Lambda_1\left(\nabla + i\frac{2\pi}{d}\delta\boldsymbol{n}\right)\psi^*\left(\nabla - i\frac{2\pi}{d}\delta\boldsymbol{n}\right)\psi, \qquad (9.48)$$

where it has been taken into account that $\nabla\phi = 2\pi/d$. The gradient term in the right-hand side of the preceding equation has the same form as the corresponding term of the Ginzburg–Landau functional for a superconductor in the presence of an applied magnetic field. Therefore, fluctuations of the director that result in distortions of the smectic phase play the same formal role as an applied magnetic field in a superconductor. Considering only the elastic energy associated with twist distortions, the free energy functional for the smectic system can be written in the form

$$\mathcal{F}_{n-s} = \mathcal{F}(T) + \int_\Omega \left[\frac{1}{2}A_s|\psi|^2 + \frac{1}{4}C_s|\psi|^4 + \frac{1}{2}\Lambda_1\left|\left(\nabla - i\frac{2\pi}{d}\delta\boldsymbol{n}\right)\psi\right|^2\right]d\boldsymbol{r}$$

$$+ \frac{1}{2}K_{22}(\hat{\boldsymbol{z}} \cdot \nabla \times \delta\boldsymbol{n})^2, \qquad (9.49)$$

where $\hat{z}$ is a unit vector in the direction of the director. Then, comparing the preceding equation with Eq. 9.31, with $B = \nabla \times A$, it is clear that δn plays the role of the vector potential A and the Frank-Oseen elastic energy the role of the magnetic energy.

The correspondence between both smectics and superconductors will not be analysed in more detail in the present book, but it is worth mentioning that it can be pushed further to show that there is even an analogue of the Abrikosov vortex lattice in liquid crystals. This is the so-called twist-grain-boundary (TGB) or smectic-A* phase. This phase was first predicted by Renn and Lubensky [334] starting from the de Gennes analogy and, soon after, it was experimentally confirmed in Ref. [335]. This TGB phase can be described as constituted of highly anisotropic molecules that are arranged in layers with their long axes normal to the layer planes, on average. Parallel to the layers, molecules display a helical ordering. It can also be viewed as an equilibrium array of screw dislocations arranged in periodically repeated low-angle grain boundaries.

9.3 The BCS theory: a microscopic explanation

The Ginzburg–Landau theory is a very powerful theory to study the transition from normal to the superconducting state, but as any phenomenological theory, it does not give insights about the microscopic mechanism which is at the origin of such a striking phenomenon. The BCS theory [336] is a microscopic theory proposed by Bardeen, Cooper and Schrieffer which posits a microscopic mechanism which permits understanding the superconducting state. The theory is based on the fact that an attractive interaction between electrons exists that, in some circumstances, can overcome the Coulomb repulsion and lead to the formation of bound pairs of electrons or Cooper pairs [337], which are bosonic entities and can condense into a superconducting state, which is also known as the BCS state. Then, while in the normal state of a metal, electrons are the electric conducting entities that move independently, in the BCS state the superconducting entities are bound pairs of electrons.

In the BCS theory, the interaction results from the virtual exchange of phonons that has an attractive character when the energy difference between the electron states involved is less than the phonon energy. In fact, this effective interaction can only be attractive near the Fermi surface. The important conclusion is that in some circumstances the Fermi liquid state comprising independent Bloch electrons can become unstable, even for a weak attraction between them, and a BCS state can form.

It is interesting to start discussing how a pair of bound electrons can form. The Cooper model considers a spherical Fermi surface of radius k_F at zero temperature with all the states with $k \leq k_F$ occupied. Then, a pair of electrons are placed outside the Fermi sphere within a shell of thickness $\hbar\omega_D$, with ω_D being the Debye frequency. These electrons, with wave vectors $\boldsymbol{k}_1$ and $\boldsymbol{k}_2$, interact through the electron–phonon interaction. Therefore, due to phonon exchange, they continually change their wave vector keeping the total momentum conserved. The minimum energy will correspond to a pair with no centre of mass motion so that the total momentum must vanish. This means that $\boldsymbol{k} = \boldsymbol{k}_1 = -\boldsymbol{k}_2$ determines the maximum strength of the attractive interaction. Therefore, the wave function of this pair of electrons should be of the general form

$$\Psi(\boldsymbol{r}_1, \sigma_1, \boldsymbol{r}_2, \sigma_2) = \varphi(\boldsymbol{r}_1 - \boldsymbol{r}_2)\phi_{\sigma_1,\sigma_2}^{spin}, \tag{9.50}$$

where $\phi_{\sigma_1,\sigma_2}^{spin}$ is the spin wave function, which can be either spin singlet with total spin $S = 0$, or triplet with $S = 1$. Almost all known superconductors have singlet pairs and, in this case, φ must be an even function to ensure that Ψ is antisymmetric. Therefore, $\varphi(\boldsymbol{r}_1-\boldsymbol{r}_2) = \varphi(\boldsymbol{r}_2-\boldsymbol{r}_1)$. Defining, $\boldsymbol{r} \equiv \boldsymbol{r}_1-\boldsymbol{r}_2$ and expanding in free electron plane wave states as, $\varphi(\boldsymbol{r}) \propto \sum_{\boldsymbol{k}} g(\boldsymbol{k})e^{i\boldsymbol{k}\cdot\boldsymbol{r}}$, where k must be within the range, $\varepsilon_F < \hbar^2 k^2/2m < \varepsilon_F + \hbar\omega_D$, with ε_F being the Fermi energy, since the interaction is restricted to the shell of thickness $\hbar\omega_D$. This restriction imposes that the coefficients $g(\boldsymbol{k})$, which are related to the probability of finding one electron in a state $\boldsymbol{k}$ and the other in the state $-\boldsymbol{k}$, vanish for $k < k_F$ and $k > \sqrt{2m(\varepsilon_F + \hbar\omega_D)/\hbar^2}$. Then, inserting this wave function in the Schrödinger equation for the two interacting electrons leads to[9]

$$\frac{\hbar^2 k^2}{m}g(\boldsymbol{k}) + \frac{1}{L^3}\sum_{\boldsymbol{k}'} g(\boldsymbol{k}')V_{\boldsymbol{k}\boldsymbol{k}'} = (\varepsilon + 2\varepsilon_F)g(\boldsymbol{k}), \tag{9.51}$$

where ε in the eigenvalue $\varepsilon + 2\varepsilon_F$, represents the gain of energy of the pair with respect to the Fermi energy and $L^3 = \Omega$ is the normalization volume. The integral, $V_{\boldsymbol{k}\boldsymbol{k}'} = \int V(r)e^{i(\boldsymbol{k}-\boldsymbol{k}')\cdot\boldsymbol{r}}d\boldsymbol{r}$, describes scattering processes of the electron pair from states $(\boldsymbol{k}, -\boldsymbol{k}) \to (\boldsymbol{k}', -\boldsymbol{k}')$ and $(\boldsymbol{k}', -\boldsymbol{k}') \to (\boldsymbol{k}, -\boldsymbol{k})$. Within the simpler approximation, it can be assumed that $V_{\boldsymbol{k}\boldsymbol{k}'} = -V_0$, with V_0 a positive constant when k is in the range between k_F and $\sqrt{2m(\varepsilon_F + \hbar\omega_D)/\hbar^2}$, and zero otherwise. Then, from the preceding Eq. 9.51

---

[9] The Schrödinger equation for the pair of electrons reads,
$\frac{\hbar^2}{2m}(\nabla_1 + \nabla_2)\varphi(\boldsymbol{r}_1, \boldsymbol{r}_2) + V(\boldsymbol{r}_1, \boldsymbol{r}_2)\varphi(\boldsymbol{r}_1, \boldsymbol{r}_2) = E\varphi(\boldsymbol{r}_1, \boldsymbol{r}_2) = (\varepsilon + \epsilon_F)\varphi(\boldsymbol{r}_1, \boldsymbol{r}_2)$. Then, the equation is obtained after multiplying by $\exp(-i\boldsymbol{k}' \cdot \boldsymbol{r})$ and integrating over the normalization volume.

it can be obtained that, $\sum_{\boldsymbol{k}}[\hbar^2 k^2/m - (\varepsilon + 2\varepsilon_F)]^{-1} = L^3/V_0$, and replacing the sum by an integral over $\boldsymbol{k}$-space,[10] and splitting the resulting integral over the whole $\boldsymbol{k}$-space into integrals over the Fermi surface and energy, it is obtained that[11]

$$\frac{V_0}{(2\pi)^3} \int \int \frac{dS_E}{|\nabla_{\boldsymbol{k}} E|} \frac{dE}{(2E - \varepsilon - 2\varepsilon_F)} = 1, \tag{9.52}$$

where $E \equiv \hbar^2 k^2/2m$. Assuming that the density of states of free electrons about the Fermi energy, $D(\varepsilon_F)$, is constant, integration leads to

$$\varepsilon = \frac{2\hbar\omega_D}{1 - \exp[2/V_0 Z(\varepsilon_F)]}, \tag{9.53}$$

that has been expressed in terms of $Z(\varepsilon_F)$, which is the half density of states at the Fermi energy, since the integration is done over pair states $(\boldsymbol{k}, -\boldsymbol{k})$. In the limit of weak interaction, $V_0 Z(\varepsilon_D) \ll 1$, it is obtained that, $\varepsilon \simeq -2\hbar\omega_D e^{-2/V_0 Z(\varepsilon_D)}$. This shows that electron-pair bound states with energy $\epsilon = \hbar^2 k^2/m - 2\varepsilon_0$ lower than that of the Fermi sea at zero Kelvin can exist. This result suggests that the ground state of non-interacting electrons can become unstable under the presence of a very weak attractive interaction between electrons. This instability should lead to the formation of Cooper pairs to reach a new lower energy ground state that should be identified with the superconducting phase. Note that, in the preceding development, the Cooper pairs consist of two bound electrons with opposite wave vectors and opposite spins (singlet state).

The BCS ground state

The energy reduction associated with the BCS ground state cannot be obtained by a simple summation of the energy reduction of a pair of electrons as obtained previously. This is because the energy reduction involved in the formation of a new pair depends on those already present, which is a consequence of the complex interaction between the electrons. To account for these effects, the minimum total energy corresponding to all possible pair configurations must be determined considering the kinetic energy of electrons and the energy reduction due to the electron–phonon interaction.

---

[10] It must be remembered that when approaching the continuum limit, $L^{-3} \sum_{\boldsymbol{k}}$ must be replaced by $\int d\boldsymbol{k}/4\pi^3$.

[11] At this step, the volume element is expressed as $d\boldsymbol{k} = dS_E dk_\perp$, where dS_E is an area element of the Fermi surface and $dk_\perp$ is a component of $d\boldsymbol{k}$ normal to the surface. Then, it is taken into account that $dE = |\nabla_{\boldsymbol{k}} E| dk_\perp$. It is important to remind that $D(E) = \frac{1}{(2\pi)^3} \left[\int \frac{dS_E}{|\nabla_{\boldsymbol{k}} E|} \right]$ defines the density of states.

The kinetic contribution[12] can be expressed as

$$E_{kin} = 2 \sum_k w_k \zeta_k, \qquad (9.54)$$

where $\zeta_k = \hbar^2 k^2 / 2m - \varepsilon_F$ and w_k is the probability that the pair $(k \uparrow, -k \downarrow)$ is occupied.

The total energy reduction due to scattering processes can be obtained from a hamiltonian that explicitly describes the scattering processes corresponding to the annihilation of a pair $(k \uparrow, -k \downarrow)$ and the simultaneous creation of a pair $(k' \uparrow, -k' \downarrow)$ with associated energy reduction $V_{kk'}$. To determine this hamiltonian, consider that the two orthogonal states, $|1\rangle_k$ and $|0\rangle_k$, represent that the state $(k \uparrow, -k \downarrow)$ is occupied or unoccupied, respectively. Then, the most general state associated with a pair should be of the form, $|\varphi_k\rangle = u_k |0\rangle_k + v_k |1\rangle_k$, with $w_k = v_k^2$ and $1 - w_k = u_k^2$ being the probabilities that the state $(k \uparrow, -k \downarrow)$ is occupied or unoccupied. Assuming that interaction between Cooper pairs can be neglected, the BCS ground state, $|\Psi_{BCS}\rangle$, can then be approximated by a product of state vectors of individual pairs. That is,

$$|\Psi_{BCS}\rangle = \prod_k \left(u_k |0\rangle_k + v_k |1\rangle_k \right). \qquad (9.55)$$

At this point, it is convenient to proceed further using a two-dimensional representation such that $|1\rangle_k = \binom{1}{0}_k$ and $|0\rangle_k = \binom{0}{1}_k$. In this representation, annihilation and creation of pairs can be described by means of the matrices, $\sigma_k^+ = [\sigma_k^{(1)} + i\sigma_k^{(2)}]/2$ and $\sigma_k^- = [\sigma_k^{(1)} - i\sigma_k^{(2)}]/2$, that combine Pauli matrices $\sigma_k^{(1)}$ and $\sigma_k^{(2)}$.[13] In terms of these two operators, it is obtained that

$$\sigma_k^+ |1\rangle_k = 0, \qquad \sigma_k^+ |0\rangle_k = |1\rangle_k, \qquad (9.56)$$

$$\sigma_k^- |1\rangle_k = |0\rangle_k, \qquad \sigma_k^- |0\rangle_k = 0. \qquad (9.57)$$

Then, the total energy reduction due to the scattering processes $(k \uparrow, -k \downarrow) \to (k' \uparrow, -k' \downarrow)$ and $(k' \uparrow, -k' \downarrow) \to (k \uparrow, -k \downarrow)$ can be described by means of the following hamiltonian:

$$\mathcal{H}_{BCS} = -\frac{V_0}{2L^3} \sum_{kk'} (\sigma_{k'}^+ \sigma_k^- + \sigma_k^+ \sigma_{k'}^-) = -\frac{V_0}{L^3} \sum_{kk'} \sigma_k^+ \sigma_{k'}^-, \qquad (9.58)$$

where the sum is restricted to wave vectors within the shell of width $\pm \hbar \omega_D$ about the Fermi energy. It is important to remark that this hamiltonian

---

[12] Note that the pairing assumes an increase of kinetic energy since an excitation above ε_F is required.

[13] These Pauli matrices are: $\sigma_k^{(1)} = \binom{0\ 1}{1\ 0}_k$ and $\sigma_k^{(2)} = \binom{0\ -i}{i\ 0}_k$.

considers only the interaction involved in Cooper pair creation/annihilation processes but neglects any other interaction between them. The energy reduction can be determined using perturbation theory as the expected value of the hamiltonian in the BCS state, $\langle \Psi_{BCS}|\mathcal{H}_{BCS}|\Psi_{BCS}\rangle$. Then, after some straightforward calculation, the energy of Cooper pairs can be obtained as

$$W_{BCS} = 2\sum_{\boldsymbol{k}} v_{\boldsymbol{k}}^2 \zeta_{\boldsymbol{k}} - \frac{V_0}{L^3}\sum_{\boldsymbol{k}\boldsymbol{k}'} v_{\boldsymbol{k}}u_{\boldsymbol{k}}v_{\boldsymbol{k}'}u_{\boldsymbol{k}'}. \tag{9.59}$$

Finally, the BCS ground-state energy will be found by minimization of W_{BCS} with respect to the probability amplitudes $u_{\boldsymbol{k}}$ and $v_{\boldsymbol{k}}$. It is worth noting that this procedure followed to solve the problem assumes a variational approach equivalent to a mean-field approximation. To proceed with the minimization, it is convenient to assume that the probability amplitudes are real[14] and to express them in terms of the occupation probabilities as, $v_{\boldsymbol{k}} = \sqrt{w_{\boldsymbol{k}}} = \cos\theta_{\boldsymbol{k}}$ and $u_{\boldsymbol{k}} = \sqrt{1-w_{\boldsymbol{k}}} = \sin\theta_{\boldsymbol{k}}$. This ensures that $u_{\boldsymbol{k}}^2 + v_{\boldsymbol{k}}^2 = \cos^2\theta_{\boldsymbol{k}} + \sin^2\theta_{\boldsymbol{k}} = 1$. Then, the minimization condition, $\frac{\partial W_{BCS}}{\partial\theta_{\boldsymbol{k}}} = 0$, leads to

$$\zeta_{\boldsymbol{k}}\tan 2\theta_{\boldsymbol{k}} = -\frac{V_0}{L^2}\sum_{\boldsymbol{k}'}\sin 2\theta_{\boldsymbol{k}'} = -\Delta_0, \tag{9.60}$$

where $\Delta_0 \equiv \frac{V_0}{L^3}\sum_{\boldsymbol{k}'} u_{\boldsymbol{k}'}v_{\boldsymbol{k}'} = \frac{V_0}{L^3}\sum_{\boldsymbol{k}'}\sin\theta_{\boldsymbol{k}'}\cos\theta_{\boldsymbol{k}'}$. Thus, the probability that the state $(\boldsymbol{k}\uparrow, -\boldsymbol{k}\downarrow)$ is occupied can be expressed as

$$w_{\boldsymbol{k}} = v_{\boldsymbol{k}}^2 = \sum_{\boldsymbol{k}}\zeta_{\boldsymbol{k}}\left(1 - \frac{\zeta_{\boldsymbol{k}}}{\sqrt{\zeta_{\boldsymbol{k}}^2 + \Delta_0^2}}\right) - L^3\frac{\Delta_0^2}{V_0}, \tag{9.61}$$

and the condensation energy is obtained by subtracting the ground-state energy of the normal phase, $W_n^0 = \sum_{|\boldsymbol{k}|\leq k_F}\zeta_{\boldsymbol{k}}$. Then, replacing the sum over $\boldsymbol{k}$ by an integral over $\boldsymbol{k}$-space, it is obtained that the condensation energy is given by

$$W_{BCS}^0 - W_n^0 = -\frac{1}{2}L^3 Z(\varepsilon_F)\Delta_0^2. \tag{9.62}$$

This equation indicates that as long as Δ_0 is a finite quantity, the BCS ground state has a lower energy than the ground state of the normal phase. The role of this parameter can be better understood by computing the energy difference between the first excited state and the ground state. The energy of the first exited state, W_{BCS}^1, can be found by considering that a given state $(\boldsymbol{k}'\uparrow, -\boldsymbol{k}'\downarrow)$ is broken. This can be easily imposed by first expressing

[14] It can be shown that the assumption that $u_{\boldsymbol{k}}$ and $v_{\boldsymbol{k}}$ are real does not influence the final result.

the BCS ground-state energy in the form $W_{BCS}^0 = -2\sum_{\mathbf{k}} v_{\mathbf{k}}^4 \sqrt{\zeta_{\mathbf{k}}^2 + \Delta_0^2}$, which is obtained after a straightforward calculation. Then, the breaking up of the state $(\mathbf{k}' \uparrow, -\mathbf{k}' \downarrow)$ is imposed by letting $v_{\mathbf{k}'}^2 = 0$, which leads to $W_{BCS}^1 = -2\sum_{\mathbf{k} \neq \mathbf{k}'} v_{\mathbf{k}}^4 \sqrt{\zeta_{\mathbf{k}}^2 + \Delta_0^2}$ and, finally, to

$$W_{BCS}^1 - W_{BCS}^0 = 2\sqrt{\zeta_{\mathbf{k}'}^2 + \Delta_0^2}. \tag{9.63}$$

The first term in the square root, that is related to the kinetic energy of the two scattered electrons scattered out of the broken Cooper pair, can be arbitrarily small since $\zeta_{\mathbf{k}'} = \hbar^2 k'^2/2m - \varepsilon_F$ and, thus, the excitation requires to supply a minimum energy of $2\Delta_0$. This result indicates that the energy spectrum of a superconducting state is characterized by a gap of energy as illustrated in Figure 9.7. Therefore, at $T = 0$ K, all electrons are condensed

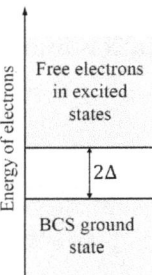

Figure 9.7 Schematic representation of the energy spectrum of single electrons in the superconducting state. At $T = 0$ all Cooper pairs are occupied and all electrons condense in the BCS ground state.

in the form of pairs in the BCS ground state that is separated from the Fermi level by a gap Δ_0. At finite temperature, in the superconducting state, some pairs are broken and the excited states begin to be populated and, well above the Fermi energy, the continuum states of a normal conductor are reached. Actually, the exponential behaviour of the electronic heat capacity (see Figure 9.4) in the low-temperature superconducting region already suggests the existence of a gap in the energy spectrum of electrons.

A convenient expression for the gap can be obtained by combining the equation, $\Delta_0 = \frac{V_0}{L^3}\sum_{\mathbf{k}'} u_{\mathbf{k}'} v_{\mathbf{k}'}$, and $2u_{\mathbf{k}} v_{\mathbf{k}} = \Delta_0 / \sqrt{\zeta_{\mathbf{k}}^2 + \Delta_0^2}$ and replacing the sum by an integral in $\mathbf{k}$-space restricted to the range $\pm\hbar\omega_D$ about the Fermi level. This leads to

$$\Delta_0 = \frac{\hbar\omega_D}{\sinh[1/V_0 Z(\varepsilon_F)]}, \tag{9.64}$$

where $Z(\varepsilon_F + \zeta) \simeq Z(\varepsilon_F)$ has been assumed. For a weak interaction, $\Delta_0 \simeq 2\hbar\omega_D e^{-1/V_0 Z(\varepsilon_F)}$, that shows that even in that case the gap is finite.

Effect of temperature

The effect of temperature on the gap, Δ, can be determined by taking into account that as temperature increases more and more Cooper pairs are destroyed and they should disappear at the critical temperature T_c, at which the gap should vanish. Since the occupation of one electron states must obey the Fermi-Dirac statistics distribution, $f_F(\zeta + \varepsilon_F, T)$, at finite temperature the gap must satisfy

$$\frac{1}{V_0 Z(\varepsilon_F)} = \int_0^{\hbar\omega_D} \frac{d\zeta}{\sqrt{\zeta^2 + \Delta^2}} \left[1 - 2f_F \left(\sqrt{\zeta^2 + \Delta^2} + \varepsilon_F, T \right) \right], \quad (9.65)$$

which is a transcendental equation for the gap that can be solved numerically. Figure 9.8 compares the temperature variation of the gap numerically obtained from the preceding equation and experimental data for type I superconductors, In, Sn and Pb.

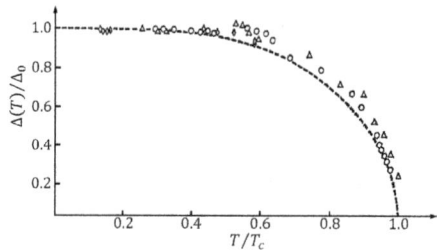

Figure 9.8 Temperature dependence of the gap in In (circles), Sn (triangles) and Pb (diamonds) obtained from tunnelling experiments. The dashed line is the numerical solution obtained from Eq. 9.65. The small deviations between experiments and theory are mainly a consequence of the fact that the interaction is assumed constant. Adapted from [338].

The critical temperature can be determined by taking into account that $\Delta \to 0$ at T_c. Then, imposing $\Delta = 0$ in the preceding equation, it can be numerically obtained that

$$T_c \simeq 1.14 \frac{\hbar\omega_D}{k} e^{-1/V_0 Z(\varepsilon_F)}. \quad (9.66)$$

Comparing the preceding expression of the critical temperature with the equation giving the gap in the weak interaction limit, it results that $\Delta_0/kT_c \simeq 1.76$. This relation is satisfied by most type I superconductors, to a reasonably good approximation. Therefore, it can be used to estimate the coupling constant $Z(\varepsilon_F)V_0$ from an experimental determination of Δ_0 and T_c.

It is worth remarking that a version of the Ginzburg–Landau theory adequate for materials with a small parameter κ ($= \lambda_L(T)/\xi(T)$) was derived from the BCS microscopic theory by Gor'kov [339]. It is found that the gap Δ plays the role of the natural order parameter for the superconducting transition and can thus be related to the order parameter ψ of the Ginzburg–Landau theory. The derivation allows a microscopic interpretation of the phenomenological parameters of the Ginzburg–Landau theory. In addition, the equivalence of both approaches shows that critical exponents in the BCS theory are the mean-field exponents characteristic of the Ginzburg–Landau theory.

It is useful to close this section by briefly discussing how the BCS theory accounts for the disappearance of the electrical resistivity and the ideal diamagnetism that are the basic properties that characterize the superconducting phase. For such a purpose, consider the BCS state in the presence of a current flux, j. In this regime, if n_s is the density of individual electrons, then $j = -n_s e v = -\frac{n_s e \hbar}{m} k$. Therefore, the existence of the current supposes an increase of the global momentum of Cooper pairs given by, $K = -\frac{2m}{n_s e \hbar} j$. In this situation, the state of Cooper pairs should be represented as, $(k + \frac{1}{2}K \uparrow, -k + \frac{1}{2}K \downarrow)$ and, thus, the positional part of their wave function should be of the form

$$\varphi(r_1, r_2) = e^{iK \cdot R}\varphi(K = 0, r), \tag{9.67}$$

where $R = (r_1 + r_2)/2$ is the centre of mass position of the pair, and $r = r_1 - r_2$. Therefore, compared with the wave function given by Eq. 9.50, the new wave function includes a phase contribution that accounts for the flux. Note, however, that $|\varphi(K \neq 0, r)|^2 = |\varphi(K = 0, r)|^2$. Moreover, the scattering matrix, $V_{kk'}$, is not affected by the current since it depends only on the separation between electrons in the Cooper pair. Consequently, all the equations that have been found beforehand remain the same in the presence of a current and, in particular, the gap in the energy spectrum remains unchanged. For instance, if a supercurrent is induced by changing the magnetic flux, the gap will continue to exist. Thus, an alteration of pair states by inelastic scattering processes will only occur if the excitation causes, at least, an increase of energy 2Δ. This increase is large and thus these kinds of excitations can be ruled out as a source of energy dissipation. On the other hand, elastic collision events can only contribute to current relaxation if they cause at least a change of one quantum of flux. These events might occur with a probability vanishingly small and the conclusion is that a superconductor can carry a current with no resistance, which supposes that the superconducting state is very stable. However, if the current is too high, the excess of energy associated with the centre of mass motion, $2\hbar^2 K^2/m$ may become larger than 2Δ and superconductivity will be destroyed. Since $K \geq k_F$, this leads to a critical current, $j_s \simeq en_s\Delta/\hbar k_F$. The associated critical field, H_c, will be

$$H_c = \lambda_L j_c \simeq \lambda_L \frac{en_s\Delta}{\hbar k_F}, \tag{9.68}$$

which shows that the critical field is proportional to the gap.

Beyond the BCS theory and high-T_c superconductors

The original results of the BCS theory consider an *s*-wave superconducting state, which is the rule among low-temperature superconductors but is not realized in many unconventional superconductors such as the *d*-wave high-temperature superconductors. Actually, extensions of the BCS theory exist to describe these cases and others. Nevertheless, so far, these extensions seem to be insufficient to completely describe the observed features of high-temperature superconductivity in cuprates that, in spite of being discovered in 1986, still remain mysterious to some extent. What is the electron pairing mechanism that sustains superconductivity up to such unexpected high temperatures? Traditionally these questions have been addressed from different viewpoints that are summarized here. A conservative point of view consists in staying within the BCS ideas but assuming a stronger coupling between electron pairs originating from the harmonic or anharmonic motion of the oxygen atoms.

Another possibility claims that magnetic features are essential to explain the condensation of electron pairs. For instance, de Gennes suggested [340] that a double-exchange mechanism[15] might induce a suitable electron–electron attractive interaction via magnetic interactions. The problem seems to be that the picture predicts ferromagnetism that has never been detected in cuprates. More recently, new experiments appear to provide strong evidence for a phenomenon called *superexchange*. This mechanism is similar to the double-exchange mechanism but has the advantage that it usually leads to antiferromagnetism and thus seems more adequate to describe superconductivity in high-T_c cuprates. The proposed mechanism suggests that virtual transitions of electrons between Cu and atoms in adjacent CuO_2 planes gen-

---

[15] The double-exchange mechanism is a type of magnetic exchange that may arise between ions in different oxidation states.

erate a spin–spin interaction that might give rise to a pairing process and thus to superconductivity [341].

It is also important to mention that another point of view considers that electrons in cuprates form a *Luttinger liquid* instead of the standard *Landau liquid*, which is in fact considered in the BCS theory. The Luttinger liquid model applies to one-dimensional chains and the interest is that it predicts the existence of a regime of attractive interaction between electrons. The existence of this regime might indicate that superconductivity will occur in a strongly interacting Luttinger liquid at high electron density.

It seems, however, that the current consensus is that high-temperature superconductivity in cuprates is associated with a pairing mechanism which involves d-waves large amplitude spin fluctuations. This mechanism is prone to occur close to the boundary between the antiferromagnetic and superconductivity *dome* in the phase diagram of high-T_c superconductors. This mechanism seems to satisfactorily explain the anisotropic properties observed in high-T_c cuprates [342].

With regard to the effect of magnetism, it is worth remarking that in some exotic U-based ferromagnetic compounds, there is some evidence that ferromagnetic fluctuations provide the indirect pairing interaction that leads to superconductivity. Strikingly, in these compounds, ferromagnetism and superconductivity seem to coexist. This might be explained taking into account that these materials are highly anisotropic itinerant ferromagnets with low magnetic ordering temperatures. Consequently, magnetic fluctuations are large and when a magnetic field is applied perpendicular to the easy-axis (c-axis of the crystal) superconductivity survives for very high fields, while it is destroyed when the field is applied along the easy-axis direction. These results provide some support to a pairing mechanism induced by spin fluctuations [343].

Exercises

9.1 Consider a superconducting thin film of thickness $2d$ limited by planes of area L^2 subjected to a magnetic field $\boldsymbol{H} = (0, 0, H_0)$ applied parallel to the planes in the z-direction. Assuming that $d \ll L$ and imposing as a boundary condition that the component of the magnetic induction $\boldsymbol{B}$ parallel to the surface is continuous across the surface, use the London theory to calculate $\boldsymbol{B}$ and the current density $\boldsymbol{j}$ as a function of position.

9.2 The two-fluid model[16] suggests the following electronic free energy function to account for superconductivity

$$F(x, T) = x^{1/2} f_n(T) + (1 - x) f_s(T),$$

where x represents the fraction of electrons that are in the *normal* fluid and $(1 - x)$ is the fraction condensed into the *superfluid*. Here f_n and f_s are chosen as:

$$f_n(T) = -\frac{1}{2} \gamma T^2,$$
$$f_s(T) = -\beta = \text{constant}, \tag{9.69}$$

where γ is the Sommerfeld coefficient and β is the condensation energy.

(a) Show that the transition temperature of the model is, $T_c = 2\sqrt{\beta/\gamma}$.

(b) Determine how x depends on temperature.

9.3 The following model is proposed to describe the heat capacities of the normal, C_n, and superconducting, C_s, phases of a type I superconductor

$$C_n(T) = \gamma T,$$
$$C_s(T) = \alpha T^3,$$

where γ is the Sommerfeld coefficient and α a parameter. Both are assumed to be constant independent of temperature, T, and magnetic field, H.

(a) Find an expression for the normal-superconducting transition temperature in the absence of an applied magnetic field, T_c, as a function of α and γ and show that at this temperature the heat capacity displays a discontinuity.

---

[16] This model was proposed by C. J. Gorter and H. G. B. Casimir [344]. In this model, there is no physical basis for the assumed expression for the free energy. In any case, it provides a fairly accurate representation of experimental results.

(*b*) Find the magnetic field dependence of the normal-superconducting phase transition.

(*c*) Compare this model with the Gorter-Casimir model considered in the preceding exercise.

9.4 Consider the BCS hamiltonian,

$$\mathcal{H}_{BCS} = -\frac{V_0}{2L^3} \sum_{kk'} (\sigma^+_{k'} \sigma^-_{k} + \sigma^+_{k} \sigma^-_{k'}) = -\frac{V_0}{L^3} \sum_{kk'} \sigma^+_{k} \sigma^-_{k'},$$

and show that the expected value of $\mathcal{H}_{BCS}$ in the BCS state can be expressed as,

$$\langle \Psi_{BCS} | \mathcal{H}_{BCS} | \Psi_{BCS} \rangle = -\frac{1}{4} \frac{V_0}{L^3} \sum_{k,k'} \sin 2\theta_k \sin 2\theta_{k'}, \qquad (9.70)$$

where $\cos^2 \theta_k = v^2_k$ and $\sin^2 \theta_k = u^2_k$ are the probabilities that the state k is occupied and unoccupied, respectively.

10

Quantum phase transitions

Quantum phase transitions occur due to quantum fluctuations instead of thermal fluctuations at the absolute zero temperature driven by the variation of a non-thermal control parameter. At this temperature, quantum fluctuations occur associated with Heisenberg's uncertainty principle and, thus, a phase transition may take place when there are competing ground state phases that are accessible for different values of certain parameters of the hamiltonian. A simple introduction to this subject can be found in Ref. [345], and a complete and pedagogical discussion in Ref. [1]

The interest in this class of phase transitions is due to the fact that they give rise to the emergence of novel quantum phenomena which can influence the properties of materials at finite temperature since their occurrence leaves clear fingerprints in extended finite temperature regions of the phase diagram. This means that significant traces of these transitions can be experimentally detected at low but still easily reachable temperatures. This is important because a thorough understanding of quantum transitions and, especially, quantum criticality, is crucial for a better understanding of certain classes of materials and, especially, some strongly correlated electron materials[1] such as high-T_c superconductors or rare earth magnetic insulators that cannot be described within the more usual independent-electron framework.

A crucial point in the study of quantum phase transitions refers to the fact that quantum fluctuations have properties quite different from those of thermal fluctuations. Consequently, the study of quantum phase transitions and quantum criticality requires novel approaches that often have no

---

[1] Strongly correlated electron materials are materials with strong electronic correlations, in which the state of a given electron depends on the positions and movements of all other electrons. Therefore, the wave function of these systems cannot be expressed as a product of simple configurations of electrons.

analogue in the standard phase transition theories. Nevertheless, the possibility of mapping d-dimensional quantum models to higher dimensional thermal models permits the use of standard statistical mechanics techniques to find solutions of quantum models.

The external parameters that can drive quantum phase transitions are, for instance, parameters such as pressure, magnetic field, or density of electrons that can be controlled by the concentration of dopant ions. Sometimes, the transition can be first order and it involves a jump in physical properties of the system. Continuous quantum transitions are more common and in this case the change of physical properties takes place gradually and quantum criticality occurs in the neighbourhood of a given value of the external parameter, which locates the quantum critical point. Far from a quantum phase transition, the system is in a state that can be described by a wave function that can be written as a product of wave functions of simple configurations of the constituents of the system. In first-order quantum phase transitions, the wave function has this structure in both sides of the transition, with just a jump from one side to the other that manifests itself in the expected values of some physical observables. In contrast, for continuous transitions, close to the critical point the wave function is very different. It comprises a complex superposition of very large sets of configurations that fluctuate at all length scales, which is the characteristic of any critical point, either quantum or classical.

There are many examples of systems exhibiting quantum criticality. Among them, high-T_c superconductors, quantum Hall systems, or ferromagnetic and antiferromagnetic materials as well as ferroelectric systems are worth mentioning. Actually, the aim of the Chapter is to briefly introduce basic concepts to understand the relevance of quantum phase transitions and to show that this is very helpful for a correct understanding of the behaviour of a number of materials, especially at low temperatures. Some selected examples will be discussed in detail in this chapter.

10.1 Thermal and quantum phase transitions

The question that must be discussed first is to what extent quantum mechanics is important for a thorough comprehension of phase transitions and what are the consequences arising from quantum mechanical effects, especially in relation to the possibility that quantum fluctuations can be dominant for a phase transition to occur. In general, it is well known that quantum mechanics is essential for an understanding of the ground state of materials undergoing phase transitions. Actually, this aspect has been considered in detail in the preceding Chapter in the case of superconductors. Nevertheless, this relevance of quantum mechanics to determine the ground state is not enough to decide whether a phase transition is a quantum transition or not. In the case of the models for superconductors that have been considered in Chapter 9, the thermal energy scale, kT, is in fact dominant and the transition is still driven by thermal fluctuations. In general, if ω is a frequency that characterizes the spectrum of excitations of the system, quantum effects are expected to be important when the energy scale, $\Delta = \hbar\omega$ becomes comparable or larger than thermal energy. In fact, Δ can be related to the lowest excitation above the ground state that can be non-zero if the spectrum has a gap or to a characteristic energy when the spectrum is gapless and there are excitations at arbitrarily low energy in the infinite lattice limit. In these situations, quantum fluctuations are important at the microscopic scale but not at the longer length scales that control the critical behaviour. This means that at any finite temperature, quantum fluctuations can be captured by a classical statistical mechanics formalism with an effective hamiltonian for the order-parameter field. This can be done, for instance, within the framework of the phenomenological Ginzburg-Landau theory. Therefore, only in the strict limit $T = 0$, a quantum phase transition can occur. In this situation, it is expected that the asymptotic critical behaviour will be different from classical behaviour due to the specific quantum features.

In this chapter, all transitions taking place at finite temperature will be considered *classical*. Only transitions that occur at zero temperature will be considered as quantum phase transitions. To better understand the differences between classical and quantum cases, it is convenient to remind that at finite temperature, thermodynamic properties near a phase transition are, in principle, derived from the partition function given by, $Q = \operatorname{Tr} \exp(-\beta\mathbf{H})$, where Tr stands for the trace and $\mathbf{H} = \mathbf{H}_k + \mathbf{H}_p$ is the hamiltonian operator that is the sum of kinetic and potential contributions. While in classical systems, these two operators commute and the partition function factorizes as $Q = Q_k Q_p$, this is not the case in the quantum case, which supposes that statics and dynamics cannot be decoupled. Therefore, the order parameter for a quantum transition will need to be a space- and time-dependent field.

To deeply analyse the consequences of this time dependence, it is worth comparing the expression of the partition function expressed in the basis of the complete set of time-independent eigenstates of $\mathbf{H}$ with the expression of a time-dependent state vector also expressed in the same basis.[2] This

---

[2] On the basis of the complete set of time-independent eigenstates of $\mathbf{H}$, which are solutions of the time-independent Schrödinger equation, $\mathbf{H}|\psi_{E_n}\rangle = E_n|\psi_{E_n}\rangle$, the partition function can be written as $Q = \sum_n \langle\psi_{E_n}|e^{-\beta\mathbf{H}}|\psi_{E_n}\rangle$. On the other hand, for a time-dependent state

comparison indicates that Q takes the form of a sum of imaginary time transition amplitudes for a system that starts in some state and returns to the same state after an imaginary time, $-i\hbar\beta$. From this point of view, the operator $\exp(-\beta\mathbf{H})$ is the same as the time-evolution operator $\exp(-i\mathbf{H}\mathcal{T}/\hbar)$ provided that the imaginary value $\mathcal{T} = i\hbar\beta$ is assigned to the time interval over which the system evolves [346, 347]. After this identification, the time evolution operator should be written as $\exp(-\mathbf{H}\mathcal{T})$. This means that the formalism involved in the calculation of thermodynamic quantities of a quantum system is the same as that corresponding to the calculation of transition amplitudes, with the total time interval fixed by the temperature of the thermodynamic system.[3]

In the language of path integrals formulation of quantum mechanics, the preceding point of view can be interpreted in the sense that the path followed by the system can be established by determining its state at a sequence of time steps. Formally, this means that the operator $e^{-\beta\mathbf{H}}$ can be written as, $[e^{-(\delta\tau/)\mathbf{H}}]^N$, with $\delta\tau$ being a real time associated with the small interval of imaginary time $i\delta\tau$, and N a large integer such that, $N\delta\tau = \hbar\beta$. It can be shown [347] that in this picture the quantum partition function of a d-dimensional system takes the form of a classical partition function for a system of $d+1$ dimensions. However, the extra dimension has a finite extent, $\hbar\beta$, at finite temperature. When $T \to 0$ the extent of this extra dimension diverges and the system can be treated as a $D \equiv (d+1)$-dimensional effective system. The conclusion is that this point of view provides a quantum-classical mapping where the temperature maps onto the inverse length of the imaginary time axis that acts as an additional dimension in the quantum system.

Suppose that at zero temperature, the system can be driven by changing a parameter g of the hamiltonian. It will be assumed that a continuous quantum phase transition occurs at a critical value g_c. This parameter can be a magnetic field, pressure, or doping among others depending on the system considered. As usual, close to a critical point, the correlation length, ξ, that determines the exponential decay of equal-time correlations in the ground state is expected to diverge such that, $\xi \sim |g - g_c|^{-\nu}$, where ν is the correlation length exponent. On the other hand, at a quantum critical point, the energy scale Δ should go to zero as, $\Delta \sim \xi^{-z}$, where z is the dynamical

---

vector $|\Psi(t)\rangle$, a solution of the general time-dependent Schrödinger equation,
$-i\hbar\frac{\partial\psi(r,t)}{\partial t} = \mathbf{H}\Psi(r,t)$, in the same basis, can be expressed as,
$|\Psi(t)\rangle = \sum_n A_n e^{-iE_n t/\hbar}|\psi_{E_n}\rangle$.
[3] The relevant aspect of this point of view is that it provides a picture of temporal propagation but the fact that the time interval is imaginary is not a central point.

exponent. Therefore, $\Delta \sim |g - g_c|^{z\nu}$. This equation suggests that close to a quantum critical point, the homogeneity condition of the singular part of the free energy density, that is given by Eq. 1.145 for a classical system, should read in the form

$$f(\epsilon, T) = b^{-(d+z)} f(\epsilon b^{1/\nu}, T b^z), \tag{10.1}$$

where in this case, $\epsilon = (g - g_c)/g_c$. The term involving T is included to account for the case of the quantum critical point being approached by lowering the temperature at $g = g_c$. The preceding equation explicitly highlights the increase of space dimension associated with time. It is worth remarking that while in general $z = 1$, in some cases other values, including fractional values, have been found. Therefore, the general form of the hyperscaling relation for a quantum critical point should read, $2 - \alpha = (d + z)\nu$, instead of Eq. 1.149 which applies to classical critical points.

In conclusion, the behaviour close to a quantum transition is determined by the relation between the correlation time, τ_c, and the extension of the imaginary time direction. A crossover from quantum to classical should occur when the correlation time becomes larger than β or, equivalently when $|\epsilon|^{\nu z} < kT$. In this situation the system realizes that it is in d dimensions and not in $d + z$. Note that when the transition is approached by lowering the temperature at $g = g_c$, both τ_c and β simultaneously diverge and quantum effects are always relevant. In the quantum critical state, the system is described by a complex, entangled, wave function. In fact, quantum effects show up in a quite wide region of the phase diagram that ends at the quantum critical point as illustrated in Figure 10.1.

In the case shown in panel (a) of Figure 10.1, a quantum critical point exists at zero temperature but the system does not show any phase transition at finite temperature. At finite temperature, three regimes can be distinguished depending on whether the behaviour is dominated by thermal or quantum fluctuations of the order parameter. In the thermally disordered region, long-range order is destroyed mainly by thermal fluctuations. Instead, in the quantum disordered region, the physics is influenced by quantum fluctuations. In between these two regions, in the region denoted as quantum critical, both types of fluctuations are relevant, but not critical. In fact, the physics, as has been discussed above, is controlled by thermal excitations of the quantum critical ground state. The boundaries of this region are determined by the condition $kT > \Delta \sim |g - g_c|^{z\nu}$. It might appear paradoxical that the width of this region increases as temperature rises. In fact, this region can only exist as long as thermal energy is lower than the characteristic energy scale of the constituent interaction.

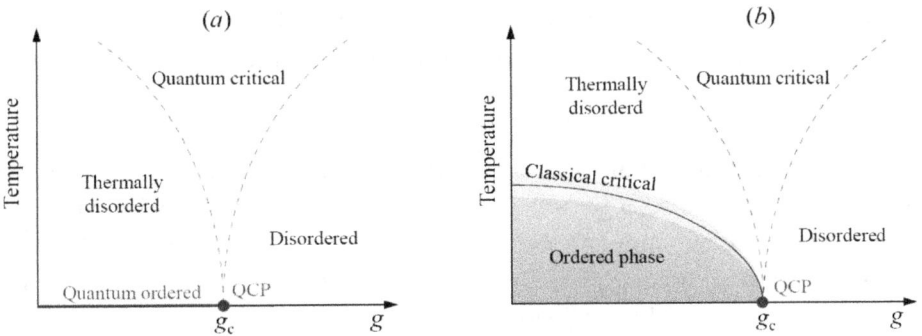

Figure 10.1 Schematic representation of the phase diagram of systems undergoing a continuous quantum phase transition. Here g is the control parameter that drives the quantum transition and quantum criticality occurs at $g = g_c$. (a) Long-range order only occurs at zero temperature. Dashed lines locate the region where quantum fluctuations arising from quantum criticality can be detected at finite temperature. (b) Long-range order can also occur at finite temperature. The classical critical line for the thermal order-to-disorder transition ends at the quantum critical point at zero temperature. The critical region along this line is indicated. Adapted from Ref. [348].

Some materials show a richer phase diagram, which is illustrated in Figure 10.1(b). This class of materials undergoes a thermal phase transition and long-range order occurs at finite temperatures. The quantum critical point corresponds to the end of the finite temperature transition temperatures. Along this line, transitions are dominated by classical fluctuations. An interesting aspect is that the corresponding classical critical region becomes narrower as temperature decreases and the quantum critical point is approached. The quantum critical region still persists, at finite temperature, above the quantum critical point.

10.1.1 Examples

Many experimental results have been reported that nicely illustrate the expected behaviour of systems undergoing quantum phase transitions and quantum criticality that have been formerly discussed. The most common example of a quantum phase transition occurs associated with magnetism and magnetic exchange originating from the spin of electrons. Notwithstanding, quantum criticality has been reported as well in other classes of materials, including superconductors and ferroelectrics. Here three magnetic materials showing quantum criticality induced by a magnetic field and pressure, respectively, and an example in a high-T_c superconductor are considered as prototypical examples of materials undergoing a quantum phase transition.

LiHoF$_4$ insulator

The insulator LiHoF$_4$ can be characterized by a net spin localized at Ho ions with low-lying magnetic excitations consisting of fluctuations between two spin states with opposite orientations along a certain crystalline axis that defines the magnetic easy axis. The existence of the easy axis is a consequence of crystal field effects.

At zero temperature, the magnetic dipolar interactions between the Ho ions cause all the spins to align in the same orientation, which gives rise to a ferromagnetic ground state. When the material is subjected to a transverse magnetic field applied perpendicular to the direction of the easy axis it induces quantum tunnelling between the two states of each Ho ion. For large enough tunnelling rate, long-range magnetic order can be destroyed and, consequently, this system is prone to display a quantum phase transition induced by the transverse field. In fact, experimental results indicate that the magnetic moment should vanish continuously at a quantum critical point [349].

At finite temperature, LiHoF$_4$ undergoes a paramagnetic-ferromagnetic phase transition at a temperature that decreases as the transverse field increases. The system behaves as a dipolar-coupled Ising ferromagnet for which the upper critical dimension is $d = 3$. Thus, mean-field criticality (with a logarithmic correction) is expected to occur. This is consistent with experimental results reported in [349] that show that the susceptibilty exponent is $\gamma = 1$ down to the lowest studied temperatures to a very good approximation. The phase diagram and the behaviour of the susceptibility of this system are shown in Figure 10.2.

At zero temperature, a quantum critical point is expected to occur at a transverse field slightly above 4.93 T. In this case, the transition should only be driven by quantum fluctuations. Moreover, it should occur from a ferromagnetic phase to a quantum paramagnetic phase. Experimental results suggest that quantum critical behaviour in this material is also mean-field-like.

CoNb$_2$O$_6$ colbalt niobate

The cobalt niobate oxide, CoNb$_2$O$_6$, is especially an interesting material [350] showing a quantum critical point. This orthorhombic material is an insulator and, thus, the electrons are tightly bound to the atomic nuclei. Therefore, only the orientations of the spins, localized at lattice sites, are

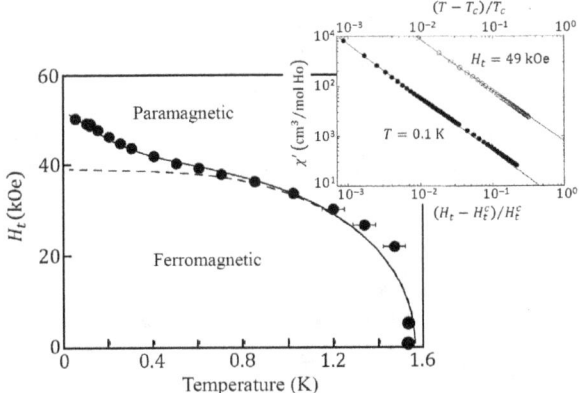

Figure 10.2 Phase diagram of LiHoF$_4$ giving the strength of the transverse field at the transition as a function of temperature of the ferromagnetic–paramagnetic transition The dashed line is the result of a mean-field model, that takes into account the electronic spin degrees for freedom only. The continuous line is an extended mean-field model that incorporates nuclear hyperfine interactions. The inset shows a log-log plot of the magnetic susceptibility (real part, χ', of the ac-susceptibility) measured at very low temperature, as a function of the reduced temperature (open symbols at $\mu_0 H_t = 4.9$ T, with $T_c = 0.114$ K) and, as a function of the reduced transverse field (solid symbols, at $T = 0.100$ K, with $\mu_0 H_t^c = 4.93$ T). A numerical fit renders a mean-field exponent value $\gamma = 1$. Adapted from Ref. [349].

relevant variables. In this material due to the spin-orbit effects, the Co^{+2} spins have a lower energy when they are either parallel or antiparallel to each other, which is a consequence of a strong easy axis anisotropy caused by crystal field effects. This means that at each Co-site j, Co^{2+}-ions can be in two different spin states, denoted as up and down states, respectively. These ions occupy the centre of CoO_6 octahedra, which are arranged into nearly-isolated zigzag chains along the c-axis as shown in Figure 10.3. Therefore, from the magnetic point of view, the material can be modelled as a quasi-1d magnetic system with a ferromagnetic ground state with all the spins parallel, pointing either up or down.

$CoNb_2O_6$ can be driven by a transverse magnetic field applied along the c-direction, which is perpendicular to the easy axis along the b-direction. For a large strength of the field, minimization of Zeeman energy favours the appearance of a new ground state with spins perpendicular to the easy axis. In contrast to the ferromagnetic ground state, this new state is invariant under the interchange of up and down states so that this change does not break the ac-plane reflection symmetry. Therefore, this phase is not ferromagnetic and is denoted as quantum paramagnetic. The strength of the transverse field plays the role of the parameter g and, at a given value, the system changes from one ground state to the other through a continuous phase transition. This means that the phase diagram displays a quantum critical point at $g = g_c$, which corresponds to a given value of the applied transverse field.

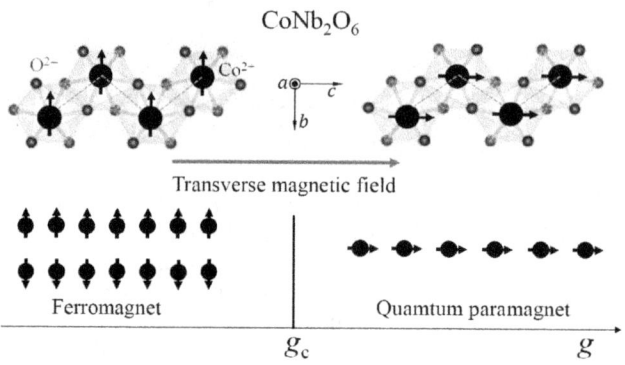

Figure 10.3 Zig-zag structure of Co^{2+} ions along the c-axis of $CoNb_2O_6$ indicated by the dashed line. The magnetic ground state can be tuned by an applied transverse magnetic field. For low magnetic fields, the material is ferromagnetic while for large enough fields magnetic moments are parallel to the field to minimize Zeeman energy and the system behaves as a quantum paramagnet. The system shows a quantum critical point at which the magnetic order changes from ferromagnetic to quantum paramagnetic. The strength of the magnetic field is the parameter g that controls the ground state order. Criticality occurs at a given value $g = g_c$.

This material does not show an ordered phase at finite temperature and, thus, corresponds to the example shown in Figure 10.1(a). The existence of the critical point was detected by means of neutron scattering experiments performed in large single crystals. Neutron scattering experiments confirm changes in spin dynamics as the transverse field increases, which corroborates the existence of a region of finite temperature dominated by quantum effects, resulting from the existence of the quantum critical point. Actually, at low field, at a finite temperature above the ordered phase, spin excitations are associated with pairs of kinks constituted of domain-wall quasi-particles. Instead, the character of the excitations changes to spin-flips above the paramagnetic quantum phase. In contrast, in between these two regions, the spin dynamics shows a fine structure with two sharp modes at low energies with a ratio consistent with the symmetry of the excitation spectrum that confirms the existence of a quantum critical region.

TlCuCl₃ antiferromagnet

Another important example of a material that displays a quantum critical point is the dimer antiferromagnet TlCuCl₃. At ambient pressure, this material is monoclinic with Cu_2Cl_6 zigzag chains extending along the *a*-axis. Cu^{2+} ions carry a net spin $1/2$ and each spin is antiferromagnetically coupled to only a single spin due to superexchange via Cl-ions. Thus, the system consists of singlet bond spin dimers as shown in Figure 10.4. However, for large pressure, the ground state changes to a Néel antiferromagnet with polarized spins forming a checkerboard pattern.

In the dimerized ground state, the coupling between dimers vanishes and the state is rotationally invariant, which is the condition that defines a quantum disordered or paramagnetic phase. Therefore, starting from this state, by increasing the applied pressure a quantum critical point is reached at a pressure p_c and the constrained system transforms to the quantum ordered Néel antiferromagnetic phase. In this system, the quantum critical point is the end of the line of Néel thermal critical points, $T_N(p)$. In this case, the parameter g can be defined as the inverse of pressure. The behaviour of this system is schematically illustrated in Figure 10.4. Similarly to the case of cobalt niobate discussed previously, quantum criticality effects extend at non-zero temperature and can be detected using neutron scattering as changes in the spin dynamics [352]. For small g or high pressure, thermal effects induce spin waves that distort the perfect antiferromagnetic arrangement of the ground state. For large g or small pressure, they break dimers and induce the formation of quasi-particles called triplons. In the region between these two behaviours, the dynamics cannot be described in terms of either triplons or spin waves, since the dynamics associated with both excitations is strongly coupled which is reminiscent of quantum criticality. These effects can be observed as long as the thermal energy remains lower than the spin-spin coupling energy. The pressure dependence of the excitations across the quantum critical point is shown in Figure 10.5.

High-T_c superconductors

The examples discussed above correspond to insulator materials that display a quantum phase transition related to the ordering of localized spins. Nevertheless, there is a lot of experimental data that corroborates the existence of quantum phase transitions in metallic compounds. Within this family, some high-T_c superconductor cuprates such as $La_{2-x}Sr_x CuO_4$ have been studied in detail [353]. These materials display an insulating ground state with Néel antiferromagnetic order in a certain range of stoichiometry. However, doping

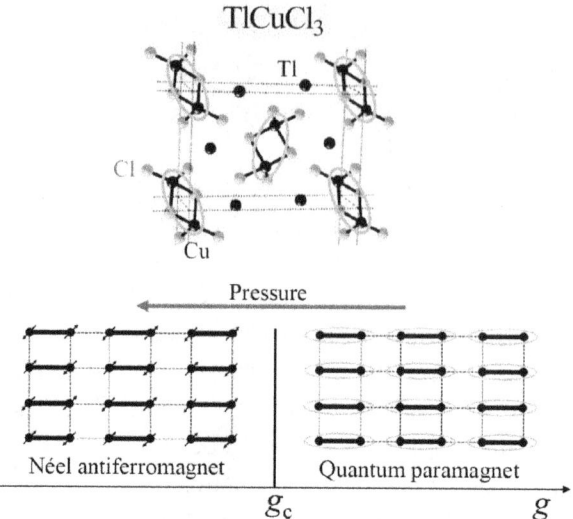

Figure 10.4 Schematic view of the crystal structure of the $TlCuCl_3$ antiferromagnet in the top panel. At low pressure, a dimerised phase is formed. In the figure, the ellipses indicate the spin dimers located at pairs of Cu^{2+} ions. At high pressure, an ordered Néel antiferromagnetic ground state occurs. The pressure-induced change of ground states is schematically represented in the bottom panel, where the existence of a quantum critical point is indicated. The parameter g is related to the inverse of the applied pressure.

the material by changing the relative amount, x, of La and Sr elements may induce the appearance of mobile charge carriers, which provide the metallic character, that allows the material to display superconductivity. Associated with this behaviour, this material is expected to show a quantum phase transition from a Néel antiferromagnetic insulator to a superconductor material controlled by doping. In reality, similar behaviour has been observed in other superconductors such as the $Pr_{2-x}Ce_xCuO_4$ cuprate [354] and even in some iron based pnictides such as the $BaFe_2(As_{1-x}P_x)_2$ compound [355]. At a first glance, the behaviour of these superconductors shares many similarities with the behaviour of the $TlCuCl_3$ antiferromagnet. For low doping or high pressure, both classes of materials show antiferromagnetic order, while at high doping or low pressure, a pairing mechanism leads to spin dimers in the insulator case or Cooper pairs when charges are mobile, which gives rise to superconductivity. In any case, new ingredients must be taken into account to understand the existence of quantum phase transitions in superconductors [356]. Due to the transition from antiferromagnetic insulator to superconductor the properties of the Fermi surface must change significantly.

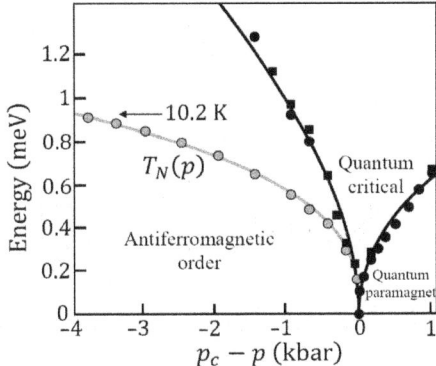

Figure 10.5 Diagram showing the pressure dependence of the excitations across the quantum critical point at $T = 1.85$ K in the TlCuCl$_3$ anti-ferromagnet. At high pressure, excitations are associated with spin waves while at low pressure they correspond to triplon quasi-particles. Continuous lines determine the quantum critical region. Symbols are measurements at selected temperatures of the pressure limits of wave (symbols on the middle curve) and quasi-particle (symbols on the right-side line) excitations, respectively. The left line locates the thermal critical points, $T_N(p)$, corresponding to transitions from paramagnetic to antiferromagetic states. This line ends at the quantum critical point at $p_c = 1.07$ kbar. The symbols on the left line are experimental data reported in Ref. [351]. Adapted from Ref. [352].

This means that a correct understanding of this quantum phase transition requires to take into account electronic effects that strongly influence the Fermi surface in addition to collective spin fluctuations associated with the pairing mechanism that, for instance, supposes a change from spin wave to triplon excitations in TlCuCl$_3$. More specifically, this supposes that models must reflect the appearance of a gap associated with the change of ground state from insulator to superconductor.

10.2 Quantum Ising systems

Most quantum phase transitions involve ordering of spins and can be described by quantum Ising models. The quantum Ising model in a transverse field is the simplest model that displays a quantum critical point. This model finds natural realizations in some of the materials discussed in the preceding section. For instance, the CoNb$_2$O$_6$ (colbalt niobiate) and the LiHoF$_4$ rare earth magnets are quite good examples of $1d$ and $3d$ realizations of this model. In both materials, the two spin states can be represented as Ising spin variables localized on Co or Ho ions, respectively. In LiHoF$_4$, the interaction has in fact a long-range dipolar nature, while the quantum Ising model takes into account only the nearest neighbour interactions between spins. In any case, as illustrated in Figure 10.2, a mean-field treatment of the model provides a reasonably good description of this material even very close to the quantum critical point at low enough temperatures. Furthermore, quantum Ising models have a much broader range of applicability since they can be easily adapted to study quantum criticality in many other classes of systems [1].

In its general form, the hamiltonian operator of the quantum Ising model can be written as

$$\mathbf{H}_{QI} = -gJ \sum_i \sigma_i^x - J \sum_{\langle ij \rangle_{nn}} \sigma_i^z \sigma_j^z, \tag{10.2}$$

where the first sum extends over all lattice sites i and the second over all i-j nearest neighbour pairs. Here J denotes the exchange coupling, which determines the microscopic energy scale of the model, and g is a dimensionless parameter that enables tuning the hamiltonian across the quantum phase transition. It is clear that the first term of the hamiltonian, proportional to gJ, can, for instance, represent an applied transverse field that disrupts the magnetic order. The second term, describes the interaction between spins. In the preceding formulation, $\sigma_i^{z,x}$ are operators localized on the i-sites of a d-dimensional lattice. These operators, that can be represented as Pauli matrices, act on the spin states. It is important to remark that since matrices localized on different lattice sites act on different spin states, matrices with $i \neq j$ must commute with each other. In the basis where σ^z is diagonal, Pauli matrices have the form,

$$\sigma^x = \begin{pmatrix} 0 & 1 \\ 1 & 0 \end{pmatrix}, \quad \sigma^y = \begin{pmatrix} 0 & -i \\ i & 0 \end{pmatrix}, \quad \sigma^z = \begin{pmatrix} 1 & 0 \\ 0 & -1 \end{pmatrix}. \tag{10.3}$$

The eigenvalues of σ_i^z are $S_i = \pm 1$. Therefore, the two spin states corresponding to these two eigenvalues can be identified with up and down orientations of the spins. They will be denoted as $| \uparrow \rangle_i$ and $| \downarrow \rangle_i$ states. It is important to note that for $g = 0$, $\mathbf{H}_{QI}$ is expressed only in terms of products of operators $\sigma_i^z \sigma_j^z$ defined on different lattice sites. Therefore, the hamiltonian will be diagonal on the basis of the eigenvalues of σ^z, which means that it can be reduced to the classical Ising model that was introduced in Chapter 1. Instead, for $g \neq 0$, the matrices σ^x are off-diagonal in that basis and, thus, may induce quantum mechanical tunnelling events which flip the spin orientation on a given site. It is, indeed, in this situation that a quantum critical point can occur due to a competition of ground states driven by the parameter g. It is then useful to analyse the ground state properties in the two limits, $g \gg 1$ and $g \ll 1$, respectively.

For $g \gg 1$ the first term of the hamiltonian 10.2 is dominant and a ground state $|0\rangle$ will be of the form,

$$|0\rangle = \prod_i | \rightarrow \rangle_i, \qquad (10.4)$$

where $| \rightarrow \rangle_i = (| \uparrow \rangle_i + | \downarrow \rangle_i)/\sqrt{2}$, and $| \leftarrow \rangle_i = (| \uparrow \rangle_i - | \downarrow \rangle_i)/\sqrt{2}$ are the two eingenstates of σ_i^x with engenvalues ± 1, respectively. When the system is in the state $|0\rangle$, the eingenstates of σ_i^z at different i-sites are totally uncorrelated, which means that $\langle 0|\sigma_i^z \sigma_j^z|0\rangle = \delta_{ij}$.

As g decreases, perturbative corrections in a power series of g^{-1} can be considered. These corrections should result in the existence of correlations between pairs of eingenstates of σ^z located in different sites. For large enough g, these two-point correlations should be short range and, thus, of the form

$$\langle 0|\sigma_i^z \sigma_j^z|0\rangle \sim e^{-|\boldsymbol{r}_i - \boldsymbol{r}_j|/\xi}, \qquad (10.5)$$

where, $\boldsymbol{r}_{i(j)}$ are the coordinates of lattice sites $i(j)$ and ξ is the correlation length.

In the opposite limit, $g \ll 1$, the second term of the hamiltonian 10.2 will be dominant and the spins must be in either up or down eingentates of σ^z. Strictly, at $g = 0$, the spins will be either in the up, $| \uparrow \rangle = \prod_i | \uparrow \rangle_i$, or the down, $| \downarrow \rangle = \prod_i | \downarrow \rangle_i$, degenerate states. For a small value of g, the majority of spins will be in one of these two orientations, while a small fraction will be in the corresponding opposite one. In any case, in an infinite system, the degeneracy will remain for any small value of g, due to the invariance of the hamiltonian under the up $\rightarrow$ down symmetry, $\sigma_i^z \rightarrow -\sigma_i^z$ with $\sigma_i^x \rightarrow \sigma_i^x$.[4] In fact, this symmetry behaviour is similar to that satisfied by the thermal Ising model at a temperature below the critical point. In both, thermal and quantum cases, a real thermodynamic system will show spontaneous symmetry breaking of the up $\rightarrow$ down symmetry, which means that the system will choose one of the two degenerate ground states due to the influence of some infinitesimal perturbation. Therefore, from perturbation theory in g, it is clear that the expectation value $\langle 0|\sigma_i^z|0\rangle = \pm m$, with m being the spontaneous magnetization per spin. Obviously, this expectation

---

[4] Formally, this invariance is due to the global Z_2 symmetry transformation of the hamiltonian operator generated by the unitary operator $\prod_i \sigma_i^x$ that maps the two ground states into each other.

value must reach the maximum value $m = 1$ when $g = 0$. Therefore, the two-point correlations for a large spin separation should be

$$\lim_{|\boldsymbol{r}_i - \boldsymbol{r}_j| \to \infty} \langle 0 | \sigma_i^z \sigma_j^z | 0 \rangle = m^2. \tag{10.6}$$

This equation indicates that the effect of quantum fluctuations reduces m^2 to a value lower than one but not to zero.

The important point is to realize that it is not possible for a system characterized by the two-point correlations given by Eq. 10.6 to analytically transform into one characterized by correlations given by Eq. 10.5 driven continuously by varying the parameter g. Actually, this means that at least a critical value $g = g_c$ should exist at which correlations in the large $|\boldsymbol{r}_i - \boldsymbol{r}_j|$ limit change from one behaviour to the other. At this value of g, a quantum phase transition should exist and, thus, at this point the two-point correlation function is expected to diverge. In fact, more than one critical point might exist but the fact that the thermal Ising model driven by temperature shows only one thermal critical point, suggests that the quantum Ising model has only one quantum critical point.

It is worth noting that the energies of thermal excited states of the model are separated, in an infinite lattice, by a finite energy gap from the ground state energy for all the states with $g \neq g_c$. Nevertheless, the gap goes to zero at $g = g_c$.

10.2.1 *Quantum-classical mapping*

It is useful to analyse the mapping of a quantum model in d-dimensions with a finite temperature model in $D = d + z$ dimensions using the quantum Ising model introduced above. Since, usually $z = 1$, it is worth considering the simplest mapping expected between the quantum $d = 0$ single site model and the classical Ising chain even though this model does not show a phase transition. The hamiltonian of the classical Ising chain can be written as

$$\mathcal{H}_I = -K \sum_{i=1}^{M} S_i S_{i+1} - \tilde{h} \sum_{i=1}^{M} S_i, \tag{10.7}$$

where the Ising spins S_i, located on the i-sites of a chain of M sites, are the eigenvalues of σ_i^z that take values ± 1. Temperature is already included in the two hamiltonian parameters, which are defined as $K \equiv \beta J$ and $\tilde{h} \equiv \beta h$, with J and h being the exchange coupling parameter and external field, respectively. Assuming periodic boundary conditions, the partition function of this model can be exactly computed using the transfer matrix method

[357]. For this model, the transfer matrix can be written as, $\mathbf{T} = \mathbf{T}_K \mathbf{T}_{\tilde{h}}$, where

$$\mathbf{T}_K = \begin{pmatrix} e^K & e^{-K} \\ e^{-K} & e^K \end{pmatrix}, \tag{10.8}$$

$$\mathbf{T}_{\tilde{h}} = \begin{pmatrix} e^{\tilde{h}} & 0 \\ 0 & e^{-\tilde{h}} \end{pmatrix}. \tag{10.9}$$

In the thermodynamic limit $M \to \infty$, the partition function is then given as

$$Q_I = \lim_{M \to \infty} \mathrm{Tr} \mathbf{T}^M = \lim_{M \to \infty} \lambda_1^M \left(1 + \frac{\lambda_2^M}{\lambda_1^M} \right) = \lambda_1^M, \tag{10.10}$$

where $\lambda_{1,2}$ are the eigenvalues of $\mathbf{T}$, which are given by

$$\lambda_{1,2} = e^K \cosh(\tilde{h}) \pm [e^{2K} \sinh^2(\tilde{h}) + e^{-2K}]^{1/2}, \tag{10.11}$$

where λ_1 is the largest eigenvalue. For $\tilde{h} = 0$, the partition function is $Q_I(\tilde{h} = 0) = 2 \cosh K$.

The transfer matrix method can also be used to obtain the two-spin correlation function. For two spins located at sites i and $j \leq i$ of the chain, it can be obtained as

$$\langle S_i S_j \rangle = \frac{1}{Q_I} \sum_{\{S_i\}} S_i S_j e^{-\mathcal{H}_I} = \frac{1}{Q_I} \mathrm{Tr} \left(\mathbf{T}^i \sigma^z \mathbf{T}^{j-i} \sigma^z \mathbf{T}^{M-j} \right), \tag{10.12}$$

where σ^z is the same Pauli matrix introduced previously. For $\tilde{h} = 0$, it can be calculated in the basis that diagonalises $\mathbf{T}$ and, in the limit $M \to \infty$, the result is [5]

$$\langle S_i S_j \rangle = (\tanh K)^{j-i}. \tag{10.13}$$

In the infinite lattice limit, translational invariance holds and the origin of coordinates can be chosen at any site. Then, suppose i is the origin of coordinates and that the spins are separated by a distance $r = aj$, where a is the lattice parameter of the chain. Then, the two-point correlation function can be written in the form

$$\langle S(r) S(0) \rangle = e^{-r/\xi}, \tag{10.14}$$

---

[5] For $h = 0$, the eingenvectors of $\mathbf{T} = \mathbf{T}_K$ are the states $| \to \rangle$ and $| \leftarrow \rangle$ introduced previously with eigenvalues $\lambda_1 = 2 \cosh K$ and $\lambda_2 = 2 \sinh K$, respectively. Then, the calculation is done taking into account that $| \to \rangle \sigma^z | \to \rangle = | \leftarrow \rangle \sigma^z | \leftarrow \rangle = 0$ and $| \to \rangle \sigma^z | \leftarrow \rangle = | \leftarrow \rangle \sigma^z | \to \rangle = 1$.

where ξ is the correlation length at $\tilde{h} = 0$ that is given by, $\xi^{-1} = \frac{1}{a}\ln\coth K$. In the limit $K \gg 1$, it is obtained that

$$\frac{\xi}{a} \simeq \frac{1}{2}e^{2K} \gg 1. \tag{10.15}$$

The point now is to show that statistical mechanics results of the Ising chain can be mapped onto the quantum mechanics results of a single Ising spin. For that purpose, it is convenient to follow the development given in Ref. [1] and rewrite the transfer matrices $\mathbf{T}_K$ and $\mathbf{T}_{\tilde{h}}$ in the form

$$\mathbf{T}_K = e^K(1 + e^{-2K}\sigma^x) \simeq e^K(1 + \frac{a}{2\xi}\sigma^x) \simeq e^{a(-E_0 + \sigma^x/2\xi)}, \tag{10.16}$$

$$\mathbf{T}_{\tilde{h}} = e^{a\tilde{h}'\sigma^z}, \tag{10.17}$$

where $E_0 = -K/a$ and $\tilde{h}' = \tilde{h}/a$. Note that, taking into account Eq. 10.15, the obtained expression of $\mathbf{T}_K$ represents a good approximation for $K \gg 1$, which is the scaling limit of interest to establish the mapping of the $d = 0$ quantum and $d = 1$ thermal systems. Therefore, the two matrices are given now in terms of the lattice parameter, a, and the correlation length, ξ, and are expressed in the form, $e^{a\mathbf{O}}$, where $\mathbf{O}$ is an operator that acts on the up and down spin states.

Defining $\tilde{T} \equiv 1/L_r$, with $L_r = Ma$ being the system size, $2gJ \equiv \xi^{-1}$, and introducing the quantum hamiltonian

$$\mathbf{H} = E_0 - gJ\sigma^x - \tilde{h}'\sigma^z, \tag{10.18}$$

in the limit $a \to 0$,[6] $\mathbf{T}_K\mathbf{T}_{\tilde{h}} = \exp(-a\mathbf{H})$, and the partition function, $Q = \mathrm{Tr}(\mathbf{T}_K\mathbf{T}_{\tilde{h}})^M$, can be written in the form

$$Q = \mathrm{Tr}\exp(-\mathbf{H}/\tilde{T}). \tag{10.19}$$

The hamiltonian $\mathbf{H}$ is the hamiltonian that describes the dynamics of a single spin subjected to a longitudinal field $\tilde{h}'$ and a transverse field gJ. Note that, apart from the constant ground state energy term E_0, this hamiltonian represents the single site version of the hamiltonian given in Eq. 10.2 with the spin coupled to an additional longitudinal field.

The expression $\mathbf{T}_K\mathbf{T}_{\tilde{h}} = \exp(-a\mathbf{H})$ shows that the transfer matrix of the classical Ising chain is the time evolution operator $e^{-\mathbf{H}\mathcal{T}}$ over an imaginary time $\mathcal{T} = a$, which is the lattice site. Therefore, this can be interpreted in the sense that the transfer from one site to the next in the classical chain is identified with the evolution of the quantum system in imaginary time.

---

[6] It must be taken into account that $e^{a\mathbf{O}_1}e^{a\mathbf{O}_2} = e^{a(\mathbf{O}_1+\mathbf{O}_2)}[1 + \mathcal{O}(a^2)]$.

Consequently, length coordinates are translated into imaginary time coordinates in the quantum model. This justifies the increase of dimensionality in passing from the quantum to the classical model. Note that results confirm that $D = d + 1$ and, consequently, that in the case discussed, the exponent $z = 1$.

Consistently with the time-length mapping discussed above, it is worth remaking that the energy scale $2gJ$ represents the energy gap, Δ, between the ground state of $\mathbf{H}$ and the excited state in the absence of an applied longitudinal field, which is given by the inverse of the correlation length of the classical Ising chain. Similarly, the partition function of the quantum single spin corresponds to an effective temperature $\tilde{T}$ which is precisely the inverse of the total length of the Ising chain. These correspondences, which are very general and apply to all models, permit estimating the effective free energy, $\tilde{F} = \tilde{T} \ln Q$, of the quantum single spin. Taking into account that the eigenvalues of $\mathbf{H}$ are $E_0 \pm \sqrt{(gJ)^2 + \tilde{h}'^2}$, the result is

$$\tilde{F} = E_0 - \tilde{T} \ln \left[\cosh \sqrt{(gJ)^2 + \tilde{h}'^2} \right], \tag{10.20}$$

that corresponds to the partition function of the Ising chain in the scaling limit.

The time-length mapping can be extended to the analysis of correlation functions. In the quantum system, the time-ordered correlation function, $\tilde{G}$, in imaginary time is defined as

$$\tilde{G}(\mathcal{T}_1, \mathcal{T}_2) = \frac{1}{Q} \mathrm{Tr} \left[e^{-\mathbf{H}/\tilde{T}} \sigma^z(\mathcal{T}_{1,2}) \sigma^z(\mathcal{T}_{2,1}) \right], \tag{10.21}$$

where $\sigma^z(\mathcal{T}) \equiv e^{\mathbf{H}/\tilde{T}} \sigma^z e^{-\mathbf{H}/\tilde{T}}$, which must be ordered as $\sigma^z(\mathcal{T}_1)\sigma^z(\mathcal{T}_2)$ for $\mathcal{T}_1 > \mathcal{T}_2$, and as $\sigma^z(\mathcal{T}_2)\sigma^z(\mathcal{T}_1)$ for $\mathcal{T}_1 < \mathcal{T}_2$, to take into account that the transfer matrix evolves from *earlier* to *later* sites. Then, at $T = 0$, considering a complete set of eingenstates, $\{|n\rangle\}$, of $\mathbf{H}$ with eigenvalues $\{E_n\}$, $\tilde{G}$ can be obtained as

$$\tilde{G}(\mathcal{T}_1, \mathcal{T}_2) = \sum_n |\langle 0|\sigma^z|n\rangle|^2 e^{-(E_n - E_0)|\mathcal{T}_1 - \mathcal{T}_2|}, \tag{10.22}$$

where the state $|0\rangle$ represents the ground state with energy E_0. It is expected that for a large time interval, $|\mathcal{T}_1 - \mathcal{T}_2|$, the sum over n be dominated by the lowest energy state with non-zero matrix elements. This leads to an exponential decay of the ordered-time correlation function over a length $\xi = (E_1 - E_0)^{-1} = \Delta^{-1}$. The result is exact in the present model for which only two states exist. In any case, it is of a more general nature and applies to more complex models.

The calculation can also be performed within the framework of the time-length mapping discussed above, which supposes proceeding as in the case of the free energy calculation. In this case, the obtained result is

$$\tilde{G}(\mathcal{T}_1, \mathcal{T}_2) = \lim_{a \to 0} \langle \sigma^z(\mathcal{T}_1) \sigma^z(\mathcal{T}_2) \rangle_{\mathcal{H}_I}, \tag{10.23}$$

where the average is taken over the classical model with hamiltonian $\mathcal{H}_I$. The calculation of $\tilde{G}$ in this case is much more involved than the preceding quantum calculation. In any case, the results of such a calculation should lead to an expression consistent with the scaling

$$\tilde{G}(\mathcal{T}_1, \mathcal{T}_2) = \Phi \left(\tilde{T}(\mathcal{T}_1 - \mathcal{T}_2), \frac{\Delta}{\tilde{T}}, \frac{\tilde{h}'}{\tilde{T}} \right), \tag{10.24}$$

where Φ is a scaling function.

All the conclusions resulting from the simple mapping that has been discussed above can be generalized to a d-dimensional transverse quantum Ising model that should be mapped to a D-dimensional classical Ising model. In particular, this means that the quantum Ising chain should be mapped to a $2d$ classical Ising model, which means that a quantum Ising chain must show a critical point. This is in agreement with the existence of a quantum critical point in the cobalt niobate system discussed previously. In general, once the mapping is done for a given quantum model, all the techniques adequate to solve thermal models can be used to find the solution of the quantum system.

Exercises

10.1 Consider a q-state ferromagnetic Potts model defined on the sites of a $1d$ chain. It is assumed that the lattice has N sites with periodic boundary conditions. The hamiltonian of the model is then written as

$$\beta\mathcal{H}_{Potts} = -K \sum_{i=1}^{N-1} \delta_{S_i,S_{i+1}} - K\delta_{S_N,S_1},$$

where variables S_i can take the q values, $S_i = 1, 2, ..., q$, and $K > 0$.

(a) For the case $q = 3$, find the eigenvectors and eigenvalues of the transfer matrix.

(b) Obtain, in the thermodynamic limit $N \to \infty$, the exact partition function and the free energy per site for an arbitrary value of q.

(c) Find the high- and low-temperature limits of the heat capacity.

10.2 Consider the XY model defined on a $1d$ chain with N sites. In each lattice site, a two-component, unit-length spin vector $\boldsymbol{S}_i = (\cos\theta_i, \sin\theta_i)$ is defined, where θ_i is the angle formed by the spin and the chain. The spins interact with their nearest neighbours through the hamiltonian

$$\beta\mathcal{H}_{XY} = -K \sum_{i=1}^{N} \boldsymbol{S}_i \cdot \boldsymbol{S}_{i+1} - \sum_{i=1}^{N} \boldsymbol{h} \cdot \boldsymbol{S}_i,$$

where $K > 0$ and $\boldsymbol{h}$ is an applied external field.

(a) Find an expression for the transfer matrix elements, $\langle\theta|\mathbf{T}|\theta'\rangle$, and show that in the absence of an applied external field $\mathbf{T}$ can be diagonalized with eigenvectors of the form $f_m(\theta) \propto \exp(im\theta)$, with m an integer.

(b) Calculate the free energy per site and discuss the behaviour of the heat capacity in the high- and low-temperature limits.

10.3 Show that in the scaling limit, which corresponds to large values of K, when θ_i is not expected to vary much from one site to the next, the XY-hamiltonian can be written, apart from a constant term, in the following continuous form

$$\mathcal{H}_{XY_c} = \int_0^{L_\tau} \left[\frac{1}{4}\xi \left(\frac{d\theta}{d\tau}\right)^2 - \tilde{h}\cos\theta \right] d\tau,$$

where the continuous coordinate $\tau = aj$ with a being the lattice spacing in the chain, $\xi = 2aK$, $\tilde{h} = h/a$, and $L_\tau = aN$ is the length of the chain.

Appendix

Landau theory and symmetry

A.1 Introduction: symmetry considerations

Symmetry is a fundamental concept in nature and physics. When a phase transition takes place in a material, some symmetry is broken. In 1937, Landau [62] realized that this broken symmetry must correspond to a change in a physical quantity which can be captured in what he termed as an order parameter. The latter is zero in the high-symmetry phase but attains a finite value in the low-symmetry phase. Landau further suggested that one can write a free energy in terms of a polynomial expansion of the order parameter which is invariant under all symmetry operations of the parent phase. This is known as the Landau free energy. In this Appendix, we will explain how to obtain the form of the Landau free energy for crystallographic phase transitions based on the symmetry of the parent and product phases as well as the nature of the order parameter [28]. The latter can be a scalar such as density, a polar vector such as polarization or shuffle modes, an axial vector such as magnetization with an implicit sense of time built-in, a second-rank polar tensor such as strain or complex valued such as the wave function in superconductivity.

Crystalline symmetry is described by a combination of translational symmetry of the underlying lattice and the rotational or point group symmetry of the unit cell. This combined symmetry is described by the space group of the crystal. There are 230 space groups and the corresponding 32 point groups in three dimensions with 14 Bravais lattices [358]. In two dimensions, there are 17 space groups and 10 point groups with 5 Bravais lattices. Quasi-two dimensional situations are described by the 80 layer or diperiodic groups, or more generally by subperiodic groups [358].

We will denote by G_0 and G the space group symmetry of the high-symmetric and low-symmetric phases, respectively. Here we are mostly concerned with phase transitions wherein G is a subgroup of G_0. This is called the first Landau-Lifshitz condition. The second Landau-Lifshitz condition (for a second order transition) states that the phase transition leads to such crystal structure changes which correspond to a single irreducible representation (or *irrep*) Γ_0 of the space group G_0. Note that Γ_0 is said to be an active representation, in other words the basis functions of Γ_0 appear in the density change of the crystal in a phase transition or more precisely in the free energy expansion. The meaning of reducible and irreducible representations will become clear in the examples discussed further.

In the reciprocal space, the crystal symmetry is captured by the Brillouin zone. The centre point of any Brillouin zone is always denoted by $k = 0$ or the Γ point (not to be confused with the irrep notation). The inverse of $k = 0$ means the size of the whole crystal, and thus the Γ-point or its associated representations correspond to strain. It is therefore clear that for strain-based or ferroelastic transitions discussed in what follows translational symmetry plays no role and the point group symmetry will be sufficient.

However, to describe unit cell size change (doubling, etc. in certain crystallographic directions) such as in the case of intra-unit cell shuffle modes, different rotations of perovskite octahedra in neighbouring unit cells or for ferroelectric transitions the symmetry change is dictated by the high-symmetry points such as M, X, R, etc. on the corner, edge or the surface of the relevant Brillouin zone. In these cases, translational symmetry is indispensable and the full space group symmetry must be taken into account. The corresponding (reducible) representations and the irreducible representations (irreps) are then labelled by these high-symmetry points. Incommensurate phase transitions (not considered in this book) are necessarily described by points inside the Brillouin zone.

For these high-symmetry point-driven transitions, an irrep, say Γ, is characterized by a unique *star* $\mathbf{k}$ of wave vectors. The latter is the full set of inequivalent wave vectors which are transformed from $\mathbf{k}$ while being acted upon by all symmetry elements of the crystal point group g_0 (associated with the space group G_0) supplemented by the inversion, that is, $\bar{g}_0$. The inversion is needed to obtain the real irreps which contain both the basis functions of $\mathbf{k}$ and $-\mathbf{k}$. Since the translations do not change the wave vector $\mathbf{k}$, the action of G_0 reduces simply to the action of g_0. We will consider these transitions in detail in Section A.3. Here we first address strain-based transitions.

A.2 Ferroelastic transitions

For pure strain-based transitions, only the unit cell shape changes and thus translation symmetry is irrelevant and point group symmetry is sufficient. First we will consider the following four $2d$ transitions [359]: (i) triangular lattice (p6mm) to centred rectangular lattice (c2mm), (ii) triangular lattice (p6mm) to oblique lattice (p2), (iii) square lattice (p4mm) to rectangular lattice (c2mm) and (iv) square lattice (p4mm) to oblique lattice (p2).

The Lagrangian symmetric strain tensor in $2d$ consists of three components: E_{xx}, E_{yy} and E_{xy}. Under the symmetry elements of G_0, the strain components will transform to E'_{xx}, E'_{yy} and E'_{xy}. The 3×3 matrices (for each symmetry element of G_0) transforming (E_{xx}, E_{yy} and E_{xy}) to (E'_{xx}, E'_{yy} and E'_{xy}) comprise the reducible representation Γ. These matrices can be block diagonalized, with each block corresponding to an irrep.

For a square lattice, the reducible representation can be decomposed into three one-dimensional irreps as $\Gamma = \Gamma^1 \oplus \Gamma^2 \oplus \Gamma^3$ (i.e., fully diagonalized) with corresponding basis functions[1]

$$e_1 = \frac{1}{2}(E_{xx} + E_{yy}), \quad e_2 = \frac{1}{2}(E_{xx} - E_{yy}), \quad e_3 = \frac{1}{2}(E_{xy} + E_{yx}). \quad \text{(A.1)}$$

Intuitively, Γ^1 is always the identity representation meaning no symmetry change with strain e_1, that is for a square it is either expansion (bigger square) or contraction (smaller square). Geometrically, Γ^2 means length change but no angle change, that is, rectangular (or deviatoric) strain e_2. Similarly, Γ^3 means angle change but no length change, that is, shear strain e_3.

For a triangular lattice, Γ can be decomposed into one one-dimensional irrep and one two-dimensional irrep, respectively, as $\Gamma = \Gamma^1 \oplus \Gamma^3$. Again, Γ^1 being the identity representation implies an equilateral triangle remains a (bigger or smaller) equilateral triangle. However, unlike the square, two ways of deforming an equilateral triangle are on the same footing meaning that e_2 and e_3 are both shears and must be treated together as $\{e_2, e_3\}$ corresponding to the two-dimensional irrep Γ^3.

Note that the free energy is a scalar quantity. Therefore, we need to construct second- and higher-order terms in the free energy that are various multiples of the strain components such that they respect the symmetry of the parent crystal structure and yet provide scalars. Since rotations, inversion and mirror operations are coordinate transformations, all we need to pay attention to is how the subscripts in various strain components change under different symmetry operations.

---

[1] Note the slight difference in the normalization of the symmetrized strains (both in $2d$ and $3d$) compared to those in Chapter 3, which we have adopted here for simplicity.

For a triangle (Figure A.1(a)), the two relevant symmetry operations are a threefold rotation (C_3) and the vertical mirror plane/line (σ_y). Under σ_y: $x \to -x$ and $y \to y$, which means $xx - yy \to xx - yy$ and $xy \to -xy$. In other words, $e_2 \to e_2$ and $e_3 \to -e_3$. Similarly, under C_3: $x \to x/2 - (\sqrt{3}/2)y$ and $y \to (\sqrt{3}/2)x + y/2$. Just looking at the subscripts of the two strain components, we find that $e_2 \to e_2/2 + (\sqrt{3}/2)e_3$ and $e_3 \to -(\sqrt{3}/2)e_2 + e_3/2$. Using this information, writing the second- and fourth-order invariant terms are straightforward, but a third-order scalar can also be created. Note that each term in the free energy is invariant under all the symmetry operations of the triangle. Note also that C_3 and σ_y are 2×2 matrices corresponding to the two-dimensional irrep Γ^3 mentioned earlier. Thus, for the triangular lattice to a centred rectangular lattice (and an oblique lattice) phase transition, we have

$$F_L = \frac{A_1}{2}e_1^2 + \frac{A}{2}(e_2^2 + e_3^2) + \frac{B}{3}(e_3^3 - 3e_2^2 e_3) + \frac{C}{4}(e_2^2 + e_3^2)^2 + C_1 e_1(e_2^2 + e_3^2). \quad \text{(A.2)}$$

Whether the transition is to a centred rectangular or an oblique lattice phase depends on the coefficients in the free energy. The latter depend on the choice of the material under consideration. The last term is the symmetry allowed coupling between different strain components. The corresponding most general strain gradient (or Ginzburg) energy is given by [359]

$$\begin{aligned}
F_G = {} & g_1(3e_{2,x}^2 + e_{2,y}^2 + e_{3,x}^2 + 3e_{3,y}^2 + e_{2,x}e_{3,y} + 2e_{2,y}e_{3,x}) \\
& + g_2(e_{2,x}^2 + e_{3,y}^2 - e_{2,y}^2 - e_{3,x}^2 + 6e_{2,x}e_{3,y} - 2e_{2,y}e_{3,x}) \\
& + g_3(e_{2,x}^2 + e_{3,y}^2 + 3e_{2,y}^2 + 3e_{3,x}^2 - 2e_{2,x}e_{3,y} - 2e_{2,y}e_{3,x}),
\end{aligned} \quad \text{(A.3)}$$

where the commas in the subscripts denote derivatives with respect to the coordinates following the commas. For simplicity, instead of the three terms in the earlier expression, only a single term $F_G = (g/2)(|\nabla e_2|^2 + |\nabla e_3|^2)$ is usually retained in the Ginzburg free energy for analysis purposes and microstructure simulations.

For a square (Figure A.1(b)), the two relevant symmetry operations are a fourfold rotation C_4 and the vertical mirror plane/line (σ_y). Under C_4, $x \to -y$ and $y \to x$, which means $xx - yy \to yy - xx$ and $xy \to -xy$. In other words, $e_2 \to -e_2$ and $e_3 \to -e_3$. Similarly, under σ_y: $x \to -x$ and $y \to y$, which means $e_2 \to e_2$ and $e_3 \to -e_3$.

Thus, for the square lattice to a rectangular lattice (and an oblique lattice) phase transition, we have

$$F_L = \frac{A_1}{2}e_1^2 + \frac{A_2}{2}e_2^2 + \frac{B_2}{4}e_2^4 + \frac{A_3}{2}e_3^2 + \frac{B_3}{4}e_3^4 + C_1 e_1 e_2^2 + C_2 e_1 e_3^2 + C_3 e_2^2 e_3^2. \quad \text{(A.4)}$$

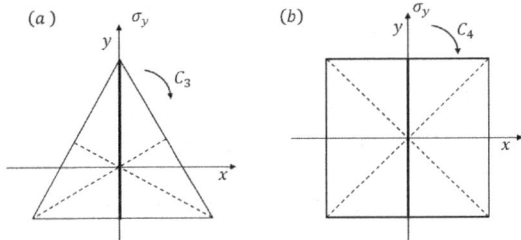

Figure A.1 (*a*) Threefold rotation operation and vertical mirror line for an equilateral triangle. (*b*) Fourfold rotation operation and vertical mirror line for a square.

For the rectangular low-symmetry phase, the order parameter is e_2 (corresponding to the one-dimensional irrep Γ^2) and thus the non-linear B_3 term should be suppressed in the free energy. In contrast, if the low-symmetry phase is oblique then the order parameter is e_3 (corresponding to the one-dimensional irrep Γ^3) and the non-linear term B_2 should be suppressed. We have also included symmetry allowed coupling terms between different strains denoted by C_1, C_2 and C_3. Again, whether the transition is to a rectangular or an oblique lattice phase depends on the coefficients in the free energy, which in turn depend on the choice of the material.

The corresponding gradient (or Ginzburg) energy is given by

$$F_G = g_1(e_{2,x}^2 + e_{2,y}^2) + g_2(e_{3,x}^2 + e_{3,y}^2) + g_3(e_{2,x}e_{3,y} + e_{2,y}e_{3,x}). \quad (A.5)$$

Again, for simplicity, only a single gradient term $(g/2)|\nabla e_2|^2$ or $(g/2)|\nabla e_3|^2$ is kept in the Ginzburg free energy depending on whether the transition is to a rectangular or an oblique lattice, respectively.

Next we describe strain-based free energy in $3d$. For that purpose, we will consider the following two sets of transitions: (i) simple cubic to tetragonal or orthorhombic and (ii) simple cubic to trigonal (i.e., rhombohedral) or monoclinic or triclinic (see Figure A.2).

The symmetric strain tensor in $3d$ comprises six components: E_{xx}, E_{yy}, E_{zz}, E_{xy}, E_{yz} and E_{xz} corresponding to the reducible representation Γ. For a simple cubic lattice, it can be decomposed as $\Gamma = \Gamma^1 \oplus \Gamma^3 \oplus \Gamma^5$. Once again Γ^1 is the identity representation meaning it leads to either a bigger or smaller cube corresponding to the (volume) strain $e_1 = (1/3)(E_{xx} + E_{yy} + E_{zz})$.

The two-dimensional irrep Γ^3 corresponds to deviatoric strains, that is, lengths change but no angles change representing respectively the tetragonal and orthorhombic strains (see Table A.1)

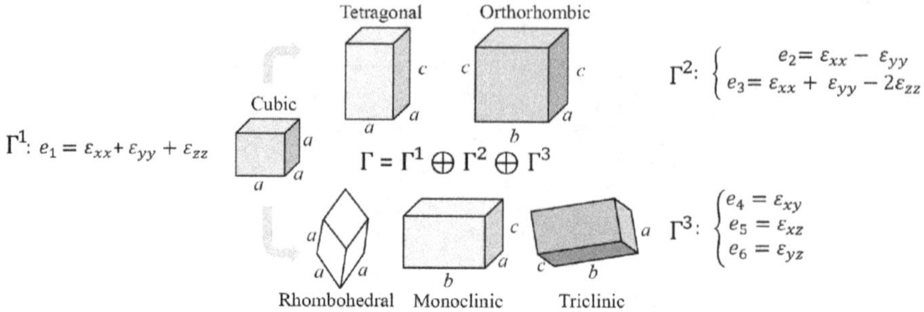

Landau theory and symmetry

Figure A.2 Two different sets of transitions to lower symmetry lattices arising from a cubic lattice.

$$e_2 = \frac{1}{6}(E_{xx} + E_{yy} - 2E_{zz}), \quad e_3 = \frac{1}{2}(E_{xx} - E_{yy}). \tag{A.6}$$

Similarly, the three-dimensional irrep Γ^5 corresponds to shear strains, that is, angles change but no lengths change (see Table A.1)

$$e_4 = E_{xy}, \quad e_5 = E_{yz}, \quad e_6 = E_{xz}. \tag{A.7}$$

If the three shear strains are equal, one gets a trigonal (or rhombohedral) phase; if two strains are equal one gets a monoclinic phase and if the three stains are different one gets a triclinic phase (see Table A.1).

The point group of a simple cube has $48 = 24 \times 2$ symmetry elements. The factor of 2 comes from the inversion symmetry. The 24 elements can be generated by just 3 symmetry operations, called generators, listed in the Table, namely a fourfold rotation about z-axis, a threefold rotation about the [111] direction (i.e., body diagonal) and a twofold rotation about the line joining two edge centres or the [110] direction. Based on the coordinate transformation under these symmetry operations, one can determine how the six different strain components transform as given in Table A.1. From the latter, it is apparent that $\{e_2, e_3\}$ transform into each other, thus irrep Γ^3 and $\{e_4, e_5, e_6\}$ transform among themselves, corresponding to Γ^5.

Thus, for the cubic to tetragonal (and orthorhombic) transitions, the order parameter is a two-component one $\{e_2, e_3\}$ and we have

$$F_L = \frac{A_1}{2}e_1^2 + \frac{A_3}{2}(e_4^2 + e_5^2 + e_6^2) + \frac{A_2}{2}(e_2^2 + e_3^2) + \frac{B}{3}(e_3^3 - 3e_2^2 e_3) + \frac{C}{4}(e_2^2 + e_3^2)^2. \tag{A.8}$$

The form of the third-order term is a consequence of how the two strain components transform under the threefold rotation about the [111] axis.

Note that the third-order term in Eq. A.8 is identical to that in Eq. A.2. The reason being the structure of $\{e_2, e_3\}$ in the two cases is similar and the cubic lattice when viewed along the [111] direction has a threefold rotational symmetry similar to a triangle. Analogous to Eq. A.5, we could write the most general symmetry-based gradient terms here but for simplicity (of analysis and simulations), it suffices that $F_G = (g/2)(|\nabla e_2|^2 + |\nabla e_3|^2)$.

Table A.1 *Transformation properties of the six symmetry-adapted strains for a cube under the three generators of the point group of a cube*

Symmetry generator	e_1	e_2, e_3	e_4, e_5, e_6
$3[111]$: $(x \rightarrow y, y \rightarrow z, z \rightarrow x)$	e_1	$-\frac{1}{2}e_2 + \frac{\sqrt{3}}{2}e_3, -\frac{\sqrt{3}}{2}e_2 - \frac{1}{2}e_3$	e_6, e_4, e_5
$4[001]$: $(x \rightarrow -y, y \rightarrow x, z \rightarrow z)$	e_1	$-e_2, e_3$	$-e_4, e_6, -e_5$
$2[110]$: $(x \rightarrow y, y \rightarrow x, z \rightarrow -z)$	e_1	$-e_2, e_3$	$e_4, -e_6, -e_5$

The transformation properties of the three strains $\{e_4, e_5, e_6\}$ are also contained in Table A.1. Thus, for the cubic to trigonal (and monoclinic and triclinic) transitions, we have

$$F_L = \frac{A_1}{2}e_1^2 + \frac{A_2}{2}(e_2^2 + e_3^2) + \frac{A_3}{2}(e_4^2 + e_5^2 + e_6^2) + \frac{B}{3}e_4 e_5 e_6$$
$$+ \frac{C_1}{2}(e_4^2 + e_5^2 + e_6^2)^2 + \frac{C_2}{2}(e_4^4 + e_5^4 + e_6^4). \tag{A.9}$$

Note that the third-order term is a consequence of the permutation of three shears under the threefold rotation about the [111] axis. Again, we could write the most general symmetry-based gradient terms here but for simplicity (of analysis and simulations) it suffices that $F_G = (g/2)(|\nabla e_4|^2 + |\nabla e_5|^2) + |\nabla e_6|^2)$.

In the above discussion, for the sake of illustration, we constructed by inspection various free energies using the symmetry properties of strain components for different lattices. However, this procedure for both the Landau and Ginzburg free energies have been automated and can be generated directly from the ISOTROPY program [360]. The same holds true for transitions corresponding to any high-symmetry point at the corners, edges, or surfaces of different Brillouin zones for the 230 space groups. Another useful resource in this context is the Bilbao Crystallographic Server [361].

A.3 Phonon-driven and ferroelectric transitions

Atomic displacement or a phonon mode relates to a polar vector, that is, a usual transformation of coordinates (x, y, z) to (x', y', z') under the symmetry operations of the high-symmetry group. A dipole or polarization vector is just charge times displacement, that is, polarization transforms as $\boldsymbol{P} = q(x, y, z)$. Since charge q is a scalar, $\boldsymbol{P}$ transforms just like the coordinates (x, y, z). Let us illustrate with a couple of examples, that of surface reconstruction on transition metals (e.g. Pt and Pd) and semiconductor (e.g. Si) vicinal surfaces and a $2d$ paraelectric to ferroelectric transition.

In surface reconstruction and in a phonon (corresponding to the high symmetry point on the boundary of the Brillouin zone)-driven transition, the size of the unit cell doubles or enlarges by an integer amount in one or more directions. For a square lattice, the Brillouin zone is shown in Figure A.3(a) with two different stars shown in Figure A.3(b). The first star is $\mathbf{k}^1 = \{\boldsymbol{k}_1 = (1/2, 0), \boldsymbol{k}_2 = (0, 1/2)\}$ at the X points whereas the second star is $\mathbf{k}^2 = (1/2, 1/2)$ at the M point. For a normalization constant A, the change in density on the square lattice surface points is given by

$$\rho_i = \frac{1}{A} \Sigma_i \exp(-\boldsymbol{k} \cdot \boldsymbol{r}) \rho_i \,. \tag{A.10}$$

Equivalently, we can write

$$\rho_i = \rho + \rho(\boldsymbol{k}_1) \cos(\boldsymbol{k}_1 \cdot \boldsymbol{r}_i) + \rho(\boldsymbol{k}_2) \cos(\boldsymbol{k}_2 \cdot \boldsymbol{r}_i) \,. \tag{A.11}$$

Since $\boldsymbol{r}_i = a(m\hat{\imath} + n\hat{\jmath})$, $\boldsymbol{k}_1 = (\pi/a)\hat{\imath}$ and $\boldsymbol{k}_2 = (\pi/a)\hat{\jmath}$, where m, n are integers and a is the lattice constant, we have

$$\rho_i = \rho + (-1)^m \rho(\boldsymbol{k}_1) + (-1)^n \rho(\boldsymbol{k}_2) \,. \tag{A.12}$$

In the general case, this density variation corresponds to the doubling of the unit cell along both directions, that is, the $p(2 \times 2)$ reconstruction. In the special case when either of the two density amplitudes $\rho(\boldsymbol{k}_1)$ or $\rho(\boldsymbol{k}_2)$ vanishes, the unit cell doubles along only one direction, that is, a $p(2 \times 1)$ reconstruction results. Based on these observations, associated with the star wave vector $\boldsymbol{k}_s$ a general free energy expression can be written as

$$F[\rho(\boldsymbol{k}_s)] = a[\rho^2(\boldsymbol{k}_1) + \rho^2(\boldsymbol{k}_2)] + b[\rho^2(\boldsymbol{k}_1) + \rho^2(\boldsymbol{k}_2)]^2 + c[\rho^4(\boldsymbol{k}_1) + \rho^4(\boldsymbol{k}_2)] \,. \tag{A.13}$$

Interestingly, from a critical phenomena perspective, this free energy form (or the universality class) is precisely that of the XY model with cubic anisotropy.

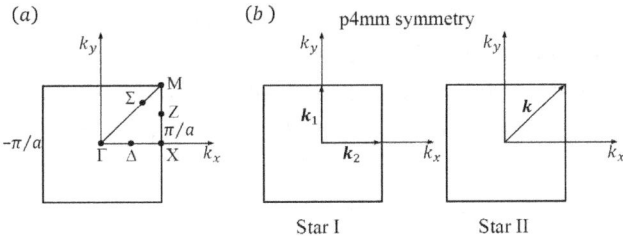

Figure A.3 (*a*) Brillouin zone of a square lattice and (*b*) two different vector stars.

Similarly, for the second star we have

$$\rho_i = \rho + \rho(\mathbf{k})\cos(\mathbf{k}\cdot\mathbf{r}_i)\,. \tag{A.14}$$

Since $\mathbf{k} = (\pi/a)(\hat{\imath} + \hat{\jmath})$ and $\mathbf{r}_i = a(m\hat{\imath} + n\hat{\jmath})$, in this case we obtain

$$\rho_{mn} = \rho + (-1)^{m+n}\rho(\mathbf{k})\,. \tag{A.15}$$

This density distribution leads to a bigger centred square or $c(2\times2)$ reconstruction. In this case, the free energy is much simpler and belongs to the Ising universality class

$$F[\rho(\mathbf{k}_s)] = a\rho^2(\mathbf{k}) + b\rho^4(\mathbf{k})\,. \tag{A.16}$$

Next, let us consider a first-order $2d$ ferroelectric transition on a square lattice driven by the M-point phonon that breaks the inversion symmetry due to the relative atomic displacement along the [11] direction. In this case, based on the above symmetry considerations, the full Ginzburg-Landau free energy in terms of the two-component polarization $\mathbf{P} = (P_x, P_y)$ and the strain accompanying the transition [362, 363] can be written as

$$\begin{aligned}
F(\mathbf{P}) = {} & a_1(P_x^2 + P_y^2) + a_{11}(P_x^4 + P_y^4) + a_{12}P_x^2P_y^2 + a_{111}(P_x^6 + P_y^6) \\
& + a_{112}(P_x^2P_y^4 + P_x^4P_y^2) + g_1(P_{x,x}^2 + P_{y,y}^2) + g_2(P_{x,y}^2 + P_{y,x}^2) \\
& + g_3P_{x,x}P_{y,y} + A_1e_1^2 + A_2e_2^2 + A_3e_3^2 + b_1e_1(P_x^2 + P_y^2) \\
& + b_2e_2(P_x^2 - P_y^2) + b_3e_3P_xP_y\,. \tag{A.17}
\end{aligned}$$

Note that, to capture a first-order transition, we expanded the free energy up to sixth order in polarization since no third-order term is symmetry allowed on a square lattice. Note also that there are two fourth-order and two sixth-order polarization invariants that are symmetry allowed. In addition, there are three gradient (or Ginzburg) terms allowed in this case which follow from the combined symmetry properties of gradients and polarization, that is, they both transform like a usual vector. We have included the strain energy to harmonic order and the lowest-order electrostriction terms (coupling strain to polarization). In the latter three terms, note that similar quantities couple to each other under coordinate transformation: e_1 has the symmetry of $xx + yy$ and thus it couples to $P_x^2 + P_y^2$, likewise e_2 has the symmetry of $xx - yy$ and thus it couples to $P_x^2 - P_y^2$. Similarly, e_3 has the symmetry of xy and it couples to $P_x P_y$.

As a quite complex example of a $3d$ phonon-driven structural transition, we next consider the ferroic transition occurring in materials with CsCl structure ($Pm\bar{3}m$) to a body-centred tetragonal structure ($I4/mmm$) that is induced by an M-point phonon mode and strain [364]. The doubly extended CsCl structure is depicted in Figure A.4. The corresponding Brillouin zone of the simple cubic lattice and the star associated with M-point are depicted in Figures A.5(a) and (b). The reducible representation of the M-point can be decomposed in irreps as $M = M_2^- \oplus M_3^- \oplus 2M_5^-$.

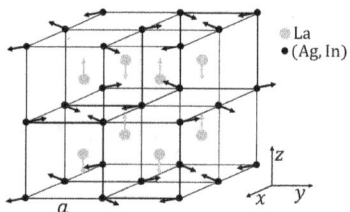

Figure A.4 Displacements associated with the simple cubic ($Pm\bar{3}m$) to body-centred tetragonal ($I4/mmm$) transition in a CsCl structure material La(Ag,In) induced by an M-point phonon mode and strain.

Brillouin zone of the SC-lattice

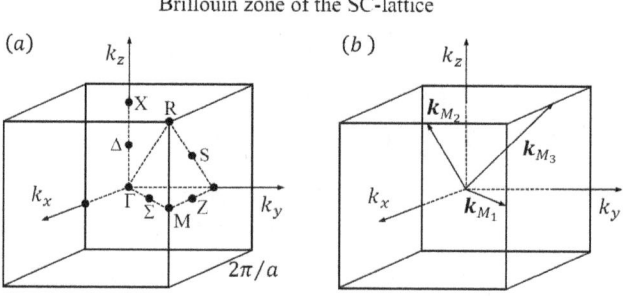

Figure A.5 (a) Brillouin zone of a simple cubic lattice and (b) the star vectors corresponding to the M point

The transition we consider is driven by the doubly degenerate irrep M_5^-, which is an acoustic phonon. Because this soft mode is doubly degenerate and since there are three distinct vectors in the star, the shuffle order parameter is a six-component one: $Q = \{Q_1, Q_2, Q_3, Q_4, Q_5, Q_6\}$. This order parameter and the corresponding Landau free energy can be obtained either using the lattice dynamics representation theory or via ISOTROPY [360], details of which can be found in [364]. Based on the symmetry analysis, we find five distinct invariants of fourth order (and ten distinct invariants of sixth order). To fourth order, the Landau free energy is given by

$$
\begin{aligned}
F_L(Q) = {} & a_1(Q_1^2 + Q_2^2 + Q_3^2 + Q_4^2 + Q_5^2 + Q_6^2) \\
& + b_1(Q_1^2 + Q_2^2 + Q_3^2 + Q_4^2 + Q_5^2 + Q_6^2)^2 \\
& + b_2(Q_1^2 Q_3^2 + Q_1^2 Q_5^2 + Q_3^2 Q_5^2 + Q_2^2 Q_4^2 + Q_2^2 Q_6^2 + Q_4^2 Q_6^2) \\
& + b_3(Q_1^2 Q_4^2 + Q_2^2 Q_5^2 + Q_3^2 Q_6^2) \\
& + b_4(Q_1^2 Q_6^2 + Q_2^2 Q_3^2 + Q_4^2 Q_5^2) \\
& + b_5(Q_1^2 Q_2^2 + Q_3^2 Q_4^2 + Q_5^2 Q_6^2).
\end{aligned}
\tag{A.18}
$$

The strain energy can be added to harmonic order: $F_L(e) = A_1 e_1^2 + A_2(e_2^2 + e_3^2) + A_3(e_4^2 + e_5^2 + e_6^2)$. For the sake of illustration, the coupling of shear strains to Q gives a third-order invariant of the form $I = e_5 e_6 Q_5 Q_6 + e_4 e_6 Q_3 Q_4 + e_4 e_5 Q_1 Q_2$. The coupling of deviatoric strain (e_2, e_3) to the shuffle order parameter Q, the 10 sixth-order invariants in Q and other details can be found in Ref. [364].

We will illustrate a second example of a (reconstructive) structural transition driven by a $3d$ phonon in Section A.5.

A.4 Colour symmetry and magnetic transitions

A current carrying loop gives rise to a magnetic moment. Current is the time derivative of charge. Thus, we cannot think of magnetization without a sense of time. In other words, magnetization M transforms as time multiplied by coordinates (or a polar vector): (xt, yt, zt). Thus, magnetization is an axial vector. Just as a broken spatial inversion symmetry leads to polarization, a broken time reversal symmetry leads to magnetization. In other words, crystals that lack a spatial inversion operation are ferroelectric and crystals that lack time reversal symmetry are ferromagnetic. Crystals that lack both these symmetries are multiferroic with an intrinsic coupling between $\boldsymbol{P}$ and $\boldsymbol{M}$. A subset of such crystals is ferrotoroidic, that is, it leads to an antisymmetric magnetoelectric contribution [365].

Time can either go forward or backward, thus it can be represented by two states: black and white at a lattice point or at a face of the unit cell. Inclusion of time in this way in crystals leads to enormous possibilities and expands the number of space groups and point groups substantially. Such groups are called colour groups or magnetic groups or Shubnikov groups [366]. In $3d$, the number of space groups expands to 1651 (from 230) and the number of point groups expands to 122 (instead of 32). The number of Bravais lattices increases to 22 (from 14).

There are three kinds of magnetic point (and space) groups [365, 366]. The usual (or crystallographic) 32 point groups are called the first kind, G_I. One obtains additional 32 grey point groups by multiplying each element of G_I by the antisymmetry operation $1'$: second kind $G_{II} = G_I 1'$. These groups are nonmagnetic because they have equal amount of black and white at every point, that is, one cannot break time reversal symmetry. However, they can be ferroelectric. There are 58 black and white groups (of third kind) which are obtained from G_I by taking a subgroup of index two (i.e., with half the symmetry elements) $(G_I)_2$ and subsequently multiplying the remaining symmetry elements by $1'$. Specifically, $G_{III} = (G_I)_2 + 1'[G_I - (G_I)_2]$. Such groups are necessarily magnetic and a subset can also be ferroelectric. The $1'$ operation was first introduced by Heesch in 1929; it changes white colour to black or vice versa. Equivalently, it reverses the direction of time or flips a spin up to spin down or vice versa. Therefore, magnetic point groups are also called Heesch-Shubnikov groups. It is clear that $122 = 32 + 32 + 58$. The corresponding number of magnetic space groups is $1651 = 230 + 230 + 1191$.

It is simpler and instructive to understand colour symmetry in $2d$. Therefore, in Figure A.6 we demonstrate how to pictorially construct magnetic point groups in $2d$. The first kind are the 10 usual point groups, out of which we depict only 5 high symmetry ones for illustration purposes, each denoted by G_0. These groups are necessarily non-ferroelectric because they have spatial inversion symmetry ($\bar{1}$). Nevertheless, their five subgroups, not shown here, can be ferroelectric. As noted previously, we can create an equal number (10) of grey groups by doubling the number of symmetry elements, $G = G_0 + 1'G_0$. To reiterate, these groups are necessarily nonmagnetic because they explicitly contain the element $1'$, that is, each point has equal amount of black and white (i.e., grey). However, they could be ferroelectric for the subgroups of the ones shown.

To create the third kind of colour groups, there are eleven distinct ways in which we can colour the crystal unit cell black and white in equal amount [365]. These are the $2d$ magnetic (or black and white) point groups, a subset of which can also be ferroelectric. It is important to note that any other black

2d Magnetic point groups

I. (White, polar, Fedorov) G_0: $1, 2, 2mm, 4mm, 3mm, 6mm$

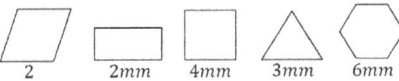

II. (Neutral, Grey) $G = G_0 + 1'G_0$: $1', 21', m1', 2mm, 1'41', 4mm1', 3\,1', 3mm\,1', 61', 6mm1'$

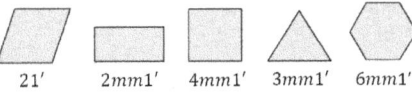

III. (Black − white, mixed polarity) $G = G_0 + 1'G_2$: $2', m', 2'm'm, 2m'm', 4', 4'm'm, 4m'm', 3m', 6', 6'm'm. \, 6m'm'$

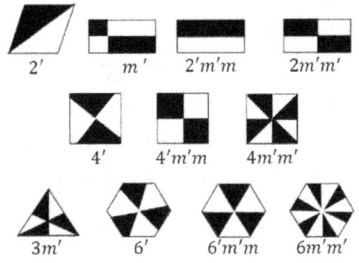

Figure A.6 Three types of magnetic groups in two dimensions. Adapted from Ref. [365]

and white pattern with equal area of black and white is necessarily equivalent to these eleven patterns. Thus there are $31 = 10 + 10 + 11$ magnetic point groups and corresponding $80 = 17 + 17 + 46$ magnetic space groups for a $2d$ crystal. It is clear that there are only 46 corresponding black and white space groups in $2d$. There are five magnetic Bravais lattices in $2d$.

Next we describe a multiferroic transition on a triangular lattice using some of the above notions. In hexagonal multiferroic materials, there can be magnetic and structural transitions between a non-ferroic parent group and a multiferroic, in particular a magnetoelectric, product phase, for example $6mm \to 3m'$. Additionally, there can be phase transitions between two different multiferroic groups, for instance $3m' \to 3$. As a representative example consider a triangular lattice undergoing a structural transition to a centred rectangular lattice with both in-plane magnetization and polarization. Note that $\boldsymbol{M}$ must appear in even powers due to the time reversal symmetry of the parent phase and $\boldsymbol{P}$ must also appear in even powers due to the spatial inversion symmetry of the parent phase. For simplicity ignoring the area change strain e_1, the symmetry allowed Landau free energy in this case [365] is thus given by

$$F(\boldsymbol{P}, \boldsymbol{M}, e) = a_1(M_1^2 + M_2^2) + b_1(M_1^2 + M_2^2)^2 + c_1(M_1^4 + M_2^4)$$
$$+ a_2(P_1^2 + P_2^2) + b_2(P_1^2 + P_2^2)^2 + c_2(P_1^4 + P_2^4)$$
$$+ a_3(e_2^2 + e_3^2) + b_3(e_2^3 - 3e_2 e_3^2) + c_3(e_2^2 + e_3^2)^2$$
$$+ A[e_2(M_1^2 - M_2^2) + e_3 M_1 M_2]$$
$$+ B[e_2(P_1^2 - P_2^2) + e_3 P_1 P_2]$$
$$+ C[(M_1^2 + M_2^2)(P_1^2 + P_2^2)].$$
(A.19)

As noted above, e_2 and e_3 are the two components of shear strain of a triangle. Here A, B, and C terms represent magnetostriction, electrostriction and magnetoelectric coefficients, respectively, of the relevant multiferroic material undergoing a phase transition.

The different order parameters can be inhomogeneous in the presence of domain walls and vortices. Then, assuming a generic variable ε for strain instead of the two strain components e_2 and e_3 and similarly for the gradient of $\boldsymbol{M}$ and $\boldsymbol{P}$, the most general gradient (or Ginzburg) energy appropriate for this transition is given by

$$F_G = g_1(\nabla\varepsilon)^2 + g_2(\nabla M)^2 + g_3(\nabla P)^2$$
$$+ g_4 M^2(\nabla \cdot P) + g_5 M \cdot [(M \cdot P)P]$$
$$+ g_6(P \cdot M)(\nabla \cdot M) + g_7 P \cdot [(M \cdot \nabla)M]$$
$$+ g_8\varepsilon(\nabla \cdot P) + g_9 P \cdot (\nabla\varepsilon).$$
(A.20)

Note that the last two terms have one strain (i.e., two coordinates), one gradient (i.e., one coordinate), and one polarization (i.e., one coordinate). Overall, there are four coordinates multiplied together, which combined are invariant under the symmetry operations. Same applies to the first seven terms also, which can be readily checked since each term has to be a scalar.

We also point out that a coupling of polarization gradient to magnetization and a coupling of magnetization gradient to polarization can occur in some unusual situations involving non-collinear dipoles or non-collinear magnetism. Specifically, in the latter case, a coupling of the form $\boldsymbol{P} \cdot [\boldsymbol{M}(\nabla \cdot \boldsymbol{M}) - (M \cdot \nabla)\boldsymbol{M}]$ can exist [159]. This coupling is discussed in Section 4.1.1.

A.5 Common subgroup method for reconstructive transitions

The Landau theory is readily applicable when the symmetry of the two phases, described by G_0 and G, has a group–subgroup relationship [28]. Such transitions (at least the structural ones) are called displacive. However, there are transitions in materials where there is no such group–subgroup relationship. An example being the *bcc* to *hcp* transition in titanium upon cooling [367] or a similar but hydrostatic pressure-induced transition in iron. Such transitions are also referred to as reconstructive transitions. Certain ways of constructing the Landau free energy for such transitions using transcendental order parameters have been proposed [96], but they are not very amenable to symmetry analysis or easily understandable. Here we illustrate an alternative way which is more transparent and uses the group–subgroup ideas by invoking the concept of a common subgroup of the groups of the two phases [368].

Once again, we first resort to 2*d* for illustrative purposes before describing the 3*d* transitions. The square lattice (*p4mm*) to triangular lattice (*p6mm*) transition is a good example. A square has a fourfold rotation axis whereas a triangle has a threefold axis. Therefore, the two lattices cannot have a group–subgroup relationship. The idea is to first find a lattice with lower symmetry that is a subgroup of both the square and the triangular lattice. A centred rectangular lattice (*c2mm*) or an oblique lattice (*p2*) can fulfill this condition (Figure A.7). In addition, we need to ascertain that after the transition to this (intermediate virtual) lattice from the two parent lattices the atoms (or the lattice points) must coincide. This is attained by the notion of a lock-in strain which couples to the primary order parameter, as we proceed to illustrate now.

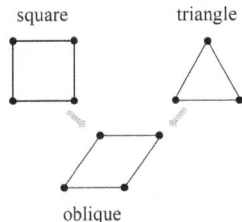

Figure A.7 Square lattice to triangular lattice transition enabled by an oblique lattice as a common subgroup.

For oblique lattice (*p2*) as a common subgroup of *p4mm* and *p6mm*, an *X*-point phonon (or shuffle mode) first displaces the atoms in opposite directions in alternating rows of a square lattice. This is the primary order parameter with two equal components (η, η). To obtain the correct shape of the triangular unit cell, two strains act as secondary order parameters that couple to the shuffle mode, namely the area strain e_1 and the deviatoric strain e_2. Thus, the Ginzburg-Landau free energy for this first order transition [368] can be written as

$$F_{GL} = a\eta^2 + b\eta^4 + c\eta^6 + g_1(\eta_x^2 + \eta_y^2) + g_2(\eta_x^2 - \eta_y^2)$$
$$+ c_1\eta^2 e_1 + c_2\eta^2 e_1^2 + c_3\eta^2 e_2^2 + a_1 e_1 + a_2 e_1^2 + b_1 e_2^2. \qquad (A.21)$$

Note that there are two gradient invariants. The elastic energy due to e_1 and e_2 is also included. The end point lock-in values from both parent lattices (square and triangle) can be obtained from ISOTROPY program [360, 368].

Next we study the *bcc* ($Im\bar{3}m$) to *hcp* ($P6_3/mmc$) transition for which an orthorhombic lattice or a monoclinic lattice is a good candidate for being a common subgroup. Indeed a base-centred orthorhombic lattice (Figure A.8) is a viable common subgroup (Cmcm) and we proceed to explore this pathway for the *bcc-hcp* phase transition. As determined using ISOTROPY, the primary order parameter is the intra-unit cell shuffle mode η. This distortional mechanism corresponds to an appropriate high symmetry point (N) along the [110] direction in the *bcc* Brillouin zone (Figure A.9(a)). Specifically, the mechanism is described by the irrep N_4^- [367] and the corresponding atomic displacements are also depicted in Figure A.9(b).

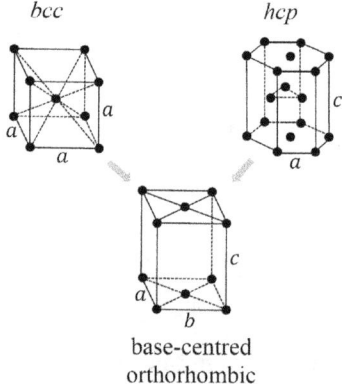

bcc

hcp

base-centred
orthorhombic

Figure A.8 *bcc* to *hcp* transition enabled by a base-centred orthorhombic lattice as a common subgroup.

The N_4^- shuffle mode (or phonon) couples to three secondary order parameter strains: volumetric (e_1), deviatoric (e_2), and shear (e_4), which lead to the exact *hcp* stacking sequence and squeeze the *bcc* octahedron to transform it into a regular *hcp* cell as well as provide an appropriate volume dilatation

Brillouin zone of the *bcc*-lattice

(*a*)

(*b*)

phonon N_4^-

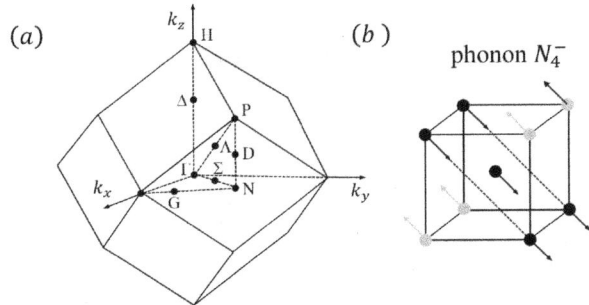

Figure A.9 (*a*) Brillouin zone of the *bcc* lattice. (*b*) Illustration of the atomic displacements caused by the N_4^- phonon.

to reach the correct end volume. The most general Landau free energy for this first-order transition [367] is then given by

$$
\begin{aligned}
F_L = {}& a_1\eta^2 + b_1\eta^4 + c_1\eta^6 + a_2\eta^2 e_2^2 + a_3\eta^2 e_4^2 \\
& + a_4 e_1^2 + a_5 e_1^3 + a_6 e_2^2 + a_7 e_2^3 + a_8 e_2^4 + a_9 e_4^2 + a_{10} e_4^4 \\
& + a_{11}\eta^2 e_4 + a_{12}\eta^2 e_2 + a_{13}\eta^2 e_1 \\
& + a_{14} e_1 e_2^2 + a_{15} e_1 e_4^2 + a_{16} e_2 e_4^2 .
\end{aligned}
\tag{A.22}
$$

The associated shuffle gradient (or Ginzburg) free energy has three distinct invariants, and is given by

$$
F_G(\eta) = g_1 \eta_x^2 + g_2(\eta_y^2 + \eta_z^2) + g_3 \eta_y \eta_z .
\tag{A.23}
$$

Akin to the 2*d* case, the end point lock-in values from both parent lattices (*bcc* and *hcp*) can be obtained from ISOTROPY program [360, 367].

References

[1] Sachdev, S., *Quantum Phase Transitions*, 2nd ed., Cambridge University Press, Cambridge, 2011.

[2] Jaeger, G., *The Ehrenfest Classification of Phase Transitions: Introduction and Evolution*, Arch. Hist. Exact Sci. **53**, 51–81 (1998).

[3] Yeoman, J. M., *Statistical Mechanics of Phase Transitions*, Oxford University Press, New York, 1992.

[4] Stanley, H. E., *Introduction to Phase Transitions and Critical Phenomena*, Oxford University Press, New York, 1971.

[5] ter Haar, D. and Wergeland, H., *Elements of Thermodynamics*, Addison-Wesley Publishing Company, Reading, 1966.

[6] Sun, W. and Powell-Palm, M. J., *Generalized Gibbs' Phase Rule*, arXiv:2105.01337v1 [math-ph].

[7] Gibbs, J. W., *The Scientific Papers of J. Willard Gibbs, Vol. 1: Thermodynamics*, Ox Bow Press, Woodbridge, CT, 1993.

[8] Tolman, R. C., *The Effect of Droplet Size on Surface Tension*, J. Chem. Phys. **17**, 333–337 (1949).

[9] Hill, T. L., *Thermodynamics of Small Systems*, J. Chem. Phys. **36**, 3182–3197 (1962).

[10] Hill, T. L., *Perspective: Nanothermodynamics*, Nano Lett. **11**, 111–112 (2001).

[11] Gunton, J. D. and Droz, M., *Introduction to the Theory of Metastable and Unstable States*, Sringer-Verlag, Berlin, 1983.

[12] Hiley, B. J. and Joyce, G. S., *Ising Models with Long Range Interactions*, Proc. Phys. Soc. **85**, 493–507 (1965).

[13] Smart, J. S., *Effective Field Theories of Magnetism*, Saunders, Philadelphia, PA, 1966.

[14] Blume, M., Emery, V. J., and Griffiths, R. B., *Ising Model for the λ Transition and Phase Separation in He^3-He^4 Mixtures*, Phys. Rev. A **4**, 1071–1077 (1971).

[15] Baxter, R. J., *Exactly Solved Models in Statistical Mechanics*, Academic Press, London, 1982.

[16] Kadanoff, L. P., *More Is the Same; Phase Transitions and Mean Field Theory*, J. Stat. Phys. **137**, 777–797 (2009).

[17] Bragg, W. L. and Williams, E. J., *The Effect of Thermal Agitation on Atomic Arrangement in Alloys*, Proc. R. Soc. Lond. **145** A, 699–730 (1934).

[18] Kinkaid, J. M. and Cohen, E. G. D., *Phase Diagrams of Liquids Helium Mixtures and Metamagnets: Experiment and Mean Field Theory*, Phys. Rep. **22**, 57–143 (1975).

[19] Weiss, P., *L hypothèse du champ moléculaire et la propriété ferromagnétique*, J. Phys. Theor. Appl. **6**, 661–690 (1907).

[20] Bethe, H., *Statistical Theory of Superlattices*, Proc. R. Soc. Lond. A **150**, 552–575 (1935).

[21] Kirkwood, J. G., *Order and Disorder in Binary Solid Solutions*, J. Chem. Phys. **6**, 70–75 (1938).

[22] Kruseman Aretz, F. E. J. and Cohen, E. G. D., *A Theory of Order-Disorder Phenomena I*, Physica **26**, 967–980 (1960).

[23] Moessner, R. and Ramirez, A. P., *Geometrical Frustration*, Phys. Today **59**, 24–29 (2006).

[24] van de Walle, A. and Ceder, G., *The Effect of Lattice Vibrations on Substitutional Alloy Thermodynamics*, Rev. Mod. Phys. **74**, 11–45 (2002).

[25] Jones, R. A. L., *Soft Condensed Matter*, Oxford University Press, Oxford, 2002.

[26] Born, M. and Huang, K., *Dynamical Theory of Crystal Lattices*, Oxford University Press, Oxford, 1954.

[27] Vives, E., Castán, T., and Planes, A., *Unified Mean-Field Study of Ferro- and Antiferromagnetic Behavior of the Ising Model with External Field*, Am. J. Phys. **65**, 907–913 (1997).

[28] Tolédano, J. C. and Tolédano, P., *The Landau Theory of Phase Transitions*, World Scientific, Singapore, 1987.

[29] Triguero, C., Porta, M., and Planes, A., *Coupling between Lattice Vibrations and Magnetism in Ising-Like Systems*, Phys. Rev. B **73**, 054401 (2006).

[30] Landau, L. D. and Lifshitz, E. M., *Statistical Physics*, 3rd ed., Part 1, Course in Theoretical Physics, Vol. 5, Pergamon Press, Oxford, 1980.

[31] Devonshire, A., *Theory of Ferroelectrics*, Adv. Phys. **3**, 85–130 (1954).

[32] Ginzburg, V. L. and Landau, L. D., *On the Theory of Superconductivity*, Zh. Eksp. Teor. Fiz. **20**, 1064–1082 (1950); English translation in: L. D. Landau, *Collected Papers*, Pergamon Press, Oxford, pp. 546–568, 1965.

[33] Salje, E. K. H., Wruck, B., and Thomas, H., *Order-Parameter Saturation and Low-Temperature Extension of Landau Theory*, Z. Phys. B **82**, 399–404 (1991).

[34] Mermin, N. D. and Wagner, H., *Absence of Ferromagnetism or Antiferromagnetism in One- or Two-Dimensional Isotropic Heisenberg Models*, Phys. Rev. Lett. **17**, 1133–1136 (1966).

[35] Chaikin, P. M. and Lubensky, T. C., *Principles of Condensed Matter Physics*, Cambridge University Press, Cambridge, 1995.

[36] Binney, J. J., Dowrick, N. J., Fisher, A. J., and Newman, M. E. J., *The Theory of Critical Phenomena*, Oxford University Press, New York, 1992.

[37] Nishimori, H. and Ortiz, G., *Elements of Phase Transitions and Critical Phenomena*, Oxford University Press, Oxford, 2011.

[38] Widom, B., *Equation of State in the Neighborhood of the Critical Point*, J. Chem. Phys. **43**, 3898–3905 (1965).

[39] Kadanoff, L. P., *Scaling Laws for Ising Models near T_c*, Phys. **2**, 263 (1966).

[40] Ho, J. T. and Litster, J. D., *Magnetic Equation of State of $CrBr_3$, Near the Critical Point*, Phys. Rev. Lett. **22**, 603–606 (1969).

[41] Maris, H. J. and Kadanoff, L. P., *Teaching the Renormalization Group*, Am. J. Phys. **46**, 653–657 (1978).

[42] Reichl, L. E., *A Modern Course in Statistical Physics*, 4th revised and updated edition, Wiley-VCH, Weinheim, 2016.

[43] Wilson, K. G., *The Renormalization Group: Critical Phenomena and the Kondo Problem*, Rev. Mod. Phys. **47**, 773–840 (1975).

[44] Gunton, J. D., San Miguel, M., and Sahni, P. S., *The Dynamics of First Order Phase Transitions*, in *Phase Transitions and Critical Phenomena*, Vol. 8, pp. 269–466, edited by Domb, C. and Lebowitz, J. L., Academic Press, London 1983.

[45] Mouritsen, O. G., *Pattern Formation in Condensed Matter*, Int. J. Mod. Phys. **B 4** 1925–1954 (1990).

[46] Mazenko, G. F., *Universal Features in Growth Kinetics: Some Experimental Tests*, Phys. Rev. B **43**, 8204–8210 (1991).

[47] Noda, Y., Nishihara, S., and Yamada, Y., *Critical Behavior and Scaling Law in Ordering Process of the First Order Phase Transition in Cu_3Au Alloy*, J. Phys. Soc. Jpn. **53**, 4241–4249 (1984).

[48] Shannon, R. F., Nagler, E. E., Harkless, C. R., and Nicklow, R. M., *Time-Resolved X-ray-Scattering Study of Ordering Kinetics in Bulk Single-Crystal Cu_3Au*, Phys. Rev. B **46**, 40–54 (1992).

[49] Gaulin, B. D., Spooner, S., and Morii, Y., *Kinetics of Phase Separation in $Mn_{0.67}Cu_{0.33}$*, Phys. Rev. Lett. **59** 668–671 (1987).

[50] Glas, R., Blaschko, O., and Rosta, L., *Structure Functions in Decomposing Au-Pt Systems*, Phys. Rev. B **46**, 5972–5981 (1992).

[51] Gunton, J. D. and Droz, M., *Introduction to the Theory of Metastable and Unstable States*, Springer-Verlag, Berlin, 1983.

[52] Binder, K., *Theory of First-Order Phase Transitions*, Rep. Prog. Phys. **50** 783–859 (1987).

[53] Hänggi, P., Talkner, P., and Borkovec, M., *Reaction-Rate Theory: Fifty Years after Kramers*, Rev. Mod. Phys. **62**, 251–341 (1990).

[54] Ortín, J. and Planes, A., *Hysteresis in Shape-Memory Materials*, in *The Science of Hysteresis*, Vol. III, pp. 467–553, edited by Bertotti, G. and Mayergoyz, I., Elsevier Inc., Amsterdam 2005.

[55] Mahato, M. C. and Shenoy, R. C., *Langevin Dynamics Simulations of Hysteresis in Field-Swept Landau Potentials*, J. Stat. Phys. **73**, 123–145 (1993).

[56] Langer, J. S., *Theory of Spinodal Decomposition in Alloys*, Ann. Phys. **65**, 53–86 (1971).

[57] Hohenberg, P. C. and Halperin, B. I., *Theory of Dynamic Critical Phenomena*, Rev. Mod. Phys. **49**, 435–479 (1977).

[58] Cahn, J. W. and Hilliard, J. E., *Free Energy of a Nonuniform System. I. Interfacial Free Energy*, J. Chem. Phys. **28**, 258–267 (1958).

[59] Cahn, J. W., *Phase Separation by Spinodal Decomposition in Isotropic Systems*, J. Chem. Phys. **42**, 93–99 (1965).

[60] Onuki, A., *Phase Transition Dynamics*, Cambridge University Press, Cambridge, 2002.

[61] Angell, A. *Supercooled Water*, Ann. Rev. Phys. Chem. **34**, 593–630 (1984).

[62] Landau, L. D. and Lifshitz, E. M., *Statistical Physics*, 3rd ed., Pergamon Press, Oxford, 1980.

[63] Lifshitz, E. M. and Pitaevskii, L. P., *Physical Kinetics*, Pergamon Press, Oxford, 1981.

[64] Becker, E. and Döring, W., *Kinetische Behandlung der Keimbildung in übersättigten Dämpfen*, Ann. Phys. **24**, 719–752 (1935).

[65] Mullins, W. W. and Sekerka, R. F., *Morphological Stability of a Particle Growing by Diffusion or Heat Flow*, J. Appl. Phys. **33**, 323–329 (1963).

[66] Balluffi, R. W., Allen, S. M., and Carter, W. C., *Kinetics of Materials*, John Wiley & Sons, Hoboken, 2005.

[67] Cahn, J. W., *Nucleation on Dislocations*, Acta Metall. **5**, 165–172 (1957).

[68] Larché, F. C., *Nucleation and Precipitation on Dislocations*, in *Dislocations in Solids*, edited by Nabarro, F. R. N., Vol. 4, pp. 137–152, North Holland, Amsterdam, 1979.

[69] Hirth, J. P. and Lothe, J., *Theory of Dislocations*, 2nd ed., John Wiley & Sons, New York, 1982.

[70] Langer, J. S., Bar-on, M., and Miller, H. D., *New Computational Method in the Theory of Spinodal Decomposition*, Phys. Rev. A **11**, 1417–1429 (1975).

[71] Mainville, J., Yang, Y. S., Elder, K. R., Sutton, M., *X-ray Scattering Study of Early Stage Spinodal Decomposition in $Al_{0.62}Zn_{0.38}$*, Phys. Rev. Lett. **78**, 2787–2790 (1997).

[72] Bray, A. J., *Theory of Phase Ordering Kinetics*, Adv. Phys. **43**, 357–459 (1994).

[73] Allen, S. M. and Cahn, J. W., *A Microscopic Theory for Antiphase Boundary Motion and Its Application to Antiphase Domain Coarsening*, Acta metall. **27**, 1085–1085 (1979).

[74] Lifshitz, I. M. and Slyozov, V. V., *The Kinetics of Precipitation from Super-Saturated Solid Solutions*, J. Phys. Chem. Solids **19**, 35–50 (1961).

[75] Wagner, C., *Theorie der Alterung von Niderschlagen durch Umlösen (Ostwald Reifung)*, Z. Electrochem. **65**, 581–591 (1961).

[76] Katano, S., Iizumi, M., Nicklov, R. M., and Child, H. R., *Scaling in the Kinetics of the Order-Disorder Transition in Ni_3Mn*, Phys. Rev. **B 38**, 2659–2663 (1988).

[77] Oki, K., Sagane, H., and Eguchi, T., *Separation and Domain Structure of α + B2 Phase in Fe-Al Alloys*, J. Phys. **38**, C7-414–417 (1977).

[78] Kikuchi, R. and Cahn, J. W., *Theory of Interphase and Antiphase Boundaries in f.c.c. Alloys*, Acta Metall. **27**, 1337–1353 (1979).

[79] Frontera, C., Vives, E., Castán, T., and Planes, A., *Monte Carlo Study of the Growth of $L1_2$-ordered domains in fcc A_3B binary alloys*, Phys. Rev. B **55**, 212–225 (1997).

[80] Castán, T. and Lindgård, P.-A., *$n = \frac{1}{4}$ Domain-Growth Univesrality Class: Crossover to the $n = \frac{1}{2}$ Class*, Phys. Rev. B **41**, 2534–2536 (1990).

[81] Aizu, K., *Possible Species of Ferromagnetic, Ferroelectric, and Ferroelastic Crystals*, Phys. Rev. B **2**, 754–772 (1970).

[82] Aizu, K., *Possible Species of "Ferroelastic" Crystals and of Simultaneously Ferroelectric and Ferroelastic Crystals*, J. Phys. Soc. Japan **27**, 387–396 (1969).

[83] Gutfleisch, O., Willard, M. A., Brück, E., Chen, C. H., Sankar, S. G., and Liu, J. P., *Magnetic Materials and Devices for the 21st Century: Stronger, Lighter, and More Energy Efficient*, Adv. Mater. **23**, 821–842 (2011).

[84] Scott, J. F, *Applications of Modern Ferroelectrics*, Science **315**, 954–959 (2007).

[85] Otsuka, K. and Wayman, C. M., eds., *Shape Memory Materials*, Cambridge University Press, Cambridge, 1998.

[86] Lehmann, J., Bortis, A., Derlet, P. M., Donnelly, C., and Leo, N., Heyderman, L. J., Fiebig, M., *Relation between Microscopic Interactions and Macroscopic Properties in Ferroics*, Nature Nanotech. **15**, 896–900 (2020).

[87] Jackson, J. D., *Classical Electrodynamics*, 3rd ed., John Wiley & Sons, Inc., 1999.

[88] Dubovik, V. M., Martsenyuk, M. A., and Saha, B., *Materials Equations for Electromagnetism with Toroidal Polarization*, Phys. Rev. E **61**, 7087–7097 (2000).

[89] Beardley, I. A., *Reconstruction of the Magnetization in a Thin Film by a Combination of Lorentz Microscopy and External Field Measurements*, IEEE Trans. Mag. **25**, 671–677 (1989).

[90] Prosandeev, S. and Bellaiche, L., *Hypertoroidal Moment in Complex Dipolar Structures*, J. Mater. Sci. **44**, 5235–5248 (2009).

[91] Shenoy, S. R., Lookman, T., and Saxena, A., *Scaled Free Energies, Power-Law Potentials, Strain Pseudospins, and Quasiuniversality for First-Order Structural Transitions*, Phys. Rev. B **82**, 144103 (2010).

[92] Onuki, A., *Phase Transition Dynamics*, Cambridge University Press, Cambridge, 2002.

[93] Porta, M. and Lookman, T., *Heterogeneity and Phase Transformation in Materials: Energy Minimization, Iterative Methods and Geometric Nonlinearity*, Acta mater. **61**, 5311–5340 (2013).

[94] Gröger, R., Lookman, T., and Saxena, A., *Defect-Induced Incompatibility of Elastic Strains: Dislocations within the Landau Theory of Martensitic Phase Transformations*, Phys. Rev. B **78**, 184101 (2008).

[95] Tolédano, P. and Tolédano, J. C., *Order-Parameter Symmetries for the Phase Transitions of Nonmagnetic Secondary and Higher-Order Ferroics*, Phys. Rev. B **16**, 386–407 (1977).

[96] Tolédano, P. and Dimitriev, V., *Reconstructive Phase Transitions in Crystals and Quasicrystals*, World Scientific, Singapur, 1996.

[97] Moriya, T. and Takahashi, Y., *Itinerant Electron Theory*, Ann. Rev. Mater. Sci. **14**, 1–25 (1984).

[98] Buschow, K. H. J. and de Boer, F. R. *Physics of Magnetism and Magnetic Materials*, Kluwer Academic Publishers, New York, 2003.

[99] Spaldin, N. A., *Magnetic Materials. Fundamentals and Applications*, 2nd ed., Cambridge University Press, Cambridge, 2003.

[100] Tayler, F., *The Magnetization-Temperature Curves of Iron, Cobalt, and Nickel*, Philos. Mag. **11**, 596–602 (1931).

[101] Aharoni, A., *Introduction to the Theory of Ferromagnetism*, Oxford University Press, New York, 1996. 1996.

[102] Osborn, J. A., *Demagnetizing Factors of the General Ellipsoid*, Phys. Rev. **67**, 351–357 (1945).

[103] Bertotti, G., *Hysteresis in Magnetism*, Academic Press, San Diego, 1998.

[104] Smart, J. S., *Effective Field Theories of Magnetism*, W. B. Saunders Company, Philadelphia, 1966.

[105] Yoshimori, A., *A New Type of Antiferromagnetic Structure in the Rutile Type Crystal* J. Phys. Soc. Japan **14**, 807–821 (1959).

[106] Borriello, I., Cantele, G., and Ninno, D., *Ab initio Investigation of Hybrid Organic-Inorganic Perovskites Based on Tin Halides*, Phys. Rev. B **77**, 235214 (2008).

[107] Bersuker, I. B., *Pseudo-Jahn-Teller Effect – A Two-State Paradigm in Formation, Deformation, and Transformation of Molecular Systems and Solids*, Chem. Rev. **113**, 1351–1390 (2013).

[108] Van Aken, B., Rivera, J. P., Schmid, H., and Fiebig, M., *Observation of Ferrotoroidic Domains*, Nature **449**, 702–705 (2007).

[109] Zimmermann, A. S., Meier, D., and Fiebig, M., *Ferroic Nature of Magnetic Toroidal Order*, Nature Comm. **5**, 4796 (2014).

[110] Tang, J., Hewitt, I., Madhu, N. T., Chastanet, G., Wernsdorfer, W., Anson, C. E., Benelli, C., Sessoli, R., and Powell, A. K., *Dysprosium Triangles Showing Single-Molecule Magnet Behavior of Thermally Excited Spin States*, Angew. Chem., Int. Eds. **45**, 1729–1733 (2006).

[111] Ungur, L., Lin, S.-Y., Tang, J., and Chibotaru, L. F., *Single-Molecule Toroidics in Ising-Type Lanthanide Molecular Clusters*, Chem. Soc. Rev. **43**, 6894–6905 (2014).

[112] Luzon, J., Bernot, K., Hewitt, I. J., Anson, C. E., and Powell, A. K., Sessoli, R., *Spin Chirality in a Molecular Dysprosium Triangle: The Archetype of the Noncollinear Ising Model*, Phys. Rev. Lett. **100**, 247205 (2008).

[113] Lehmann, J., Donnelly, C., Derlet, P. M., Heyderman, L. J., and Fiebig, M., *Poling of an Artificial Magneto-Toroidal Crystal*, Nature Nanotech. **14**, 141–144 (2019).

[114] Schmid, H., *On Ferrotoroidics and Electrotoroidics, Magnetotoroidics and Piezotoroidics Effects*, Ferroelectrics **252**, 41–50 (2001).

[115] Spaldin, N. A., Fiebig, M., and Mostovoy, M., *The Toroidal Moment in Condensed-Matter Physics and Its Relation to the Magnetoelectric Effect*, J. Phys. Condens. Matter. **20**, 434203 (2008).

[116] Baum, M., Schmalzl, K., Steffens, P., Hiess, A., Regnault, L. P., Meven, M., Becker, P., Bohatý, L., and Braden, M., *Controlling Toroidal Moments by Crossed Electric and Magnetic Field*, Phys. Rev. B **88**, 024414 (2013).

[117] Yu, S.-Y., Mao, Schmid, H., Triscone, G., and Muller, J., *Spontaneous Magnetization and Magnetic Susceptibility of a Ferroelectric/Ferromagnetic/Ferroelastic Single Domain Crystal of Nickel Bromine Boracite $Ni_3B_7O_{13}Br$*, J. Mag. Mag. Mater. **195**, 65–75 (1999).

[118] Popov Y. F., Kadomtseva, A. M., Vorob'ev, Tomofeeva, V. A., Ustinin, D. M., Zvezdin, A. K., and Tegeranvhi, M. M., *Magnetoelectric Effect and Toroidal Ordering in $Ga_{2-x}Fe_xO_3$*, JEPT **87**, 146–151 (1998).

[119] Popov, Y. F., Kadomtseva, A. M., Vorob'ev, G. P., and Zvezdin, A. K., *Magnetic-Field Induced Toroidal Moment in the Magnetoelectric Cr_2O_3*, JEPT Lett. **69**, 300–335 (1999).

[120] Sannikov, D. G., *Ferrotoroidics*, Ferroelectrics **354**, 39–43 (2007).

[121] Ederer, C. and Spaldin, N. A., *Towards a Microscopic Theory of Toroidal Moments in Bulk Periodic Crystals*, Phys. Rev. B **76**, 214404 (2007).

[122] Eshelby, J. D., *The Determination of the Elastic Field of an Ellipsoidal Inclusion, and Related Problems*, Proc. R. Soc. A **241** 376–396 (1957); *The Elastic Field Outside an Ellipsoidal Inclusion*, Proc. R. Soc. A **252**, 561–569 (1959).

[123] Wechsler, M. S., Lieberman, D. S., and Read, T. A., *Theory of the Formation of Martensite*, Trans. AIME **197**, 1503–1515 (1953).

[124] Bowles, J. S. and Mackenzie, J. K., *The Crystallography of Martensite Transformations I*, Acta Metal. **2**, 129–137 (1954).

[125] Khachaturyan, A. G., *The Theory of Structural Transformations in Solids*, John Wiley & Sons, New York, 1983.

[126] Ball, J. M. and James, R. D., *Fine Phase Mixtures as Minimizers of Energy*, Arch. Ration Mech. Anal. **100**, 13–52 (1987).

[127] Bhattacharya, K., *Microstructure of Martensite*, 1st ed., Oxford University Press, New York, 2003.

[128] Porta, M., Castán, T., Lloveras. P., Lookman, T., Saxena, A., and Shenoy, S. R., *Interfaces in Ferroelastics: Fringing Fields, Microstructure, and Size and Shape Effects*, Phys. Rev. B **79**, 214117 (2009).

[129] Kittel, C., *Physical Theory of Ferromagnetic Domains*, Rev. Mod. Phys. **21**, 541–583 (1949).

[130] Bales, G. S. and Gooding, R. J., *Interfacial Dynamics at a First-Order Phase Transition Involving Strain: Dynamical Twin Formation*, Phys. Rev. Lett. **67**, 3412–3415 (1991).

[131] Lookman, T., Shenoy, S. R., Rasmussen, K. Ø., Saxena, A., and Bishop, A. R., *Ferroelastic Dynamics and Strain Compatibility*, Phys. Rev. B **67**, 024114 (2003).

[132] Otsuka, K. and Wayman, C. M., eds., *Shape Memory Materials*, Cambridge University Press, Cambridge, 1998.

[133] Kaspar, G., Ravoo, B. J., van der Wiel, W. G., Wegner, S. V., and Pernice, W. H. P., *The Rise of Intelligent Matter*, Nature **594**, 345–355 (2021).

[134] Schmid, H., *Multi-ferroic Magnetoelectrics*, Ferroelectrics **162**, 317–338 (1994).

[135] Eerenstein, W., Mathur, N. D., and Scott, J., *Multiferroic and Magnetoelectric Materials*, Nature **442**, 759–765 (2006).

[136] Martin, L. W., Crane, S. P., Chu, Y.-H., Holcomb, M. B., Gajek, M., Huijben, M., Yang, C.-H., Balke, N., and Ramesh, R., *Multiferroics and Magnetoelectrics: Thin Films and Nanostructures*, J. Phys.: Condens. Matter **20**, 434220 (2008).

[137] Spaldin, N. A. and and Ramesh, R., *Advances in Magnetoelectric Multiferroics*, Nature Mater. **18**, 203–212 (2019).

[138] Forsbergh, P. W., *Domain Structures and Phase Transitions in Barium Titanate*, Phys. Rev. **76**, 1187–1201 (1949).

[139] Liebermann, H. H. and Graham, C. D., *Plastic and Magnetoplastic Deformation of Dy Single Crystals*, Acta Metall. **25**, 715–720 (1977).

[140] Ullakko, K., Huang, J. K., Kantner, C., O'Handley, R. C., and Kokorin, V. V., *Large Magnetic-Field-Induced Strains in* Ni_2MnGa *Single Crystals*, Appl. Phys. Lett. **69**, 1966–1968 (1996).

[141] Schmid, H., *Some Symmetry Aspects of Ferroics and Single Phase Multiferroics*, J. Phys. Condens. Matter **20**, 434201 (2008).

[142] Hill, N. A., *Why Are There So Few Magnetic Ferroelectrics?* J. Phys. Chem. B **104**, 6694–6709 (2000).

[143] Ascher, E., Rieder, H., Schmid. H., and Stössel, H., *Some Properties of Ferromagnetoelectric Nickel-Iodine Boracite*, $Ni_3B_7O_{13}I$, J. Appl. Phys. **37**, 1404–1405 (1966).

[144] Kiselev, S. V., Ozerov, R. P., and Zhdanov, G. S., *Detection of Magnetic Order in Ferroelectric* $BiFeO_3$ *by Neutron Diffraction*, Sov. Phys. Dokl. **7**, 742–744 (1963).

[145] Bertaut, E. F., Forrat, F., and Fang, P., *Les Manganites de Terres Rares et d'Ytrium: Une Nouvelle Classe de Ferroélectriques*, Comptes Rendus Acad. Sci. **256**, 1958–1961 (1963).

[146] Scott, J. F. and Blinc, R., *Multiferroic Magnetoelectric Fluorides: Why Are There So Many Magnetic Ferroelectrics?*, J. Phys. Condens. Matter **23**, 113202 (2011).

[147] Khomskii, D., *Classifying Multiferroics: Mechanisms and Effects*, Physics **2**, 20 (2009).

[148] Smolenskii, G. A. and Chupis, I. E., *Ferroelectromagnets*, Sov. Phys. Usp. **25**, 475–493 (1982).

[149] Cheong, S.-W. and Mostovoy, M., *Multiferroics: A Magnetic Twist for Ferroelectricity*, Nature Mater. **6**, 13–20 (2007).

[150] Efremov, D. V., Van den Brink, J., and Khomskii, D. I., *Bond-versus Site-Centred Ordering and Possible Ferroelectricity in Manganites*, Nature Mater. **3**, 853–856 (2004).

[151] Ikeda, N., Ohsumi, H., Ohwada, K., Ishii, K., Inami, T., Kakurai, K., Murakami, Y., Yoshii, K., Mori, S., Horibe, Y., and Kito, H., *Ferroelectricity from Iron Valence Ordering in the Charge-Frustrated System* $LuFe_2O_4$, Nature **436**, 1136–1138 (2005).

[152] van Aken, B. B., Palastra, T. T. M., Filippetti, A., and Spaldin, N. A., *The Origin of Ferroelectricity in Magnetoelectric* $YMnO_3$, Nature Mater. **3**, 164–170 (2004).

[153] Newnham, R. E., Kramer, J. J. Schulze, W. A., and Cross, L. E., *Magneto-ferroelectricity in* Cr_2BeO_4, J. Appl. Phys. **49**, 6088–6091 (1978).

[154] Kimuta, T., Goto, T., Shintani, H., Ishizaka, K., Arima, T., and Tokura, Y., *Magnetic Control of Ferroelectric Polarization*, Nature **426**, 55–58 (2003).

[155] Choi, Y. J., Yi, H. T., Lee, S., Huang, Q., Kiryukhin, V., and Cheong, S.-W., *Ferroelectricity in an Ising Chain Magnet*, Phys. Rev. Lett. **100**, 047601 (2008).

[156] Tokura, Y., Seki, S., and Nagaosa, N., *Multiferroics of Spin Origin*, Rep. Progress Phys. **77**, 076501 (2014).

[157] Bulaevskii, L. N., Batista, C. D., Mostovoy, M. V., and Khomskii, D. I., *Electronic Orbital Currents and Polarization in Mott Insulators*, Phys. Rev. B **78**, 024402 (2008).

[158] Cherifi, R. O., Ivanovskaya, V., Phillips, L. C., Zobelli, A., Infante, I. C., Jacquet, E., Garcia, V., Fusil, S., Briddon, P. R., Guiblin, N., Mougin, A., Ünal, A. A., Kronast, F., Valencia, S., Dkhil, B., Barthélémy, A., and Bibes, M., *Electric-Field Control of Magnetic Order above Room Temperature*, Nature Mater. **13**, 345–351 (2014).

[159] Mostovoy, M., *Ferroelectricity in Spiral Magnets*, Phys. Rev. Lett. **96**, 067601 (2006).

[160] Kenzelmann, M., Harris, A. B., Jonas, S., Broholm, C., Schefer, J., Kim, S. B., Zhang, C. L., Cheong, S.-W., Vajk, O. P., and Lynn, J. W., *Magnetic Inversion Symmetry Breaking and Ferroelectricity in* $TbMnO_3$, Phys. Rev. Lett. **95**, 087206 (2005).

[161] Kimura, T. and Tokura, Y., *Magnetoelectric Phase Control in a Magnetic System Showing Cycloidal/Conical Spin Order*, J. Phys. Condens. Matter **20**, 434204 (2008).

[162] Ge, Y., Heczko, O., Söderberg, O., and Lindroos, V. K., *Various Magnetic Domain Structures in a Ni–Mn–Ga Martensite Exhibiting Magnetic Shape Memory Effect*, J. Appl. Phys. **96**, 2159 (2004).

[163] Merz, W. J., *Domain Formation and Domain Wall Motions in Ferroelectric* $BaTiO_3$ *Single Crystals*, Phys. Rev. **95**, 690–698 (1954).

[164] Sozinov, A., Likhachev, A. A., Lanska, N., and Ullakko, K., *Giant Magnetic-Field-Induced Strain in NiMnGa Seven-Layered Martensitic Phase*, Appl. Phys. Lett. **80**, 1746–1748 (2002).

[165] Gebbia, J. F., Lloveras, P., Castán, T., Saxena, A., and Planes, A., *Modelling Shape-Memory Effects in Ferromagnetic Alloys*, Shap. Mem. Superelasticity **1**, 347–358 (2015).

[166] James, R. D. and Wuttig, M., *Magnetostriction of Martensite*, Philos. Mag. A **77**, 1273–1299 (1998).

[167] Lavrov, A. N., Komiya, S., and Ando, Y., *Magnetic Shape-Memory Effects in a Crystal*, Nature **418**, 385–386 (2002).

[168] Buchelnikov, V. D., Entel, P., Taskaev, S. V., Sokolovskiy, V. V., Hucht, A., Ogura, M., Akai, H., Gruner, M. E., and Nayak, S. K., *Monte Carlo Study of the Influence of Antiferromagnetic Exchange Interactions on the Phase Transitions of Ferromagnetic Ni-Mn-X alloys (X = In,Sn,Sb)*, Phys. Rev. B **78**, 184427 (2008).

[169] Goldschmidt, V. M., *Crystal Structure and Chemical Constitution*, Trans. Faraday Soc. **25**, 253–283 (1929).

[170] Liu, W. and Ren, X., *Large Piezoelectric Effect in Pb-Free Ceramics*, Phys. Rev. Lett. **103**, 257602 (2009).

[171] Yang, S., Yang, S., Bao, H., Zhou, C., Wang, Y., Ren, X., Matsushita, Y., Katsuya, Y., Tanaka, M., Kobayashi, K., Song, X., and Gao, J., *Large Magnetostriction from Morphotropic Phase Boundary in Ferromagnets*, Phys. Rev. Lett. **104**, 197201 (2010).

[172] Rossetti, G. A., Khachaturyan, A. G., Akcay, G., and Ni, Y., *Ferroelectric Solid Solutions with Morphotropic Boundaries: Vanishing Polarization Anisotropy, Adaptive, Polar Glass, and Two-Phase States*, J. Appl. Phys. **103**, 114113 (2008).

[173] Porta, M. and Lookman, T., *Effects of Tricritical Points and Morphotropic Phase Boundaries on the Piezoelectric Properties of Ferroelectrics*, Phys. Rev. B **83**, 174108 (2011).

[174] Tagantsev, A. K., *Susceptibility Anomaly in Films with Bilinear Coupling between Order Parameter and Strain*, Phys. Rev. Lett. **94**, 247603 (2005).

[175] Joule, J. P., *On Some Thermo-Dynamic Properties of Solids*, Phil. Trans. R. Soc. Lond. **149**, 91–131 (1859).

[176] Weiss, P. and Piccard, A., *Les Phenomènes Magnétocaloriques*, J. Phys. Theor. Appl. **7**, 103–109 (1917).

[177] Debye, P., *Einige Bemerkungen zur Magnetisierung bei tiefer Temperatur*, Ann. Phys. **81**, 1154–1160 (1926).

[178] Giauque, W. F., *A Thermodynamic Treatment of Certain Magnetic Effects. A Proposed Method of Producing Temperatures Considerably Below 1° Absolute*, J. Am. Chem. Soc. **49**, 1864–1870 (1927).

[179] Giauque, W. F. and MacDougall, D. P., *Attainment of Temperatures Below 1° Absolute by Demagnetization of $Gd_2(SO_4)_3 \cdot 8H_2O$*, Phys. Rev. **43** 768 (1933).

[180] Kobeko, P. and Kurtschatov, J., *Dielektrische Eigenschaften der Seignettesalzkristalle*, Z. Phys. **66**, 192–205 (1930).

[181] Brown, G. V. *Magnetic Heat Pumping Near Room Temperature*. J. Appl. Phys. 47, 3673–3680 (1976).

[182] Pecharsky, V. K. and Gschneider, K. A., *Giant Magnetocaloric Effect in $Gd_5(Si_2Ge_2)$* Phys. Rev. Lett. **78**, 4494–4497 (1997).

[183] Mischenko, A. S., Zhang, Q., Scott, J. F., Whatmore, R. W., Mathur, N. D., *Giant Electrocaloric Effect in Thin-Film* $PbZr_{0.95}Ti_{0.05}O_3$, Science **311**, 1270–1271 (2006).

[184] Bonnot, E., Romero, R., Vives, E., Mañosa, L., and Planes, A., *Elastocaloric Effect Associated with the Martensitic Transition in Shape-Memory Alloys*, Phys. Rev. Lett. **100**, 125901 (2008).

[185] Moya, X., Kar-Narayan, S., and Mathur, N. D., *Caloric Materials Near Ferroic Phase Transitions*, Nature Mater. **13**, 439–450 (2014).

[186] Castán, T., Planes, A., and Saxena, A., *Thermodynamics of Ferrotoroidic Materials: Toroidocaloric Effect*, Phys. Rev. B **85**, 144429 (2012).

[187] Mathon, J. and Wohlfarth, E. P., *Thermodynamic Properties of Nickel Near the Curie Temperature*, J. Phys. C: Solid St. Phys. **2**, 1647–1652 (1969).

[188] Gottschall, T., Ku'zmin, M. D., Skokov, K. P., Skourski, Y., Fries, M., Gutfleish, O., Ghorbani Zavareh, M., Schlagel, D. L., Mudryk, Y., Pecharsky, V., and Wosnitza, J., *Magnetocaloric Effect of Gadolinium in High Magnetic Fields*, Phys. Rev. B **99**, 134429 (2019).

[189] Srinath, S. and Kaul, S. N., *Static Universality Class for Gadolinium*, Phys. Rev. B **60**, 12176 (1999).

[190] Fujita, A., Fujieda, S., Fukachimi, K., Mitamura, H., and Goto, T., *Itinerant-Electron Metamagnetic Transition and Large Magnetovolume Effects in* $La(Fe_xSi_{1-x})_{13}$ *Compounds*, Phys. Rev. B, **65**, 014410 (2001).

[191] Fujita, A., Fujieda, S., Hasegawa, Y., and Fukachimi, K., *Itinerant-Electron Metamagnetic Transition and Large Magnetocaloric Effects in* $La(Fe_xSi_{1-x})_{13}$ *Compounds and their Hydrides*, Phys. Rev. B **67**, 104416 (2003).

[192] Bean, C. P. and Rodbell, D. S., *Magnetic Disorder as a First-Order Phase Transformation*, Phys. Rev. **126**, 104–115 (1962).

[193] Triguero, C., Porta, M., Planes, A., *Magnetocaloric Effect in Metamagnetic Systems*, Phys. Rev. B **76**, 094415 (2007).

[194] Planes, A., Castán, T., and Saxena, A., *Thermodynamics of Multicaloric Effects in Multiferroics*, Philos. Mag. **94**, 1893–1908 (2014).

[195] Stern-Taulats, E., Castán, T., Planes, A., Lewis, L. H., Barua, R., Pramanick, S., Majumdar, S., and Mañosa, L., *Giant Multicaloric Response of Bulk* $Fe_{49}Rh_{51}$, Phys. Rev. B **95**, 104424 (2017).

[196] Stephen, M. J. and Straley, J. P., *Physics of Liquid Crystals*, Rev. Mod. Phys. **46**, 617–704.

[197] Mitov, M., *A Brief History of Liquid Crystals Can Be Found*, in *Liquid-Crystal Science from 1888 to 1922: Building a Revolution*, Chem. Phys. Chem. **15**, 1245–1250 (2014).

[198] *Liquid Crystals – Applications and Uses*, Vol. III, edited by B. Bahadur, World Scientific, Singapur, 1992.

[199] Kast, W., *Landolt-Bornstein Tables*, Vol. 2, Part 2a, Springer-Verlag, Berlin, 1969.

[200] Chandrasekhar, S., *Liquid Crystals*, 2nd ed., Cambridge University Press, Cambridge, 1992.

[201] Jákli, A., Lavrentovich, O. D., and Selinger, J. V., *Physics of Liquid Crystals of Bent-Shaped Molecules*, Rev. Mod. Phys. **90**, 045004 (2018).

[202] Bushbya, R. J. and Kawata, K., *Liquid Crystals That Affected the World: Discotic Liquid Crystals*, Liquid Crystals **38**, 1415–1426 (2011).

[203] Maier, W. and Saupe, A., *Eine Einfache Molekulare Theorie des Nematischen Kristallinflüssigen Zustandes*, Z. Naturforsch. A **13a**, 564–566 (1958); *Eine Einfache Molekular-Statistische Theorie der Nematischen Kristallinflüssigen Phase. Teil 1*, Z. Naturforsch. A **14** 882–889 (1959); *Eine Einfache Molekular-Statistische Theorie der Nematischen Kristallinflüssigen Phase. Teil 2*, Z. Naturforsch. A **15**, 287–292 (1960).

[204] de Gennes, P. G., *Short Range Order Effects in the Isotropic Phase of Nematics and Cholesterics*, Mol. Cryst. Liq. Cryst. **12**, 193–214 (1971).

[205] Gramsbergen, E. F., Longa, L., and Jeu, W. H., *Landau Theory of the Nematic-Isotropic Phase Transition*, Phys. Rep. **135**, 195–257 (1986).

[206] Chaikin, P. M. and Lubesnsky, T. C., *Principles of Condensed Matter Physics*, Cambridge University Press, New York, 1995.

[207] Luckhurst, G. R. and Zannoni, C., *Why Is the Maier-Saupe Theory of Nematic Liquid Crystals So Successful?*, Nature **267**, 412–414 (1977).

[208] Gelbart, W. M., *Molecular Theory of Nematic Liquid Crystals*, J. Chem. Phys. **86**, 4298–4307 (1986).

[209] Saupe, A., *Biaxial Nematic Phases in Amphiphilic Systems*, J. Chem. Phys. **80**, 7–13 (1983).

[210] Anisimov, M. A., Garber, S. R., Esipov, V. S., Mannitskiĭ, V. M., Ovodov, G. I., Smolenko, L. A., and Sorkin, E. L., *Anomaly of the Specific Heat and the Nature of the Phase Transition from an Isotropic Liquid to a Nematic Liquid Crystal*, Sov. Phys. JETP **45**, 1042–1047 (1977).

[211] Oseen, C. W., *The Theory of Liquid Crystals*, Trans. Faraday Soc. **29**, 883–899 (1933).

[212] Frank, F. C. I., *I Liquid Crystals. On the Theory of Liquid Crystals*, Discuss. Faraday Soc. **25**, 19–28 (1958).

[213] Mori, H., Gartland Jr., E. C., Kelly, J. R., and Bos, P. J., *Multidimensional Director Modeling Using the Q Tensor Representation in a Liquid Crystal Cell and Its Application to the Cell with Patterned Electrodes*, Jpn. J. Appl. Phys. **38**, 135–146 (1999).

[214] Binder, K., Egorov, S. A., Milchev, A., and Nikoubashman, A., *Understanding the Properties of Liquid-Crystalline Polymers by Computational Modeling*, J. Phys. Mater. **3**, 032008 (2020).

[215] Zink, H. and De Jeu, W. H., *A Light-Scattering Study of Pretransitional Behavior Around the Isotropic-Nematic Phase Transition in Alkyl-cyanobiphenyls*, Mol. Cryst. Liq. Cryst. **124**, 287–304 (1985).

[216] Helfrich, W., *Effect of Electric Fields on the Temperature of Phase Transitions of Liquid Crystals*, Phys. Rev. Lett. **24**, 201–203.

[217] Rosenblatt, C., *Magnetic Field Dependence of the Nematic-Isotropic Transition Temperature*, Phys. Rev. A **24**, 2236–2238 (1981).

[218] Nicastro, A. J. and Heyes, P. H., *Electric-Field-Induced Critical Phenomena at the Nematic-Isotropic Transition and the Nematic-Isotropic Critical Point*, Phys. Rev. A **30**, 3156–3160 (1984).

[219] Fréedericksz, V. and Zolina, V., *Forces Causing the Orientation of an Anisotropic Liquid*, Trans. Faraday Soc. **29**, 919–930 (1933).

[220] Mukherjee, P. K., Pleiner, H., and Brand, H. R., *A Simple Landau Model for the Smectic-A-Isotropic Phase Transition*, Eur. Phys. J. **E4**, 293–297 (2001).

[221] Brinkman, W. F. and Cladis, P. E., *Defects in Liquid Crystals*, Phys. Today **35**, 48–54 (1982).

[222] Kleman, M. and Friedel, J., *Disclinations, Dislocations, and Continuous Defects: A Reappraisal*, Rev. Mod. Phys. **80**, 61–115 (2008).

[223] Caroli, C. and Dubois-Violette, E., *Energy of a Disinclination Line in an Anisotropic Cholesteric Liquid Crystal*, Solid State Commun. **7**, 799 (1969).

[224] Kamien, R. D. and Selinger, J. V., *Order and Frustration in Chiral Liquid Crystals*, J. Phys. Condens. Matter **13**, R1–R22 (2001).

[225] Duzgun, A., Selinger, J. V., and Saxena, A., *Comparing Skyrmions and Merons in Liquid Crystals and Magnets*, Phys. Rev. B **97**, 062706 (2018).

[226] Mühlbauer, S., Binz, B., Jonietz, F., Pfleiderer, C., Rosch, A., Neubauer, A., Georgii, R., and Böni, P., *Skyrmion Lattice in a Chiral Magnet*, Science **323**, 915–919 (2019).

[227] Bogdanov, A. N. and Rößler, U. K., *Chiral Symmetry Breaking in Magnetic Thin Films and Multilayers*, Phys. Rev. Lett. **87**, 037203 (2001).

[228] Imry, Y. and Ma, S.-K, *Random-Field Instability of the Ordered State of Continuous Symmetry*, Phys. Rev. Lett. **35**, 1399–1401 (1975).

[229] Imry, Y. and Wortis, M., *Influence of Quenched Impurities on First-Order Phase Transitions*, Phys. Rev. B **19**, 3580–3585 (1979).

[230] Lubchenko, V. and Wolynes, P. G., *Theory of Structural Glasses and Supercooled Liquids*, Ann. Rev. Phys. Chem. **58**, 235–266 (2007).

[231] Biroli, G and Garrahan, J. P., *Perspective: The Glass Transition*, J. Chem. Phys. **138**, 12A301 (2013).

[232] Mydosh, J. A., *Spin Glasses: An Experimental Introduction*, Taylor & Francis, London, 1993.

[233] Cowley, R. A., Gvasaliyac†, S. N., Lushnikov, S. G., Roesslicand, B., and Rotaruc, G. M., *Relaxing with Relaxors: A Review of Relaxor Ferroelectrics*, Adv. Phys. **60**, 229–327 (2011).

[234] Sharma, P. A., Kim, S. B., Koo, T. Y., Guha, S., and Cheong, S. -W., *Reentrant Charge Ordering Transition in the Manganites as Experimental Evidence for Strain Glass*, Phys. Rev. B **71**, 224416 (2005).

[235] Tolédano, P. and Manchon, D., *Structural Mechanism Leading to a Ferroelastic Strain Glass State: Interpretation of Amophization under Pressure*, Phys. Rev. B **71**, 024210 (2005).

[236] Sarkar, S., Ren, X., Otsuka, K., *Evidence for Strain Glass in the Ferroelastic-Martensitic System* $Ti_{50+x}Ni_{50-x}$, Phys. Rev. Lett. **95**, 205702 (2005).

[237] Lookman, T. and Ren, X., eds., *Frustrated Materials and Ferroic Glasses*, Springer-Verlag, Cham, Switzerland, 2018.

[238] Yamaguchi, Y. and Kimura, T., *Magnetoelectric Control of Frozen State in a Toroidal Glass*, Nature Comm. **4**, 2063 (2013).

[239] Hurd, C. M., *Varieties of Magnetic Order in Solids*, Contemp. Phys. **23**, 469–493 (1982).

[240] Bedanta, S. and Kleemann, W., *Supermagnetism*, J. Phys. D Appl. Phys. **42**, 013001 (2009).

[241] Ashcroft, N. W. and Mermin, N. D., *Solid State Physics*, Holt, Rinehart and Winston, Philadelphia, 1976.

[242] Ruderman, M. A. and Kittel, C., *Indirect Exchange Coupling of Nuclear Magnetic Moments by Conduction Electrons*, Phys. Rev. **96**, 99 (1954).

[243] Kasuya, T. *A Theory of Metallic Ferro- and Antiferromagnetism on Zener's Model*, Prog. Theor. Phys. **16**, 45 (1956).

[244] Yoshida, K., *Magnetic Properties of Cu-Mn Alloys*, Phys. Rev. **106**, 893 (1957).

[245] Cole, R. B., Sarkissian, B. V. B., Taylor, R. H., *The Role of Finite Magnetic Clusters in Au-Fe Alloys Near the Percolation Concentration*, Philos. Mag. B **37**, 489–498 (1978).

[246] Maletta, H. and Convert, P., *Onset of Ferromagnetism in $Eu_xSr_{1-x}S$ Near x = 0.5*, Phys. Rev. Lett. **42**, 108–111 (1979).

[247] Kasuya, T., *Exchange Mechanisms in Europium Chalcogenides*, IBM J. Res. Dev. **14**, 214–223 (1970).

[248] Börgermann, F. -J., Maletta, H., Zinn, W., *$Eu_xSr_{1-x}Te$: Spin-Glass Behavior in a Diluted Antiferromagnet*, Phys. Rev. B **35**, 8454–8461 (1986).

[249] Fischer, K. H. and Hertz, J. A., *Spin Glasses*, Cambridge University Press, Cambridge, 1991.

[250] Nagata, S., Keesom, P. H., Harrison, H. R., *Low-dc-field Susceptibility of CuMn Spin Glass*, Phys. Rev. B **19**, 1633–1638 (1979).

[251] Suzuki, M., *Phenomenological Theory of Spin-Glasses and Some Rigorous Results*, Prog. Ther. Phys. **58**, 1151–1165 (1981).

[252] Barbara, B., Malozernoff, A. I., and Imry, I., *Scaling of Nonlinear Susceptibility in MnCu and GdAl Spin-Glasses*, Phys. Rev. Lett. **47**, 1852–1855 (1981).

[253] Mulder, C. A. M., van Duyneveldt, A. J., Mydosh, J. A., *Susceptibility of the CuMn Spin-Glass: Frequency and Field Dependences*, Phys. Rev. **23**, 1384–1396 (1981).

[254] Hohenberg, P. C. and Halperin, B. I., *Theory of Dynamic Critical Phenomena*, Rev. Mod. Phys. **49**, 435–479 (1977).

[255] Shtrikman, S. and Wohlfarth, E. P., *The Theory of the Vogel-Fulcher Law of Spin Glasses*, Phys. Lett. A **85**, 467–470 (1981).

[256] Bohn, H. G., Zinn, W., Dorner, D., and Kollmar, A., *Neutron Scattering Study of Spin Waves and Exchange Interactions in Ferromagnetic EuS*, Phys. Rev. B **22**, 5447–5452 (1980).

[257] Edwards, S. F. and Anderson, P. W., *Theory of Spin Glasses*, J. Phys. F: Metal Phys. **5**, 965–974 (1975).

[258] Nattermann, T., *Theory of the Random Field Ising Model*, in *Spin Glasses and Random Fields*, edited by Young, A. P., Series on Directions in Condensed Matter Physics, World Scientific, pp. 227–298 (1997).

[259] Southern, B. W., *Effective-Field Approximations for Disordered Magnets*, J. Phys. C: Solid Stat. Phys, **9**, 4011–4020 (1975).

[260] Sherrington, D. and Kirkpatrick, S., *Solvable Model of a Spin-Glass*, Phys. Rev. Lett. **35**, 1792–1796 (1975).

[261] de Almeida, J. R. L. and Thouless, D. J., *Stability of the Sherrington-Kirkpatrick Solution of a Spin Glass Model*, J. Phys. A: Math. Gen. **11**, 983–990 (1978).

[262] Parisi, G., *Infinite Number of Order Parameters for Spin-Glasses*, Phys. Rev. Lett. **43**, 1754–1756 (1979).

[263] Kutnjak, Z., Pirc, R., Levstik, A., Levstik, I., Filipic, C., and Blinc, R., *Observation of the Freezing Line in a Deuteron Glass*, Phys. Rev. B **50**, 12421–12428 (1994).

[264] Pirc, R., Tadić, B., and Blinc, R., *Random-Field Smearing of the Proton-Glass Transition*, Phys. Rev. B. **36**, 8607–8615 (1987).

[265] Cowley, R. A., Gvasaliyac, S. N., Lushnikov, S. G., Roessli, B., and Rotaru, G. M., *Relaxing with Relaxors: A Review of Relaxor Ferroelectrics*, Adv. Phys. **60**, 229–327 (2011).

[266] Cross, L. E., *Relaxor Ferroelectrics*, Ferroelectrics **76**, 241–267 (1987).

[267] Samara, G. A., *Ferroelectricity Revisited – Advances in Materials and Physics*, Sol. State Phys. **56**, 239–458 (2001).

[268] Westphal, V., Kleeman, W., and Glinchuk, M. D., *Diffuse Phase Transitions and Random-Field-Induced Domain States of the "Relaxor" Ferroelectric* $PbMg_{1/3}Nb_{2/3}O_3$, Phys. Rev. Lett. **68**, 847– (1992).

[269] Pirc, R. and Blinc, R., *Spherical Random-Bond–Random-Field Model of Relaxor Ferroelectric*, Phys. Rev. B **60**, 13470–13478 (1999).

[270] Kutnjak, Z., Filipič, C., Pirc, R., Levstik, A., Farhi, R., and El Marssi, M., *Slow Dynamics and Ergodicity Breaking in a Lanthanum-Modified Lead Zirconate Titanate Relaxor System*, Phys. Rev. B **59**, 294–301 (1999).

[271] Ren, X., Wang, Y., Zhou, Y., Zhang, Z., Wang, D., Fan, G., Otsuka, K., Suzuki, T., Ji, Y., Zhang, J., Tian, Y., Hou, S., and Ding, X., *Strain Glass in Ferroelastic Systems: Premartensitic Tweed versus Strain Glass*, Philos. Mag. **90**, 141–157 (2010).

[272] Lloveras, P., Castán, T., Porta, M., Planes, A., and Saxena, A., *Influence of Anisotropy on Structural Nanoscale Textures*, Phys. Rev. Lett. **100**, 165707 (2008).

[273] Planes, A., Lloveras, P., Castán, T., Saxena, A., and Porta, M., *Ginzburg-Landau Modelling of Precusror Nanoscale Textures in Ferroeleastic Materials*, Continnum Mech. Thermodym. **24**, 619–627 (2012).

[274] Vasseur, R. and Lookman, T., *Effects of Disorder in Ferroelastics: A Spin Model for Strain Glass*, Phys. Rev. B **81**, 094107 (2010).

[275] Chakrabarti, B. K. and Acharyya, M., *Dynamic Transitions and Hysteresis*, Rev. Mod. Phys. **71**, 847–859 (1999).

[276] Sethna, J. P., Dahmen, K. A., and Myers, C. R., *Crackling Noise*, Nature **410**, 242–250 (2001).

[277] Koshelev, A. E. and Vinokur, V. M., *Dynamic Melting of the Vortex Lattice*, Phys. Rev. Lett. **73**, 3580–3583 (1994).

[278] Kes, P. H., Kokubo, N., and Besseling, R., *Vortex Matter Driven through Mesoscopic Channels*, Physica C **408–410**, 478 (2004).

[279] Jiang, Q., Yang, H.-N., and Wang, G.-C., *Scaling and Dynamics of Low-Frequency Hysteresis Loops in Ultrathin Co Films on a Cu(001) Surface*, Phys. Rev. B **52**, 14911 (1995).

[280] Ogawa, N., Murakami, Y., and Miyano, K., *Charge-Density-Wave Phase Reconstruction in the Photoinduced Dynamic Phase Transition in* $K_{0.3}MoO_3$, Phys. Rev. B **65**, 155107 (2002).

[281] Geng, Z., Peters, K. J. H., Trichet, A. A. P., Malmir, K., Kolkowski, R., Smith, J. M., and Rodriguez, S. R. K., *Universal Scaling in the Dynamic Hysteresis, and Non-Markovian Dynamics, of a Tunable Optical Cavity*, Phys. Rev. Lett. **124**, 153603 (2020).

[282] Tomé, T. and de Oliveira, M. J., *Dynamic Phase Transition in the Kinetic Ising Model under a Time-Dependent Oscillating Field*, Phys. Rev. A **41**, 4251–4254 (1990).

[283] Bader, S. D. and Moog, E. R., *Magnetic Properties of Novel Epitaxial Films*, J. Appl. Phys. **61**, 3729 (1987).

[284] He, Y.-L. and Wang, G.-C., *Observation of Dynamic Scaling of Magnetic Hysteresis in Ultrathin Ferromagnetic Fe/Au(001) Films*, Phys. Rev. Lett. **70**, 2336–2339 (1993).

[285] Suen J.-S and Erskine, J. L., *Magnetic Hysteresis Dynamics: Thin* $p(1 \times 1)$ *Fe Films on Flat and Stepped W(110)*, Phys. Rev. Lett. **78**, 3567 (1997).

[286] Gallardo, R., Idigoras, O., Landeros, P., and Berger, A., *Analytical Derivation of Critical Exponents of the Dynamic Phase Transition in the Mean-Field Approximation*, Phys. Rev. E **86**, 051101 (2012).

[287] Sethna, J. P., Dahmen, K., Kartha, S., Krumhansl, J. A., Roberts, B. W., and Shore, J. D., *Hysteresis and Hierarchies: Dynamics of Disorder-Driven First-Order Phase Transformations*, Phys. Rev. Lett. **70**, 3347–3350 (1993).

[288] Kumar, S. K., Biroli, G., and Tarjus, G., *Spinodals with Disorder: From Avalanches in Random Magnets to Glassy Dynamics*, Phys. Rev. Lett **116**, 145701 (2016).

[289] McClure, J. C. and Schröder, K., *The Magnetic Barkhausen Effect*, Crit. Rev. Solid State Sci. **6**, 45–83 (1976).

[290] Durin, G. and Zapperi, S., *The Role of Stationarity in Magnetic Crackling Noise*, J. Stat. Mech.: Theory and Experiment **2006**, P01002.

[291] Zapperi, S., Cizeau, P., Durin, G., and Stanley, H. E., *Dynamics of a Ferromagnetic Domain Wall: Avalanches, Depinning Transition, and Barkhausen Effect*, Phys. Rev. B **58**, 6353–6366 (1998).

[292] da Silveira, R., *An Introduction to Breakdown Phenomena in Disorderd Systems*, Am. J. Phys. **67**, 1177–1188 (1999).

[293] Perković, O., Dahmen, K., and Sethna, J. P., *Avalanches, Barkhausen Noise, and Plain Old Criticality*, Phys. Rev. Lett. **75**, 4528–4531 (1995).

[294] Berger, A., Inomata, J. S., Jiang, J. S., Pearson, J. E., and Bader, S. D., *Experimental Observation of Disorder-Driven Hysteresis-Loop Criticality*, Phys. Rev. Lett. **85**, 4176–4179 (2000).

[295] Marcos, J., Vives, E., Mañosa, L., Acet, M., Duman, E., Morin, M., Novák, V., and Planes, A., *Disorder Induced Non-Equilibrium Phase Transition in Magnetically Glassy Cu-Al-Mn*, Phys. Rev. B **67**, 224406 (2003).

[296] Pérez-Reche, F. J., *Experimentos y modelos en sistemas que presentan transiciones de fase de primer orden con dinámica de avalanchas*, PhD Thesis, University of Barcelona, 2005, http://hdl.handle.net/2445/35492.

[297] Lieneweg, U. and Grosse-Nobis, W., *Distribution of Size and Duration of Barkhausen Pulses and Energy Spectrum of Barkhausen Noise Investigated on 81% Nickel-Iron after Heat Treatment*, Int. J. Magn. **3**, 11–16 (1972).

[298] Durin, G. and Zapperi, S., *The Barkhausen Effect* in *The Science of Hysteresis Vol. II, Chap. 3*, edited by Bertotti, G. and Mayergoyz, I., pp. 181–267, Academic Press, Oxford, 2006.

[299] Schwarz, A., Liebmann, M., Kaiser, U., Wiesendanger, R., Noh, T. W., and Kim, D. W., *Visualization of the Barkhausen Effect by Magnetic Force Microscopy*, Phys. Rev. Lett. **92**, 077206 (2004).

[300] Perković, O., Dahmen, K., and Sethna, J. P. *Disorder-Induced Critical Phenomena in Hysteresis: A Numerical Scaling Analysis*, arXiv:cond-mat/9609072v1 6 Sep 1996, unpublished.

[301] Sethna, J. P., Dahmen, K., and Perković, *Random-Field Ising Models of Hysteresis* in The Science of Hysteresis Vol. II, Chap. 2, edited by Bertotti, G. and Mayergoyz, I., pp. 107–179, Academic Press, Oxford, 2006.

[302] Tan, C. D., Flannigan, C., Gardner, J., Morrison, F. D., Salje, E. K. H., and Scott, J. F., *Electrical Studies of Barkhausen Switching Noise in Ferroelectric PZT: Critical Exponents and Temperature Dependence*, Phys. Rev. Mat. **3**, 034402 (2019).

[303] Salje, E. K. H., Xue, D., Ding, D., Dahmen, K., and Scott, J. F., *Ferroelectric Switching and Scale Invariant Avalanches in $BaTiO_3$*, Phys. Rev. Mat. **3**, 014415 (2019).

[304] Pérez-Reche, F. J., Stipcich, M., Vives, E., Mañosa, L., Planes, A., and Morin, M., *Kinetics of Martensitic Transitions in Cu-Al-Mn under Thermal Cycling: Analysis at Multiple Length Scales*, Phys. Rev. B **69**, 0641001 (2004).

[305] Pérez-Reche, F. J., Truskinovsky, L., and Zanzotto, G., *Training-Induced Criticality in Martensites*, Phys. Rev. Lett. **99**, 075501 (2006).

[306] Porta, M., Castán, T., Saxena, A., and Planes, A., *Influence of the Number of Orientational Domains on Avalanche Criticality in Ferroelastic Transitions*, Phys. Rev. E **100**, 062115 (2019).

[307] Xu, Y., Xue, D., Zhou, Y., Su, T., Ding, X., Sun, J., and Salje, E. K. H., *Avalanche Dynamics of Ferroelectric Phase Transitions in $BaTiO_3$ and $0.7Pb(Mg_{2/3}Nb_{1/3})O3$-$0.3PbTiO_3$ Single Crystals*, Appl. Phys. Lett. **115**, 022901 (2019).

[308] Bonnot, E., Mañosa, L., Planes, A., Soto-Parra, D., and Vives, E., *Acoustic Emission in the fcc-fct Martensitic Transition of $Fe_{68.8}Pd_{31.2}$*, Phys. Rev. B **78**, 184103 (2008).

[309] Bertotti, G., *Hysteresis in Magnetism*, Academic Press, San Diego, 1998.

[310] Barker, J. A., Schreiber, D. E., Huth, B. G., and Everett, D. H., *Magnetic Hysteresis and Minor Loops: Models and Experiments*, Proc. R. Soc. Lond. A **386**, 251–261 (1983).

[311] Gutenberg, R. and Richter, C. F., *Seismicity of the Earth and Associated Phenomena*, Princeton University Press, Princeton, NJ, 1949.

[312] Baró, J., Corral, A., Illa, X., Planes, A., Salje, E. K. H., Schranz, W., Soto-Parra, E., and Vives, E., *Statistical Similarity between the Compression of a Porous Material and Earthquakes*, Phys. Rev. Lett. **110**, 088702 (2013).

[313] Bonamy, D. and Bouchaud, E., *Failure of Heterogeneous Materials: A Dynamic Phase Transition?*, Phys. Rep. **498**, 1–44 (2011).

[314] Van Delft, D. and Kes, P., *The Discovery of Superconductivity*, Phys. Today **63**, 38–42 (2010).

[315] Drude, P., *Zur Elektronentheorie der Metalle*, Annalen der Physik **306**, 566–613 (1900).

[316] Kamerlingh Onnes, H., *The Resistance of Pure Mercury at Liquid Helium*, Commun. Phys. Lab. Univ. Leiden, **120b**, 1479–1481 (1911); *The Disappearance of the Resistance of Mercury, ibid.*, **122b**, 81–83 (2011); *On the Sudden Change in the Rate at Which the Resistance of Mercury Disappears, ibid.*, **124c**, 799–801 (2011).

[317] Meissner, W. and Ochsenfeld, R., *Ein neuer Effekt bei Eintritt der Supraleitfähigkeit*, Naturwissenschaften **21**, 787–788 (1933).

[318] McMillan, W. L., *Transition Temperature of Strong-Coupled Superconductors*, Phys. Rev. **167**, 331–344 (1968).

[319] Bednorz, J. G. and Müller, K. A., *Possible High Tc Superconductivity in the Ba-La-Cu-O System*, Z. Phys. B **64**, 189–193 (1986).

[320] Kleinert, H., *Order of Superconductive Phase Transition*, Condensed Matter Phys. **8**, 75–86 (2005).

[321] Gottlieb, U., Lasjaunias, J. C., Tholence, J. L., Laborde, O., Thomas, O., and Madar, R., *Superconductivity in $TaSi_2$ Single Crystals*, Phys. Rev. B **45**, 4803–4806 (1992).

[322] London, F. and London, H., *The Electromagnetic Equations of the Supracon-ductor*, Proc. Roy. Soc. A **149**, 71–88 (1935).

[323] Abrikosov, A. A., *On the Magnetic Properties of Superconductors of the Second Group*. J. Phys. Chem. Solids **5**, 1174–1182 (1957).

[324] Abramowitz, M. and Stegun, I. A., eds., *Handbook of Mathematical Functions*, National Bureau of Standards, Applied Mathematical Series, Washington DC, 1964.

[325] Phillips, N. E., *Heat Capacity of Aluminum between 0.1 °K and 4.0 °K*, Phys. Rev. **114**, 676–685 (1959).

[326] Ginzburg, V. L. and Landau, L. D., *On the Theory of Superconductivity*, Zh. Eksp. Teor. Fiz. **20**, 1064–1082 (1950). English translation in: Landau, L. D *Collected Papers*, edited by Ter Haar, D., Pergamon Press, Oxford, 1965, p. 546–568.

[327] Annett, J. F., *Superconductivity, Superfluids and Condensates*, Oxford University Press, Oxford, 2004.

[328] Abrikosov, A. A., *On the Magnetic Properties of Superconductors of Second Group*, Sov. Phys. JEPT **5**, 1174–1182 (1957).

[329] Hess, H. F., Robinson, R. B., Dynes, R. C., Valles, Jr., J. M., and Waszczak, J. V., *Scanning-Tunneling-Microscope Observation of the Abrikosov Flux Lattice and the Density of States near and inside a Fluxoid*, Phys. Rev. Lett. **62**, 214–216 (1989).

[330] Riseman, T. M., Kealey, P. G., Forgan, E. M., Mackenzie, A. P., Galvin, L. M., Tyler, A. W., Lee, L. S., Ager, C., McK. Paul, D., Aegerter, C. M., Cubittk, R., Mao, Z. Q., Akima, T., and Maeno, Y., *Observation of a Square Flux-Line Lattice in the Unconventional Superconductor Sr_2RuO_4*, Nature **396**, 242–245 (1998).

[331] Altshuler,. E. and Johansen, T. H., *Colloquium: Experiments in Vortex Avalanches*, Rev. Mod. Phys. **76**, 471–487 (2004).

[332] Field, S., Witt, J., Nori, F., and Ling, X., *Superconducting Vortex Avalanches*, Phys. Rev. Lett. **74**, 1206–1209 (1995).

[333] de Gennes, P. G., *An Analogy between Superconductors and Smectics A*, Solid State Comm. **10**, 753–756 (1972).

[334] Renn, S. R. and Lubensky, T., *Abrikosov Dislocation Lattice in a Model of the Cholesteric-to-Smectic-A Transition*, Phys. Rev. A **38**, 2132–2147 (1988).

[335] Goodby, J. W., Waugh, M. A., Stein, S. M., Chin, E., Pindak, R., and Patel, J. S., *Characterization of a New Helical Smectic Liquid Crystal*, Nature, **337**, 449–452 (1989).

[336] Bardeen, J., Cooper, L. N., and Schrieffer, J. R., *Microscopic Theory of Superconductivity*, Phys. Rev. **106**, 162–164 (1957).

[337] Cooper, L, *Bound Electron Pairs in a Degenerate Fermi Gas*, Phys. Rev. **104**, 1189–1190 (1956).

[338] Giaever, I. and Megerle, K., *Study of Superconductors by Electron Tunneling*, Phys. Rev. **122**, 1101–1111 (1961).

[339] Gor'kov, L. P., *Microscopic Derivation of the Ginzburg-Landau Equations in the Theory of Superconductivity*, Sov. Phys. JETP **36**, 1364–1367 (1959).

[340] de Gennes, P. G., *Role of Double Exchange in Copper Oxides of Mixed Valency*, Comptes Rendus Acad. Sci. (Paris) **305**, 345–348 (1987).

[341] O'Mahony, S. M., Ren, W., Chen, W., and Séamus Davis, J. C., *On the Electron Pairing Mechanism of Copper-Oxide High Temperature Superconductivity*, PNAS **119**, e2207449119 (2022).

[342] Moriya, T. and Ueda, K., *Spin Fluctuations and High Temperature Superconductivity*, Adv. Phys. **49**, 555–606 (2010).

[343] Hattori, T., Ihara, Y., Nakai, Y., Ishida, K., Tada, Y., Fujimoto, S., Kawakami, N., Osaki, E., Deguchi, K., Sato, N. K., and Satoh, I., *Superconductivity Induced by Longitudinal Ferromagnetic Fluctuations in UCoGe*, Phys. Rev. Lett. **108**, 066403 (2012).

[344] Gorter, C. J. and Casimir, H. G. B., *On Supraconductivity I*, Physica **1**, 306–320 (1934); Gorter, C. J., *The Two Fluid Model for Superconductors and Helium II*, in *Progress in Low Temperature Physics*, Vol. I, edited by Gorter, C. J., North-Hollad Publishing Company, Amsterdam, 1955, pp. 1–16.

[345] Sachdev, S., *Quantum Phase Transitions*, Phys. World, **12**, 33–38 (1999); Sachdev, S. and Keimer, B., *Quantum Criticality*, Phys. Today, **64**, 29–35 (2011).

[346] Hertz, J. A., *Quantum Critical Phenomena*, Phys. Rev. B **14**, 1166–1184 (1976).

[347] Sondhi, S. L., Girvin, S. M., Carin, J. P., and Shahar, D., *Continuous Quantum Phase Transitions*, Rev. Mod. Phys. **69**, 315–333 (1997).

[348] Vojta, M., *Quantum Phase Transitions*, Rep. Prog. Phys. **66**, 2069–2110 (2003).

[349] Bitko, D., Rosenbaum, T. F., and Aeppli, G., *Quantum Critical Behavior for a Model Magnet*, Phys. Rev. Lett. **77**, 940–943 (1996).

[350] Coldea, R., Tennant, D. A., Wheeler, E. M., Wawrzynska, E., Prabhakaran, D., Telling, M., Habicht, K., Smeibidl, P., and Kiefe, K., *Quantum Criticality in an Ising Chain: Experimental Evidence for Emergent E8 Symmetry*, Science **327**, 177–180 (2010).

[351] Rüegg, C., Furrer, A., Sheptyakov, D., Strässle, T., Krämer, K. W., Güdel, H. -U., and Mélési, *Pressure-Induced Quantum Phase Transition in the Spin-Liquid TlCuCl*, Phys. Rev. Lett. **93**, 257201 (2004).

[352] Rüegg, C., Normand, B., Matsumoto, M., Furrer, A., McMorrow, D. F., Krämer, K. W., Güdel, H. -U., Gvasaliya, S. N., Mutka, H., and Boehm, M., *Quantum Magnets under Pressure: Controlling Elementary Excitations in TlCuCl₃*, Phys. Rev. Lett. **100**, 205701 (2008).

[353] Ando, Y., Komiya, S., Segawa, K., Ono, S., and Kurita, Y., *Electronic Phase Diagram of High-Tc Cuprate Superconductors from a Mapping of the In-Plane Resistivity Curvature*, Phys. Rev. Lett. **93**, 267001 (2004).

[354] Armitage, N. P., Fournier, P., and Green, R. L., *Progress and Perspectives on Electron-Doped Cuprates*, Rev. Mod. Phys. **82**, 2421–2487 (2010).

[355] Kasahara, S., Shibauchi, T., Hashimoto, K., Ikada, K., Tonegawa, S., Okazaki, R., Shishido, H., Ikeda, H., Takeya, H., Hirata, K., Terashima, and T., Matsuda, Y., *Evolution from Non-Fermi- to Fermi-Liquid Transport via Isovalent Doping in $BaFe_2(As_{1-x}P_x)_2$ Superconductors*, Phys. Rev. B **81**, 184519 (2010).

[356] Sachdev, S., *Where Is the Quantum Critical Point in the Cuprate Superconductors?*, Phys. Stat. Sol. B **247**, 537–543 (2010).

[357] Thompson, C. J., *Classical Equilibrium Statistical Mechanics*, Clarendon Press, Oxford, 1988.

[358] *International Tables for Crystallography*, Vols. A to I, Wiley, 2019 (online version available at, https://it.iucr.org/).

[359] Hatch, D. M., Lookman, T., Saxena, A., and Shenoy, S. R., *Proper Ferroelastic Transitions in Two Dimensions: Anisotropic Long-Range Kernels, Domain Wall Orientations, and Microstructure*, Phys. Rev. B **68**, 104105 (2003).

[360] Stokes, H. T. and Hatch, D. M., *Isotropy Subgroups of the 230 Crystallographic Space Groups*, World Scientific, Singapore, 1988. The software package ISOTROPY is available at www.physics.byu.edu/stokesh/isotropy .html.

[361] Bilbao Crystallographic Server: www.cryst.ehu.es/.

[362] Hu, H. L. and Chen, L. Q., *Computer Simulation of 90° Ferroelectric Domain Formation in Two-Dimensions*, Mater. Sci. Eng. A **238**, 182–191 (1997).

[363] Ahluwalia, R. and Cao, W., *Influence of Dipolar Defects on Switching Behavior in Ferroelectrics*, Phys. Rev. B **63**, 012103 (2001).

[364] Saxena, A., Barsch, G. R., and Hatch, D. M., *Lattice Dynamics Representation Theory versus Isotropy Subgroup Method with Application to M_5^- Mode Instability in CsCl Structure*, Phase Trans. **46**, 89–142 (1994).

[365] Saxena, A. and Lookman, T., *Magnetic Symmetry of Low-Dimensional Multiferroics and Ferroelastics*, Phase Trans. **84**, 421–437 (2011).

[366] Birss, R. R., *Symmetry and Magnetism*, North-Holland, Amsterdam, 1964.

[367] Srinivasan, S. G., Hatch, D. M., Stokes, H. T., Saxena, A., Albers, R. C., and Lookman, T., *Mechanism for BCC to HCP Transformation: Generalization of the Burgers Model*, arXiv:cond-mat/0209530.

[368] Hatch, D. M., Lookman, T., Saxena, A., and Stokes, H. T., *Systematics of Group-Nonsubgroup Transitions: Square to Triangle Transition*, Phys. Rev. B **64**, 060104 (2001).

Materials Index

General Index

For EU product safety concerns, contact us at Calle de José Abascal, 56–1°,
28003 Madrid, Spain or eugpsr@cambridge.org.

www.ingramcontent.com/pod-product-compliance
Lightning Source LLC
Chambersburg PA
CBHW080442130525
26598CB00005B/52